SEMA SIMAI Springer Series

Volume 22

As of 2013, the SIMAI Springer Series opens to SEMA in order to publish a joint series aiming to publish advanced textbooks, research-level monographs and collected works that focus on applications of mathematics to social and industrial problems, including biology, medicine, engineering, environment and finance. Mathematical and numerical modeling is playing a crucial role in the solution of the complex and interrelated problems faced nowadays not only by researchers operating in the field of basic sciences, but also in more directly applied and industrial sectors. This series is meant to host selected contributions focusing on the relevance of mathematics in real life applications and to provide useful reference material to students, academic and industrial researchers at an international level. Interdisciplinary contributions, showing a fruitful collaboration of mathematicians with researchers of other fields to address complex applications, are welcomed in this series.

THE SERIES IS INDEXED IN SCOPUS

More information about this series at http://www.springer.com/series/10532

Francisco Marcellán • Edmundo J. Huertas
Editors

Orthogonal Polynomials: Current Trends and Applications

Proceedings of the 7th EIBPOA Conference

Springer

Editors
Francisco Marcellán
Departamento de Matemáticas
Universidad Carlos III de Madrid
Leganés, Spain

Edmundo J. Huertas
Departamento de Física y Matemáticas
Universidad de Alcalá
Alcalá de Henares
Madrid, Spain

ISSN 2199-3041 ISSN 2199-305X (electronic)
SEMA SIMAI Springer Series
ISBN 978-3-030-56189-5 ISBN 978-3-030-56190-1 (eBook)
https://doi.org/10.1007/978-3-030-56190-1

This Springer imprint is published by the registered company Springer Nature Switzerland AG.
The registered company address is: Gewerbestrasse 11, 6330 Cham, Switzerland

Preface

The present volume contains the Proceedings of the Seventh Iberoamerican Workshop in Orthogonal Polynomials and Applications (EIBPOA, which stands for Encuentros Iberoamericanos de Polinomios Ortogonales y Aplicaciones, in Spanish), held at the Universidad Carlos III de Madrid, Leganés, Spain, from July 3 to July 6, 2018.

The conference, as well as this volume, was warmly dedicated to celebrate the 70th birthday of Professor Guillermo López Lagomasino, professor of Applied Mathematics in the University Carlos III de Madrid. He is one of the most relevant and active mathematicians of the Spanish school of orthogonal polynomials. His contributions to orthogonal polynomials in different frameworks (the real line, the unit circle, multiple orthogonality, and Sobolev orthogonality) as well as in approximation theory (Padé and Hermite–Padé rational approximants) have been recognized worldwide. His leadership in the above fields and his commitment with young researchers constitute a sample of his scientific excellence.

This is the seventh workshop of a series that began in 2011 at the National University of Colombia in Bogotá organized by Professor Herbert Dueñas Ruíz, continued in 2012 at Universidad de Colima, México, organized by Professor Luis E. Garza Gaona and, until we reach this last edition, has also been held in São José do Rio Preto, Brazil (Universidade Estadual Paulista UNESP 2013), Bogotá, Colombia (Universidad Nacional de Colombia 2014), México D.F. (Universidad Nacional Autónoma de México, 2015), and Uberaba, Brazil (Universidade Federal do Triângulo Mineiro UFTM 2017).

These meetings on orthogonal polynomials were mainly focused to encourage research in the fields of approximation theory, special functions, orthogonal polynomials and their applications among graduate students as well as young researchers from Latin America, Spain, and Portugal. To support this idea, the core of these workshops is based on two academic courses in orthogonal polynomials theory in the morning sessions and plenary talks and short communications during the evenings. Both talks and courses are taught at postgraduate level and have a marked international character, since professors and students come from all around the world, and surprisingly enough, not only from the Latin-American region.

The EIBPOA logo reflects the spirit of this event. The Latin-American region is united through mathematics, specifically by the graph of the first seven Jacobi orthogonal polynomials.

In this occasion, we were very happy to have the two EIBPOA Courses taught by Professor Guillermo López Lagomasino (UC3M) and Professor Manuel Mañas-Baena (UCM). We had nothing less than ten plenary distinguished speakers: María José Cantero (Universidad de Zaragoza, España), Alfredo Deaño (University of Kent, United Kingdom), Diego Dominici (State University of New York at New Paltz, USA), Ulises Fidalgo (Case Western Reserve University, USA), Ana Foulquié-Moreno (Universidade de Aveiro, Portugal), Judit Mínguez Ceniceros (Universidad de La Rioja, España), Maria das Neves Rebocho (Universidade da Beira Interior, Portugal), Javier Segura (Universidad de Cantabria, España), Jan Felipe van Diejen (Universidad de Talca, Chile), and Luis Verde Star (Universidad Autónoma Metropolitana, México). Their very valuable contributions dealt with issues of absolute currency in orthogonal polynomials theory: from the role of Darboux transformations, Painlevé equations, incomplete interpolatory and Gaussian quadrature rules, new systems of discrete orthogonal polynomial families, Fourier series, Laguerre–Hahn, q-Hahn families of the q-Askey scheme, and new insights on the Riemann–Hilbert problem for orthogonal polynomials. The event was rounded off with eleven short communications and a poster contest whose winners were two young researchers: Fátima Lizarte López (Universidad de Granada) and Juan Francisco Mañas Mañas (Universidad de Almería).

We would like to thank all the members of the organizing and the local committees for the nice organization of this 7th EIBPOA, as well as the members of the Scientific Committee who helped us to make an attractive and genuine conference of the highest scientific quality. We also express our gratitude to all the participants of the workshop, to the contributors of this volume, and to the editors of the SEMA/SIMAI series for their support in the production of these proceedings.

In this edition, the workshop was co-organized by Universidad Politécnica de Madrid (UPM) and Universidad Carlos III de Madrid (UC3M). The event was partially supported by PROGRAMA PROPIO I+D+I DE LA UNIVERSIDAD POLITÉCNICA DE MADRID, through CONVOCATORIA DE AYUDAS DIRIGIDAS A PDI E INVESTIGADORES DOCTORES PARA LA ORGANIZACIÓN DE EVENTOS as well as partially supported by the network ORTHONET, project MTM2017-90694-REDT granted by Agencia Estatal de Investigación, Ministerio de Ciencia, Innovación y Universidades of Spain.

Leganés, Spain — Francisco Marcellán
Alcalá de Henares, Spain — Edmundo J. Huertas

Contents

Riemann–Hilbert Problem and Matrix Biorthogonal Polynomials

Amílcar Branquinho, Ana Foulquié-Moreno, and Manuel Mañas-Baena

Abstract Recently the Riemann–Hilbert problem with jumps supported on appropriate curves in the complex plane has been presented for matrix biorthogonal polynomials, in particular non–Abelian Hermite matrix biorthogonal polynomials in the real line, understood as those whose matrix of weights is a solution of a Sylvester type Pearson equation with coefficients first order matrix polynomials. We will explore this discussion, present some achievements and consider some new examples of weights for matrix biorthogonal polynomials.

Keywords Riemann-Hilbert problems · Matrix Pearson equations · Markov functions · Matrix biorthogonal polynomials

1 Introduction

Back in 1949, in the seminal papers [36, 37], Krein discussed matrix extensions of real orthogonal polynomials. Afterwards, this matrix extension of the standard orthogonality was studied only sporadically until the last decade of the XX century,

A. Branquinho
CMUC and Department of Mathematics, University of Coimbra (FCTUC), Coimbra, Portugal
e-mail: ajplb@mat.uc.pt

A. Foulquié-Moreno (✉)
Departamento de Matemática, CIDMA, Universidade de Aveiro, Campus de Santiago, Aveiro, Portugal
e-mail: foulquie@ua.pt

M. Mañas-Baena
Departamento de Física Teórica, Universidad Complutense de Madrid, Madrid, Spain

M. Mañas
Instituto de Ciencias Matemáticas (ICMAT), Campus de Cantoblanco UAM, Madrid, Spain
e-mail: manuel.manas@ucm.es

F. Marcellán, E. J. Huertas (eds.), *Orthogonal Polynomials: Current Trends and Applications*, SEMA SIMAI Springer Series 22,
https://doi.org/10.1007/978-3-030-56190-1_1

see [2, 33] and [1]. In[1], for a kind of discrete Sturm–Liouville operators, the authors solved the corresponding scattering problem and found a matrix version of Favard's theorem, polynomials that satisfy the three term relation

$$xP_k(x) = A_k P_{k+1}(x) + B_k P_k(x) + A_{k-1}^* P_{k-1}(x), \qquad k = 0, 1, \dots,$$

are orthogonal with respect to a positive definite matrix of measures.

Along the last decade, a number of basic results on the theory of scalar orthogonal polynomials, such as Favard's theorem [15, 16, 27, 30], quadrature formulae [17, 21, 29] and asymptotic properties (Markov's theorem [17], ratio [18, 19] weak [20] and zero asymptotics [28]), have been extended to the matrix scenario. The search of families of matrix orthogonal polynomials that satisfy second order differential equations with coefficients independent of n, can be found in [22–26]. This can be considered as a matrix extension of the classical orthogonal polynomial sequences of Hermite, Laguerre and Jacobi.

Fokas, Its and Kitaev, when discussing 2D quantum gravity, discovered that certain Riemann–Hilbert problem was solved in terms of orthogonal polynomials in the real line (OPRL), [31]. Namely, it was found that the solution of a 2×2 Riemann–Hilbert problem can be expressed in terms of orthogonal polynomials in the real line and its Cauchy transforms. Later on, Deift and Zhou combined these ideas with a non–linear steepest descent analysis in a series of papers [10, 11, 13, 14] which was the seed for a large activity in the field. Relevant results to be mentioned here are the study of strong asymptotic with applications in random matrix theory, [10, 12], the analysis of determinantal point processes [7, 8, 38, 39], applications to orthogonal Laurent polynomials [40, 41] and Painlevé equations [9, 35].

The Riemann–Hilbert problem characterization is a powerful tool that allows one to prove algebraic and analytic properties of orthogonal polynomials. The Riemann–Hilbert problem for this matrix situation and the appearance of non–Abelian discrete versions of Painlevé I (mdPI) were explored in [4], and the appearance of singularity confinement was shown in [6]. The analysis was extended further [5] for the matrix Szegő type orthogonal polynomials in the unit circle and corresponding non–Abelian versions discrete Painlevé II equations. For an alternative discussion of the use of Riemann–Hilbert problem for MOPRL see [34], were the authors focus on the algebraic aspects of the problem, obtaining difference and differential relations satisfied by the corresponding orthogonal polynomials.

In [3] we have studied a Hermite–type biorthogonal matrix polynomial system from a Riemann–Hilbert problem. In this case the matrix measure extends to an entire function on the complex plane. We have considered three types of matrix weights, W, obtained from the solution of a generalized Pearson differential or Sylvester differential equation, i.e.

$$W' = h^{\mathsf{L}}(z)W(z) + W(z)h^{\mathsf{R}}(z)$$

where h^{L} and h^{R} are polynomials of the first, second or third degrees.

In that paper we showed that for these weights the matrix solutions of the Riemann–Hilbert problem

$$Y_n^{\mathsf{L}} = \begin{bmatrix} P_n^{\mathsf{L}} & Q_n^{\mathsf{L}} \\ -C_{n-1}P_{n-1}^{\mathsf{L}} & -C_{n-1}Q_{n-1}^{\mathsf{L}} \end{bmatrix}, \quad Y_n^{\mathsf{R}} = \begin{bmatrix} P_n^{\mathsf{R}} & -C_{n-1}Q_{n-1}^{\mathsf{R}} \\ -C_{n-1}P_{n-1}^{\mathsf{R}} & Q_n^{\mathsf{R}} \end{bmatrix}$$

satisfies

$$\Big(Y_n^{\mathsf{L}} \exp\Big(\int_0^z C_{\mathsf{L}}'(t)C_{\mathsf{L}}^{-1}(t)\,dt\Big)\Big)' = M_n^{\mathsf{L}}\Big(Y_n^{\mathsf{L}} \exp\Big(\int_0^z C_{\mathsf{L}}'(t)C_{\mathsf{L}}^{-1}(t)\,dt\Big)\Big)$$

$$\Big(\exp\Big(\int_0^z C_{\mathsf{R}}'(t)C_{\mathsf{R}}^{-1}(t)\,dt\Big)Y_n^{\mathsf{R}}\Big)' = \Big(\exp\Big(\int_0^z C_{\mathsf{R}}'(t)C_{\mathsf{R}}^{-1}(t)\,dt\Big)Y_n^{\mathsf{R}}\Big)M_n^{\mathsf{R}}$$

with $C_{\mathsf{L}}'(t)C_{\mathsf{L}}^{-1}(t) = \begin{bmatrix} h^{\mathsf{L}} & 0_N \\ 0_N & -h^{\mathsf{R}} \end{bmatrix}$, $C_{\mathsf{R}}^{-1}(t)C_{\mathsf{R}}'(t) = \begin{bmatrix} h^{\mathsf{R}} & 0_N \\ 0_N & -h^{\mathsf{L}} \end{bmatrix}$, where M_n^{L} and M_n^{R} are defined by

$$(T_n^{\mathsf{L}})' = M_{n+1}^{\mathsf{L}}T_n^{\mathsf{L}} - T_n^{\mathsf{L}}M_n^{\mathsf{L}}, \qquad (T_n^{\mathsf{R}})' = T_n^{\mathsf{R}}M_{n+1}^{\mathsf{R}} - M_n^{\mathsf{R}}T_n^{\mathsf{R}}$$

and $\{T_n^{\mathsf{L}}\}$, $\{T_n^{\mathsf{R}}\}$ are sequences of transfer matrices, i.e.

$$Y_{n+1}^{\mathsf{L}} = T_n^{\mathsf{L}}Y_n^{\mathsf{L}}, \qquad Y_{n+1}^{\mathsf{R}} = Y_n^{\mathsf{R}}T_n^{\mathsf{R}}, \tag{1}$$

with

$$T_n^{\mathsf{L}}(z) = \begin{bmatrix} zI_N - \beta_n^{\mathsf{L}} & C_n^{-1} \\ -C_n & 0_N \end{bmatrix}, \qquad T_n^{\mathsf{R}}(z) = \begin{bmatrix} zI_N - \beta_n^{\mathsf{R}} & -C_n \\ C_n^{-1} & 0_N \end{bmatrix}. \tag{2}$$

In this work we extend the Riemann–Hilbert characterization for a more general class of measures, a matrix extension of the classical scalar Laguerre measures, and we find that for this class of measures appears a power logarithmic type singularities at the end point of the support of the measure.

This will open a window for the study of the differential properties of the sequences of matrix polynomials associated to this matrix measures.

2 Riemann–Hilbert Problem for Matrix Biorthogonal Polynomials

2.1 Matrix Biorthogonal Polynomials

Let

$$W = \begin{bmatrix} W^{(1,1)} & \dots & W^{(1,N)} \\ \vdots & \ddots & \vdots \\ W^{(N,1)} & \dots & W^{(N,N)} \end{bmatrix} \in \mathbb{C}^{N \times N}$$

be a $N \times N$ matrix of weights with support on a smooth oriented non self–intersecting curve γ in the complex plane $\mathbb{C}$, i.e. $W^{(j,k)}$ is, for each $j, k \in \{1, \dots, N\}$, a complex weight with support on γ. We define the *moment of order* n associated with W as

$$W_n = \frac{1}{2\pi\,\mathrm{i}} \int_\gamma z^n W(z)\,\mathrm{d}\,z, \qquad n \in \mathbb{N} := \{0, 1, 2, \dots\}.$$

We say that W is *regular* if $\det\left[W_{j+k}\right]_{j,k=0,\dots,n} \neq 0$, $n \in \mathbb{N}$. In this way, we define a *sequence of matrix monic polynomials*, $\left\{P_n^{\mathsf{L}}(z)\right\}_{n\in\mathbb{N}}$, *left orthogonal* and *right orthogonal*, $\left\{P_n^{\mathsf{R}}(z)\right\}_{n\in\mathbb{N}}$ with respect to a regular matrix measure W, by the conditions,

$$\frac{1}{2\pi\,\mathrm{i}} \int_\gamma P_n^{\mathsf{L}}(z) W(z) z^k\,\mathrm{d}\,z = \delta_{n,k} C_n^{-1}, \tag{3}$$

$$\frac{1}{2\pi\,\mathrm{i}} \int_\gamma z^k W(z) P_n^{\mathsf{R}}(z)\,\mathrm{d}\,z = \delta_{n,k} C_n^{-1}, \tag{4}$$

for $k = 0, 1, \dots, n$ and $n \in \mathbb{N}$, where C_n is an nonsingular matrix.

Notice that neither the matrix of weights is requested to be Hermitian nor the curve γ to be the real line, i.e., we are dealing, in principle with nonstandard orthogonality and, consequently, with biorthogonal matrix polynomials instead of orthogonal matrix polynomials.

The matrix of weights induce a sesquilinear form in the set of matrix polynomials $\mathbb{C}^{N\times N}[z]$ given by

$$\langle P, Q \rangle_W := \frac{1}{2\pi\,\mathrm{i}} \int_\gamma P(z) W(z) Q(z)\,\mathrm{d}\,z. \tag{5}$$

Moreover, we say that $\{P_n^{\mathsf{L}}(z)\}_{n\in\mathbb{N}}$ and $\{P_n^{\mathsf{R}}(z)\}_{n\in\mathbb{N}}$ are biorthogonal with respect to a matrix weight functions W if

$$\langle P_n^{\mathsf{L}}, P_m^{\mathsf{R}}\rangle_W = \delta_{n,m} C_n^{-1}, \qquad n, m \in \mathbb{N}. \tag{6}$$

As the polynomials are chosen to be monic, we can write

$$P_n^{\mathsf{L}}(z) = I_N z^n + p_{\mathsf{L},n}^1 z^{n-1} + p_{\mathsf{L},n}^2 z^{n-2} + \cdots + p_{\mathsf{L},n}^n,$$

$$P_n^{\mathsf{R}}(z) = I_N z^n + p_{\mathsf{R},n}^1 z^{n-1} + p_{\mathsf{R},n}^2 z^{n-2} + \cdots + p_{\mathsf{R},n}^n,$$

with matrix coefficients $p_{\mathsf{L},n}^k, p_{\mathsf{R},n}^k \in \mathbb{C}^{N\times N}$, $k = 0, \ldots, n$ and $n \in \mathbb{N}$ (imposing that $p_{\mathsf{L},n}^0 = p_{\mathsf{R},n}^0 = I$, $n \in \mathbb{N}$). Here $I \in \mathbb{C}^{N\times N}$ denotes the identity matrix.

We define the *sequence of second kind matrix functions* by

$$Q_n^{\mathsf{L}}(z) := \frac{1}{2\pi\,\mathrm{i}} \int_\gamma \frac{P_n^{\mathsf{L}}(z')}{z'-z} W(z')\,\mathrm{d}z', \tag{7}$$

$$Q_n^{\mathsf{R}}(z) := \frac{1}{2\pi\,\mathrm{i}} \int_\gamma W(z') \frac{P_n^{\mathsf{R}}(z')}{z'-z}\,\mathrm{d}z', \tag{8}$$

for $n \in \mathbb{N}$. From the orthogonality conditions (3) and (4) we have, for all $n \in \mathbb{N}$, the following asymptotic expansion near infinity for the sequence of functions of the second kind

$$Q_n^{\mathsf{L}}(z) = -C_n^{-1}\big(I_N z^{-n-1} + q_{\mathsf{L},n}^1 z^{-n-2} + \cdots\big), \tag{9}$$

$$Q_n^{\mathsf{R}}(z) = -\big(I_N z^{-n-1} + q_{\mathsf{R},n}^1 z^{-n-2} + \cdots\big) C_n^{-1}. \tag{10}$$

Assuming that the measures $W^{(j,k)}$, $j, k \in \{1, \ldots, N\}$ are Hölder continuous we obtain, by the Plemelj's formula applied to (7) and (8), the following fundamental jump identities

$$\big(Q_n^{\mathsf{L}}(z)\big)_+ - \big(Q_n(z)^{\mathsf{L}}\big)_- = P_n^{\mathsf{L}}(z) W(z), \tag{11}$$

$$\big(Q_n^{\mathsf{R}}(z)\big)_+ - \big(Q_n^{\mathsf{R}}(z)\big)_- = W(z) P_n^{\mathsf{R}}(z), \tag{12}$$

$z \in \gamma$, where, $\big(f(z)\big)_\pm = \lim_{\epsilon\to 0^\pm} f(z + i\epsilon)$; here $\pm$ indicates the positive/negative region according to the orientation of the curve γ.

2.2 Reductions: From Biorthogonality to Orthogonality

We consider two possible reductions for the matrix of weights, the symmetric reduction and the Hermitian reduction.

(i) A matrix of weights $W(z)$ with support on γ is said to be symmetric if

$$(W(z))^{\top} = W(z), \qquad z \in \gamma.$$

(ii) A matrix of weights $W(x)$ with support on $\mathbb{R}$ is said to be Hermitian if

$$(W(x))^{\dagger} = W(x), \qquad x \in \mathbb{R}.$$

These two reductions leads to orthogonal polynomials, as the two biorthogonal families are identified; i.e., for the symmetric case

$$P_n^{\mathsf{R}}(z) = \big(P_n^{\mathsf{L}}(z)\big)^{\top}, \qquad Q_n^{\mathsf{R}}(z) = \big(Q_n^{\mathsf{L}}(z)\big)^{\top}, \qquad z \in \mathbb{C},$$

and for the Hermitian case, with $\gamma = \mathbb{R}$,

$$P_n^{\mathsf{R}}(z) = \big(P_n^{\mathsf{L}}(\bar{z})\big)^{\dagger}, \qquad Q_n^{\mathsf{R}}(z) = \big(Q_n^{\mathsf{L}}(\bar{z})\big)^{\dagger}, \qquad z \in \mathbb{C}.$$

In both cases biorthogonality collapses into orthogonality, that for the symmetric case reads as

$$\frac{1}{2\pi\,\mathrm{i}}\int_{\gamma} P_n(z)W(z)\big(P_m(z)\big)^{\top}\,\mathrm{d}z = \delta_{n,m}C_n^{-1}, \qquad n, m \in \mathbb{N},$$

while for the Hermitian case can be written as follows

$$\frac{1}{2\pi\,\mathrm{i}}\int_{\mathbb{R}} P_n(x)W(x)\big(P_m(x)\big)^{\dagger}\,\mathrm{d}x = \delta_{n,m}C_n^{-1}, \qquad n, m \in \mathbb{N},$$

where $P_n = P_n^{\mathsf{L}}$.

2.3 The Riemann–Hilbert Problem

Let us consider the particular case when the $N \times N$ matrix of weights with support on a smooth oriented non self–intersecting curve γ has entrywise power logarithmic type singularities at the end points of the support of the measure, that is the entries $W^{(j,k)}$ of the matrix measure W can be described as

$$W^{j,k}(z) = \sum_{m \in I_{j,k}} h_m(z)(z-c)^{\alpha_m}\log^{p_m}(z),$$

where $I_{j,k}$ denotes a finite set of indexes, $\alpha_m > -1$, $p_m \in \mathbb{N}$ and $h_m(x)$ is Hölder continuous, bounded and non–vanishing on γ.

The biorthogonality can be characterized in terms of a left and right Riemann–Hilbert formulation,

Theorem 1

(i) The matrix function

$$Y_n^{\mathsf{L}}(z) := \begin{bmatrix} P_n^{\mathsf{L}}(z) & Q_n^{\mathsf{L}}(z) \\ -C_{n-1}P_{n-1}^{\mathsf{L}}(z) & -C_{n-1}Q_{n-1}^{\mathsf{L}}(z) \end{bmatrix}$$

is, for each $n \in \mathbb{N}$, the unique solution of the Riemann–Hilbert problem; which consists in the determination of a $2N \times 2N$ complex matrix function such that:

(RHL1): $Y_n^{\mathsf{L}}(z)$ *is holomorphic in* $\mathbb{C} \setminus \gamma$*;*
(RHL2): *has the following asymptotic behaviour near infinity,*

$$Y_n^{\mathsf{L}}(z) = \Big(I_N + \sum_{j=1}^{\infty}(z^{-j})Y_n^{j,\mathsf{L}}\Big)\begin{bmatrix} I_N z^n & 0_N \\ 0_N & I_N z^{-n} \end{bmatrix};$$

(RHL3): *satisfies the jump condition*

$$\big(Y_n^{\mathsf{L}}(z)\big)_+ = \big(Y_n^{\mathsf{L}}(z)\big)_- \begin{bmatrix} I_N & W(z) \\ 0_N & I_N \end{bmatrix}, \qquad z \in \gamma.$$

(RHL4): $$Y_n^{\mathsf{L}}(z) = \begin{bmatrix} \mathrm{O}(1) & \mathrm{O}(s_1^{\mathsf{L}}(z)) \\ \mathrm{O}(1) & \mathrm{O}(s_2^{\mathsf{L}}(z)) \end{bmatrix}, \quad as \quad z \to c,$$

where c denotes any of the end points of the curve γ if they exists,

$$\lim_{z\to c}(z-c)s_j^{\mathsf{L}}(z) = 0_N, \qquad j = 1, 2,$$

and the O *conditions are understood entrywise.*

(iii) The matrix function

$$Y_n^{\mathsf{R}}(z) := \begin{bmatrix} P_n^{\mathsf{R}}(z) & -P_{n-1}^{\mathsf{R}}(z)C_{n-1} \\ Q_n^{\mathsf{R}}(z) & -Q_{n-1}^{\mathsf{R}}(z)C_{n-1} \end{bmatrix}$$

is, for each $n \in \mathbb{N}$, the unique solution of the Riemann–Hilbert problem; which consists in the determination of a $2N \times 2N$ complex matrix function such that:

(RHR1): $Y_n^{\mathsf{R}}(z)$ *is holomorphic in* $\mathbb{C} \setminus \gamma$*;*

(RHR2): *has the following asymptotic behaviour near infinity,*

$$Y_n^{\mathsf{R}}(z) = \begin{bmatrix} I_N z^n & 0_N \\ 0_N & I_N z^{-n} \end{bmatrix} \big(I_N + \sum_{j=1}^{\infty} (z^{-j}) Y_n^{j,\mathsf{R}}\big);$$

(RHR3): *satisfies the jump condition*

$$\big(Y_n^{\mathsf{R}}(z)\big)_+ = \begin{bmatrix} I_N & 0_N \\ W(z) & I_N \end{bmatrix} \big(Y_n^{\mathsf{R}}(z)\big)_-, \qquad z \in \gamma.$$

(RHR4): $Y_n^{\mathsf{R}}(z) = \begin{bmatrix} \mathrm{O}(1) & \mathrm{O}(1) \\ \mathrm{O}(s_1^{\mathsf{R}}(z)) & \mathrm{O}(s_2^{\mathsf{R}}(z)) \end{bmatrix}$, *as* $z \to c$,

where c denotes any of the end points of the curve γ *if they exists,*

$$\lim_{z\to c} (z-c) s_j^{\mathsf{R}}(z) = 0_N, \qquad j = 1, 2,$$

and the O *conditions are understood entrywise.*

(ii) The determinant of $Y_n^{\mathsf{L}}(z)$ *and* $Y_n^{\mathsf{L}}(z)$ *are both equal to* 1, *for every* $z \in \mathbb{C}$.

Proof Using the standard calculations from the scalar case it follows that the matrices Y_n^{L} and Y_n^{R} satisfy (RHL1) – (RHL3) and (RHR1) – (RHR3) respectively. The entries $W^{(j,k)}$ of the matrix measure W can be described as

$$W^{j,k}(z) = \sum_{m \in I_{j,k}} h_m(z)(z-c)^{\alpha_m} \log^{p_m}(z),$$

where $I_{j,k}$ denotes a finite set of indexes, $\alpha_m > -1$, $p_m \in \mathbb{N}$ and $h_m(x)$ is Hölder continuous, bounded and non–vanishing on γ. At the boundary values of the curve γ if they exist and are denoted by c, for $z \to c$. It holds [32] that in a neighbourhood of the point c, the Cauchy transform of the function

$$\phi_m(z) = \frac{1}{2\pi i} \int_\gamma \frac{p(z') h_m(z') (z'-c)^{\alpha_m} \log^{p_m}(z')}{z'-z} dz',$$

where $p(z')$ denotes any polynomial in z', verifies

$$\lim_{z\to c} (z-c)\phi_m(z) = 0,$$

and the condition (RHL4), is fulfilled for the matrix Y_n^{L} and respectively the condition (RHR4), is fulfilled for the matrix Y_n^{R}. Now let us consider

$$G(z) = Y_n^{\mathsf{L}}(z) \begin{bmatrix} 0_N & I_N \\ -I_N & 0_N \end{bmatrix} Y_n^{\mathsf{R}}(z) \begin{bmatrix} 0_N & -I_N \\ I_N & 0_N \end{bmatrix}.$$

It can easily be proved that G has no jump on the curve γ. In a neighborhood of the point c

$$G(z) = \begin{bmatrix} \mathrm{O}(s_1^{\mathsf{L}}(z)) + \mathrm{O}(s_2^{\mathsf{R}}(z)) & \mathrm{O}(s_1^{\mathsf{L}}(z)) + \mathrm{O}(s_1^{\mathsf{R}}(z)) \\ \mathrm{O}(s_2^{\mathsf{L}}(z)) + \mathrm{O}(s_2^{\mathsf{R}}(z)) & \mathrm{O}(s_2^{\mathsf{L}}(z)) + \mathrm{O}(s_1^{\mathsf{R}}(z)) \end{bmatrix},$$

so $\lim_{z\to c}(z-c)G(z) = 0$, and at the point c the singularity is removable. Now using the behaviour for $z \to \infty$,

$$G(z) = \begin{bmatrix} I_N z^n & 0_N \\ 0_N & I_N z^{-n} \end{bmatrix} \begin{bmatrix} 0_N & I_N \\ -I_N & 0_N \end{bmatrix} \begin{bmatrix} I_N z^n & 0_N \\ 0_N & I_N z^{-n} \end{bmatrix} \begin{bmatrix} 0_N & -I_N \\ I_N & 0_N \end{bmatrix} = \begin{bmatrix} I_N & 0_N \\ 0_N & I_N \end{bmatrix},$$

and using Liouville's Theorem it holds that $G(z) = I$, the identity matrix. From this follows the unicity of the solution of each of the Riemann–Hilbert problems stated in this theorem.

Again using the standard arguments as in the scalar case we can conclude that $\det Y_n^{\mathsf{L}}(z)$ and $\det Y_n^{\mathsf{R}}(z)$ are both equal to 1. □

We recover a representation for the inverse matrix $\left(Y_n^{\mathsf{L}}\right)^{-1}$ given by the following result

Corollary 1 *It holds that*

$$\left(Y_n^{\mathsf{L}}\right)^{-1}(z) = \begin{bmatrix} 0_N & I_N \\ -I_N & 0_N \end{bmatrix} Y_n^{\mathsf{R}}(z) \begin{bmatrix} 0_N & -I_N \\ I_N & 0_N \end{bmatrix}. \tag{13}$$

Corollary 2 *In the conditions of Theorem 1 we have that for all $n \in \mathbb{N}$,*

$$Q_n^{\mathsf{L}}(z)P_{n-1}^{\mathsf{R}}(z) - P_n^{\mathsf{L}}(z)Q_{n-1}^{\mathsf{R}}(z) = C_{n-1}^{-1}, \tag{14}$$

$$P_{n-1}^{\mathsf{L}}(z)Q_n^{\mathsf{R}}(z) - Q_{n-1}^{\mathsf{L}}(z)P_n^{\mathsf{R}}(z) = C_{n-1}^{-1}, \tag{15}$$

$$Q_n^{\mathsf{L}}(z)P_n^{\mathsf{R}}(z) - P_n^{\mathsf{L}}(z)Q_n^{\mathsf{R}}(z) = 0. \tag{16}$$

Proof As we have already proven the matrix

$$\begin{bmatrix} -Q_{n-1}^{\mathsf{R}}(z)C_{n-1} & -Q_n^{\mathsf{R}}(z) \\ P_{n-1}^{\mathsf{R}}(z)C_{n-1} & P_n^{\mathsf{R}}(z) \end{bmatrix},$$

is the inverse of $Y_n^{\mathsf{L}}(z)$, i.e.

$$Y_n^{\mathsf{L}}(z) \begin{bmatrix} -Q_{n-1}^{\mathsf{R}}(z)C_{n-1} & -Q_n^{\mathsf{R}}(z) \\ P_{n-1}^{\mathsf{R}}(z)C_{n-1} & P_n^{\mathsf{R}}(z) \end{bmatrix} = I;$$

and multiplying the two matrices we get the result. □

2.4 Three Term Recurrence Relation

Following the standard arguments from the Riemann–Hilbert formulation we can prove

$$Y^{\mathsf{L}}_{n+1}(z) = T^{\mathsf{L}}_n(z) Y^{\mathsf{L}}_n(z), \qquad n \in \mathbb{N},$$

where T^{L}_n is given in (2). For the right orthogonality, we similarly obtain from (1) that

$$Y^{\mathsf{R}}_{n+1}(z) = Y^{\mathsf{R}}_n(z) T^{\mathsf{R}}_n(z), \qquad n \in \mathbb{N},$$

where T^{R}_n is given in (2). Hence, we conclude that the sequence of monic polynomials $\{P^{\mathsf{L}}_n(z)\}_{n\in\mathbb{N}}$ satisfies the three term recurrence relations

$$zP^{\mathsf{L}}_n(z) = P^{\mathsf{L}}_{n+1}(z) + \beta^{\mathsf{L}}_n P^{\mathsf{L}}_n(z) + \gamma^{\mathsf{L}}_n P^{\mathsf{L}}_{n-1}(z), \qquad n \in \mathbb{N}, \tag{17}$$

with recursion coefficients $\beta^{\mathsf{L}}_n := p^1_{\mathsf{L},n} - p^1_{\mathsf{L},n+1}$, $\gamma^{\mathsf{L}}_n := C^{-1}_n C_{n-1}$, with initial conditions, $P^{\mathsf{L}}_{-1} = 0_N$ and $P^{\mathsf{L}}_0 = I_N$. We can also assert that

$$zP^{\mathsf{R}}_n(z) = P^{\mathsf{R}}_{n+1}(z) + P^{\mathsf{R}}_n(z)\beta^{\mathsf{R}}_n + P^{\mathsf{R}}_{n-1}(z)\gamma^{\mathsf{R}}_n, \qquad n \in \mathbb{N}, \tag{18}$$

where $\beta^{\mathsf{R}}_n := C_n \beta^{\mathsf{L}}_n C^{-1}_n$, $\gamma^{\mathsf{R}}_n := C_n \gamma^{\mathsf{L}}_n C^{-1}_n = C_{n-1} C^{-1}_n$.

3 Matrix Weights Supported on a Curve γ on the Complex Plane that Connects the Point 0 to the Point ∞: Laguerre Weights

Motivated by different attempts that appear in the literature we try to consider some classes of weights with the aim to use the Riemann–Hilbert formulation. In this matrix case it is not so obvious which are the conditions we should to impose in order to guarantee the integrability of the matrix measure that we want to consider.

3.1 *Matrix Weights Supported on a Curve γ with One Finite End Point: $W(z) = z^A H(z)$*

We begin considering the weight $W(z) = z^A H(z)$ supported on a curve γ on the complex plane that connects the point 0 to the point ∞, where

(i) The function z^A is defined as $z^A = e^{A\log z}$, where γ is the branch cut of the logarithmic function, and we define por $t \in \gamma$, the $t^A := (z^A)_+$, where $(z^A)_+$ is the non–tangential limit as $z \to t$, from the left side of the oriented curve γ.
(ii) The constant matrix A is such that the minimum of the real part of its eigenvalues is greater than -1.
(iii) The factor $H(t)$ is the restriction to the curve γ of $H(z)$, a matrix of entire functions, $z \in \mathbb{C}$ such that $H(z)$ is invertible for all $z \in \mathbb{C}$.
(iv) The left logarithmic derivative $h(z) := \big(H(z)\big)^{-1}\big(H(z)\big)'$ is an entire function.

It is necessary, in other to consider the Riemann–Hilbert problem related to the weight function $W(z)$, to clarify the behaviour of this weight function $W(z)$ on a neighborhood of the point $z = 0$.

If we consider the Jordan decomposition of the matrix A, it holds that there exists an invertible matrix P such that $A = PJP^{-1}$, where $J = D + N$, where D is the diagonal matrix formed whose entries are the eigenvalues of the matrix A and N is a nilpotent matrix that commutes with the matrix D. This commutation enables us to obtain

$$z^A = z^{PJP^{-1}} = Pz^J P^{-1} = Pz^D z^N P^{-1}$$

where z^N is a polynomial in the variable $\log z$. The matrix z^D is a diagonal matrix whose entries are of the form $z^{\alpha_j + i\beta_j}$, where $\alpha_j + i\beta_j$ is a eigenvalue of the matrix A. Let us consider as an example the matrix

$$A = \begin{bmatrix} -\frac{1}{2} & 1 & 0 \\ 0 & -\frac{1}{2} & 0 \\ 0 & 0 & 1 \end{bmatrix}$$

In this case

$$z^A = \begin{bmatrix} z^{-\frac{1}{2}} & 0 & 0 \\ 0 & z^{-\frac{1}{2}} & 0 \\ 0 & 0 & z \end{bmatrix} \begin{bmatrix} 1 & \log z & 0 \\ 0 & 1 & 0 \\ 0 & 0 & 1 \end{bmatrix}$$

So we observe that the matrix A is not diagonalizable, as it appears a factor in the behaviour near 0, such as $z^{-\frac{1}{2}} \log z$, hence we are in presence of a power logarithmic type singularity. To assure the integrability of this kind of measure it is enough to ask that $\alpha > -1$, where is the minimum of the real part of the eigenvalues of the matrix A. So in this case for this weight we are in the conditions of the Theorem 1.

It is also valuable to comment about the factor $H(z)$ of the measure $W(z)$. In other also to have integrability of this matrix weight function we should be careful. If for example we consider $H(z) = e^{Bz}$, then it is clear that, by reasoning similarly as before, that we should impose that the real part of the eigenvalues of the matrix B are negative. If we consider $h(z) := \big(H(z)\big)^{-1}\big(H(z)\big)'$ to be a matrix polynomial $h(z) = B_0+B_1z+\cdots+B_mz^m$, it should be enough in order to guarantee integrability of the measure, to impose that the real part of the eigenvalues of the matrix B_m to be negative.

3.2 *Matrix Weights Supported on a Curve* γ *with One Finite End Point:* $W(z) = z^{\alpha}H(z)G(z)z^{B}$

In [23] appears different examples of Laguerre matrix weights for the matrix orthogonal polynomials on the real line. This motivates to consider the matrix weight $W(z) = z^{\alpha}H(z)G(z)z^{B}$ supported on a curve γ on the complex plane that connects the point 0 to the point ∞, with similar considerations as in the case treated before. Nevertheless when we try to apply the general methods from the Riemann–Hilbert formulation we find a lot of difficulties, derived from the non–commutativity of the matrix product and we should impose important restrictions.

This kind of matrix weights can be treated in a more general context. Let us consider that instead of a given matrix of weights we are provided with two matrices, say $h^{\mathsf{L}}(z)$ and $h^{\mathsf{R}}(z)$, of entire functions such that the following two matrix Pearson equations are satisfied

$$z\frac{\mathrm{d}\,W^{\mathsf{L}}}{\mathrm{d}\,z} = h^{\mathsf{L}}(z)W^{\mathsf{L}}(z), \qquad z\frac{\mathrm{d}\,W^{\mathsf{R}}}{\mathrm{d}\,z} = W^{\mathsf{R}}(z)h^{\mathsf{R}}(z); \tag{19}$$

and given solutions to them we construct the corresponding matrix of weights $W = W^{\mathsf{L}}W^{\mathsf{R}}$. Moreover, this matrix of weights is also characterized by a Pearson equation.

Proposition 1 (Pearson Sylvester Differential Equation) *Given two matrices of entire functions* $h^{\mathsf{L}}(z)$ *and* $h^{\mathsf{R}}(z)$, *any solution of the Sylvester type matrix differential equation, which we call Pearson equation for the weight,*

$$z\frac{\mathrm{d}\,W}{\mathrm{d}\,z} = h^{\mathsf{L}}(z)W(z) + W(z)h^{\mathsf{R}}(z) \tag{20}$$

is of the form $W = W^{\mathsf{L}}W^{\mathsf{R}}$ *where the factor matrices* W^{L} *and* W^{R} *are solutions of* (19), *respectively.*

Proof Given solutions W^{L} and W^{R} of (19), respectively, it follows intermediately, just using the Leibniz law for derivatives, that $W = W^{\mathsf{L}}W^{\mathsf{R}}$ fulfills (20). Moreover, given a solution W of (20) we pick a solution W^{L} of the first equation in (19), then it is easy to see that $(W^{\mathsf{L}})^{-1}W$ satisfies the second equation in (19). □

Acknowledgments Amílcar Branquinho was partially supported by the Centre for Mathematics of the University of Coimbra–UIDB/00324/2020, funded by the Portuguese Government through FCT/MCTES.

Ana Foulquié acknowledges the Center for Research & Development in Mathematics and Applications through the Portuguese Foundation for Science and Technology (FCT–Fundação para a Ciência e a Tecnologia), references UIDB/04106/2020 and UIDP/04106/2020.

Manuel Mañas Baena thanks economical support from the Spanish Ministerio de Economía y Competitividad research project [MTM2015-65888-C4-2-P], Ortogonalidad, teoría de la aproximación y aplicaciones en física matemática and Spanish Agencia Estatal de Investigación research project [PGC2018-096504-B-C33], Ortogonalidad y Aproximación: Teoría y Aplicaciones en Física Matemática.

References

1. Aptekarev, A.I., Nikishin, E.M.: The scattering problem for a discrete Sturm–Liouville operator. Math. USSR, Sb. **49**, 325–355 (1984)
2. Berezanskii, J.M.: Expansions in Eigenfunctions of Selfadjoint Operators. Translations of Mathematical Monographs, vol. 17. American Mathematical Society, Providence (1968)
3. Branquinho, A., Foulquié Moreno, A., Mañas, M.: Matrix biorthogonal polynomials: eigenvalue problems and non-abelian discrete Painlevé equations, J. Math. Anal. Appl. **494**(2), 124605 (2021)
4. Cassatella–Contra, G.A., Mañas, M.: Riemann–Hilbert problems, matrix orthogonal polynomials and discrete matrix equations with singularity confinement. Stud. Appl. Math. **128**, 252–274 (2011)
5. Cassatella–Contra, G.A., Mañas, M.: Matrix biorthogonal polynomials in the unit circle: Riemann–Hilbert problem and matrix discrete Painleve II system (2016). `arXiv:1601.07236 [math.CA]`
6. Cassatella–Contra, G.A., Mañas, M., Tempesta, P.: Singularity confinement for matrix discrete Painlevé equations. Nonlinearity **27**, 2321–2335 (2014)
7. Daems, E., Kuijlaars, A.B.J.: Multiple orthogonal polynomials of mixed type and non–intersecting Brownian motions. J. Approx. Theory **146**, 91–114 (2007)
8. Daems, E., Kuijlaars, A.B.J., Veys, W.: Asymptotics of non–intersecting Brownian motions and a 4×4 *Riemann–Hilbert problem*. J. Approx. Theory **153**, 225–256 (2008)
9. Dai, D., Kuijlaars, A.B.J.: Painlevé IV asymptotics for orthogonal polynomials with respect to a modified Laguerre weight. Stud. Appl. Math. **122**, 29–83 (2009)
10. Deift, P.A.: Orthogonal Polynomials and Random Matrices: A Riemann–Hilbert Approach. Courant Lecture Notes, vol. 3. American Mathematical Society, Providence (2000)
11. Deift, P.A.: Riemann–Hilbert methods in the theory of orthogonal polynomials. In: Spectral Theory and Mathematical Physics: A Festschrift in Honor of Barry Simon's 60th Birthday. Proceedings of Symposia in Pure Mathematics, vol. 76, pp. 715–740. American Mathematical Society, Providence (2007)
12. Deift, P.A., Gioev, D.: Random Matrix Theory: Invariant Ensembles and Universality. Courant Lecture Notes in Mathematics, vol. 18. American Mathematical Society, Providence (2009)

13. Deift, P.A., Zhou, X.: A steepest descent method for oscillatory Riemann–Hilbert problems. Asymptotics for the MKdV equation. Ann. Math. **137**, 295–368 (1993)
14. Deift, P.A., Zhou, X.: Long–time asymptotics for solutions of the NLS equation with initial data in a weighted Sobolev space. Commun. Pure Appl. Math. **56**, 1029–1077 (2003)
15. Durán, A.J.: A generalization of Favard's theorem for polynomials satisfying a recurrence relation. J. Approx. Theory **74**, 260–275 (1994)
16. Durán, A.J.: On orthogonal polynomials with respect to a positive definite matrix of measures. Can. J. Math. **47**, 88–112 (1995)
17. Durán, A.J.: Markov's theorem for orthogonal matrix polynomials. Can. J. Math. **48**, 1180–1195 (1996)
18. Durán, A.J.: Ratio asymptotics for orthogonal matrix polynomials. J. Approx. Theory **100**, 304–344 (1999)
19. Durán, A.J., Daneri–Vias, E.: Ratio asymptotics for orthogonal matrix polynomials with unbounded recurrence coefficients. J. Approx. Theory **110**, 1–17 (2001)
20. Durán, A.J., Daneri–Vias, E.: Weak convergence for orthogonal matrix polynomials. Indag. Math. **13**, 47–62 (2002)
21. Durán, A.J., Defez, E.: Orthogonal matrix polynomials and quadrature formulas. Linear Algebra Appl. **345**, 71–84 (2002)
22. Durán, A.J., de la Iglesia, M.D.: Second order differential operators having several families of orthogonal matrix polynomials as eigenfunctions. Int. Math. Res. Notices **2008**, 24 (2008)
23. Durán, A.J., Grünbaum, F.A.: Orthogonal matrix polynomials satisfying second order differential equations. Int. Math. Res. Notices **10**, 461–484 (2004)
24. Durán, A.J., Grünbaum, F.J.: Structural formulas for orthogonal matrix polynomials satisfying second order differential equations, I, Constr. Approx. **22**, 255–271 (2005)
25. Durán, A.J., Grünbaum, F.A.: Orthogonal matrix polynomials, scalar–type Rodrigues' formulas and Pearson equations. J. Approx. Theory **134**, 267–280 (2005)
26. Durán, A.J., Grünbaum, F.A.: Structural formulas for orthogonal matrix polynomials satisfying second order differential equations I. Constr. Approx. **22**, 255–271 (2005)
27. Durán, A.J., Lopez–Rodriguez, P.: Orthogonal matrix polynomials: zeros and Blumenthal's theorem. J. Approx. Theory **84**, 96–118 (1996)
28. Durán, A.J., Lopez–Rodriguez, P., Saff, E.B.: Zero asymptotic behaviour for orthogonal matrix polynomials. J. d'Analyse Math. **78**, 37–60 (1999)
29. Durán, A.J., Polo, B.: Gauss quadrature formulae for orthogonal matrix polynomials. Linear Algeb. Appl. **355**, 119–146 (2002)
30. Durán, A.J., Van Assche, W.: Orthogonal matrix polynomials and higher order recurrence relations. Linear Algeb. Appl. **219**, 261–280 (1995)
31. Fokas, A.S., Its, A.R., Kitaev, A.V.: The isomonodromy approach to matrix models in 2D quantum gravity. Commun. Math. Phys. **147**, 395–430 (1992)
32. Gakhov, F.D.: Boundary Value Problems (Translated from the Russian. Reprint of the 1966 translation). Dover Publications, New York (1990)
33. Geronimo, J.S.: Scattering theory and matrix orthogonal polynomials on the real line. Circ. Syst. Signal Proc. **1**, 471–495 (1982)
34. Grünbaum, F.A., de la Iglesia, M.D., Martínez–Finkelshtein, A.: Properties of matrix orthogonal polynomials via their Riemann–Hilbert characterization. SIGMA Symmetry Integr. Geom. Method Appl. **7**, 098 (2011)
35. Its, A.R., Kuijlaars, A.B.J., Östensson, J.: Asymptotics for a special solution of the thirty fourth Painlevé equation. Nonlinearity **22**, 1523–1558 (2009)
36. Krein, M.G.: Infinite J–matrices and a matrix moment problem. Dokl. Akad. Nauk SSSR **69**(2), 125–128 (1949)
37. Krein, M.G.: Fundamental Aspects of the Representation Theory of Hermitian Operators with Deficiency Index (m, m). AMS Translations, Series 2, vol. 97, pp. 75–143. American Mathematical Society, Providence (1971)

38. Kuijlaars, A.B.J.: Multiple orthogonal polynomial ensembles. In: Recent Trends in Orthogonal Polynomials and Approximation Theory. Contemporary Mathematics, vol. 507, pp. 155–176. American Mathematical Society, Providence (2010)
39. Kuijlaars, A.B.J., Martínez–Finkelshtein, A., Wielonsky, F.: Non–intersecting squared Bessel paths and multiple orthogonal polynomials for modified Bessel weights. Commun. Math. Phys. **286**, 217–275 (2009)
40. McLaughlin, K.T.–R., Vartanian, A.H., Zhou, X.: Asymptotics of Laurent polynomials of even degree orthogonal with respect to varying exponential weights. Int. Math. Res. Notices **2006**, 62815 (2006)
41. McLaughlin, K.T.–R., Vartanian, A.H., Zhou, X.: Asymptotics of Laurent polynomials of odd degree orthogonal with respect to varying exponential weights. Constr. Approx. **27**, 149–202 (2008)

The Symmetrization Problem for Multiple Orthogonal Polynomials

Amílcar Branquinho and Edmundo J. Huertas

Abstract We analyze the effect of symmetrization in the theory of multiple orthogonal polynomials. For a symmetric sequence of type II multiple orthogonal polynomials satisfying a high-term recurrence relation, we fully characterize the Weyl function associated to the corresponding block Jacobi matrix as well as the Stieltjes matrix function. Next, from an arbitrary sequence of type II multiple orthogonal polynomials with respect to a set of d linear functionals, we obtain a total of $d + 1$ sequences of type II multiple orthogonal polynomials, which can be used to construct a new sequence of symmetric type II multiple orthogonal polynomials. We also prove a Favard-type result for certain sequences of matrix multiple orthogonal polynomials satisfying a matrix four-term recurrence relation with matrix coefficients. Finally, we provide an example concerning multiple Hermite polynomials.

Keywords Darboux transformations · Multiple orthogonal polynomials · Matrix orthogonal polynomials · Linear functionals · Recurrence relations · Operator theory · Full Kostant-Toda systems · Symmetrization problem

AMS Subject Classification (2000) Primary 33C45; Secondary 39B42

A. Branquinho
CMUC and Department of Mathematics, University of Coimbra (FCTUC), Coimbra, Portugal
e-mail: ajplb@mat.uc.pt

E. J. Huertas (✉)
Departamento de Física y Matemáticas, Universidad de Alcalá, Madrid, Spain
e-mail: edmundo.huertas@uah.es

F. Marcellán, E. J. Huertas (eds.), *Orthogonal Polynomials: Current Trends and Applications*, SEMA SIMAI Springer Series 22,
https://doi.org/10.1007/978-3-030-56190-1_2

1 Introduction

In recent years an increasing attention has been paid to the notion of multiple orthogonality (see, for instance [1]). Multiple orthogonal polynomials are a generalization of orthogonal polynomials [13], satisfying orthogonality conditions with respect to a number of measures, instead of just one measure. There exists a vast literature on this subject, e.g. the classical works [2, 4], [24, Ch. 23] and [29] among others. A characterization through a vector functional equation, where the authors call them *d-orthogonal polynomials* instead of multiple orthogonal polynomials, was done in [17]. Their asymptotic behavior have been studied in [5], also continued in [14], and properties for their zeros have been analyzed in [22].

The Bäcklund transformation approach is one of the most effective methods for constructing explicit solutions of partial differential equations which are called integrable systems and play important roles in mechanics, physics and differential geometry. Roughly speaking, it consists in generating new solutions for a given partial differential equation by starting with a "seed" solution to the same (or a different) PDE and solving an auxiliary system of ODEs. Bäcklund transformations resulted from the symmetrization process in the usual (standard) orthogonality, which allows one to jump, from one hierarchy to another, in the whole Toda lattices hierarchy (see [21]). That is, they allow reinterpretations inside the hierarchy. In [9], the authors have found certain Bäcklund-type transformations (also known as Miura-type transformations) which allow to reduce problems in a given full Konstant-Toda hierarchy to another. Also, in [6], where Peherstorfer's work [30] is extended to the whole Toda hierarchy, it is shown how this system can be described with the evolution of only one parameter instead of two, using exactly this kind of transformations. Other application to the Toda systems appear in [3, 7], and [8], where the authors studied Bogoyavlenskii systems which were modeled by certain symmetric multiple orthogonal polynomials.

The symmetrization of linear functionals concerning just one variable is well known in the literature [13, I.8, p. 40]. Broadly speaking, given a quasi-definite, symmetric linear functional $\mathcal{S}$ defined in the linear space of polynomials with real coefficients $\mathbb{P}$, and the linear functional $\mathcal{L}$ defined by $\mathcal{L}[x^n] := \mathcal{S}[x^{2n}]$, if $\{S_n\}_{n\geq 0}$ is the monic orthogonal polynomial sequence associated to $\mathcal{S}$, then one has $S_n(-x) = (-1)^n S_n(x)$. Moreover, there exist two monic polynomial sequences $\{P_n\}_{n\geq 0}$ and $\{Q_n\}_{n\geq 0}$ such that

$$S_{2m}(x) = P_m(x), \quad S_{2m+1}(x) = xQ_m(x^2),$$

satisfying

$$\mathcal{L}[P_m(x)P_n(x)] = \mathcal{S}[S_{2m}(x)S_{2n}(x)]$$

and

$$\mathcal{L}^*[Q_m(x)Q_n(x)] = \mathcal{L}[xP_m(x)P_n(x)] = \mathcal{S}[S_{2m+1}(x)S_{2n+1}(x)].$$

For every polynomial $\pi(x) \in \mathbb{P}$, and any real or complex number κ, the linear functional $\mathcal{L}^*$ is defined by

$$\mathcal{L}_\kappa^*[\pi(x)] := \mathcal{L}[(x-\kappa)\pi(x)],$$

and $\{Q_n\}_{n\geq 0}$ is the so called monic kernel orthogonal polynomial sequence of parameter κ, associated to $\{P_n\}_{n\geq 0}$. Our aim here is to extend the above ideas to the framework of multiple orthogonality, i.e., in this work we analyze the effect of symmetrization in systems of multiple orthogonality measures. Our viewpoint seeds some new light on the subject, and we prove that the symmetrization process in multiple orthogonality is a model to define the aforementioned Bäcklund-type transformations, as happens in the scalar case with the Bäcklund transformations (see [13, 26, 30]). Furthermore, we solve the so called *symmetrization problem* in the theory of multiple orthogonal polynomials. We apply certain *Darboux transformations*, already described in [9], to a $(d+2)$-banded matrix, associated to a $(d+2)$-term recurrence relation satisfied by an arbitrary sequence of type II multiple orthogonal polynomials, to obtain a total of $d+1$ sequences of not necessarily symmetric multiple orthogonal polynomials, which we use to construct a new sequence of symmetric multiple orthogonal polynomials.

On the other hand, following the ideas in [26] (and the references therein) for standard sequences of orthogonal polynomials, in [28] (see also [16]) the authors provide a cubic decomposition for sequences of polynomials, multiple orthogonal with respect to a two different linear functionals. Concerning the symmetric case, in [27] this cubic decomposition is analyzed for a 2-symmetric sequence of polynomials, which is called a *diagonal cubic decomposition (CD)* by the authors. Here, we also extend this notion of diagonal decomposition to a more general case, considering symmetric sequences of polynomials multiple orthogonal with respect to $d > 3$ linear functionals.

The structure of the manuscript is as follows. In Sect. 2 we summarize without proofs the relevant material about multiple orthogonal polynomials, and a basic background about the matrix interpretation of the type II multi-orthogonality conditions with respect to the a regular system of d linear functionals $\{u^1, \ldots, u^d\}$ and diagonal multi-indices. In Sect. 3 we fully characterize the Weyl function $\mathcal{R}_J$ and the Stieltjes matrix function $\mathcal{F}$ associated to the block Jacobi matrix J corresponding to a $(d+2)$-term recurrence relation satisfied by a symmetric sequence of type II multiple orthogonal polynomials. In Sect. 4, starting from an arbitrary sequence of type II multiple polynomials satisfying a $(d+2)$-term recurrence relation, we state the conditions to find a total of $d+1$ sequences of type II multiple orthogonal polynomials, in general non-symmetric, which can be used to construct a new sequence of *symmetric* type II multiple orthogonal polynomials. Moreover, we also deal with the converse problem, i.e., we propose a decomposition of a given *symmetric* type II multiple orthogonal polynomial sequence, which allows us to find a set of other (in general non-symmetric) $d+1$ sequences of type II multiple orthogonal polynomials, satisfying in turn $(d+2)$-term recurrence relations. Section 5 is devoted to present a Favard-type result, showing that certain 3×3 matrix decomposition of a type II multiple 2-orthogonal polynomials, satisfy a *matrix four-term recurrence relation*, and therefore it is type II multiple 2-orthogonal (in a matrix sense) with respect to a certain system of matrix measures. Finally, in Sect. 6, we provide an example of the methods developed throughout the paper. We define a particular sequence of type II multiple Hermite orthogonal polynomials and we construct their corresponding matrix type II multiple 2-orthogonal polynomial sequence satisfying a *matrix four-term recurrence relation*. Furthermore, following the techniques of Sect. 4, we construct the corresponding sequence of *symmetric* type II multiple orthogonal polynomials.

2 Definitions and Matrix Interpretation of Multiple Orthogonality

Let $\boldsymbol{n} = (n_1, \ldots, n_d) \in \mathbb{N}^d$ be a multi-index with length $|\mathbf{n}| := n_1 + \cdots + n_d$ and let $\{u^j\}_{j=1}^d$ be a set of linear functionals, i.e. $u^j : \mathbb{P} \to \mathbb{C}$. Let $\{P_{\mathbf{n}}\}$ be a sequence of polynomials, with deg $P_{\mathbf{n}}$ is at most $|\mathbf{n}|$. $\{P_{\mathbf{n}}\}$ is said to be *type II multiple orthogonal* with respect to the set of linear functionals $\{u_j\}_{j=1}^d$ and multi-index $\boldsymbol{n}$ if

$$u^j(x^k P_{\mathbf{n}}) = 0\,, \quad k = 0, 1, \ldots, n_j - 1\,, \quad j = 1, \ldots, d\,. \tag{1}$$

A multi-index $\boldsymbol{n}$ is said to be *normal* for the set of linear functionals $\{u_j\}_{j=1}^d$, if the degree of $P_{\mathbf{n}}$ is exactly $|\mathbf{n}| = n$. When all the multi-indices of a given family are normal, we say that the set of linear functionals $\{u_j\}_{j=1}^d$ is *regular*. In the present work, we will restrict our attention to the so called *diagonal multi-indices* $\boldsymbol{n} = (n_1, \ldots, n_d) \in \mathcal{I}$, where

$$\mathcal{I} = \{(0, 0, \ldots, 0), (1, 0, \ldots, 0), \ldots, (1, 1, \ldots, 1), (2, 1, \ldots, 1), \ldots, (2, 2, \ldots, 2), \ldots\}.$$

Notice that there exists a one to one correspondence, $\mathbf{i}$, between the above set of diagonal multi-indices $\mathcal{I} \subset \mathbb{N}^d$ and $\mathbb{N}$, given by $\mathbf{i}(\mathbb{N}^d) = |\mathbf{n}| = n$. Therefore, to simplify the notation, we write in the sequel $P_{\mathbf{n}} \equiv P_{|\mathbf{n}|} = P_n$. The left-multiplication of a linear functional $u : \mathbb{P} \to \mathbb{C}$ by a polynomial $p \in \mathbb{P}$ is given by the new linear functional $p\,u : \mathbb{P} \to \mathbb{C}$ such that

$$p\,u(x^k) = u(p(x)x^k)\,, \quad k \in \mathbb{N}\,.$$

Next, we briefly review a matrix interpretation of type II multiple orthogonal polynomials with respect to a system of d regular linear functionals and a family of diagonal multi-indices. Throughout this work, we will use this matrix interpretation as a useful tool to obtain some of the main results of the manuscript. For a recent and deeper account of the theory (in a more general framework, considering quasi-diagonal multi-indices) we refer the reader to [11].

Let us consider the family of vector polynomials

$$\mathbb{P}^d = \{\left[P_1 \cdots P_d\right]^T, \ d \in \mathbb{N}, \ P_j \in \mathbb{P}\},$$

and $\mathcal{M}_{d\times d}$ the set of $d \times d$ matrices with entries in $\mathbb{C}$. Let $\{\mathcal{X}_j\}$ be the family of vector polynomials $\mathcal{X}_j \in \mathbb{P}^d$ defined by

$$\mathcal{X}_j = \left[x^{jd} \ldots x^{(j+1)d-1}\right]^T, \ j \in \mathbb{N}, \tag{2}$$

where $\mathcal{X}_0 = \left[1 \cdots x^{d-1}\right]^T$. By means of the shift $n \to nd$, associated with $\{P_n\}$, we define the sequence of vector polynomials $\{\mathcal{P}_n\}$, with

$$\mathcal{P}_n = \left[P_{nd}(x) \cdots P_{(n+1)d-1}(x)\right]^T, \ n \in \mathbb{N}, \ \mathcal{P}_n \in \mathbb{P}^d. \tag{3}$$

Let $u^j : \mathbb{P} \to \mathbb{C}$ with $j = 1, \ldots, d$ a system of linear functionals as in (1). From now on, we define the *vector of functionals* $\mathcal{U} = \left[u^1 \cdots u^d\right]^T$ acting in $\mathbb{P}^d \to \mathcal{M}_{d\times d}$, by

$$\mathcal{U}(\mathcal{P}) = \left(\mathcal{U}\cdot\mathcal{P}^T\right)^T = \begin{bmatrix} u^1(P_1) & \cdots & u^d(P_1) \\ \vdots & \ddots & \vdots \\ u^1(P_d) & \cdots & u^d(P_d) \end{bmatrix},$$

where "·" means the symbolic product of vectors $\mathcal{U}$ and $\mathcal{P}^T$. Let

$$A_l(x) = \sum_{k=0}^{l} A_k^l x^k,$$

be a matrix polynomial of degree l, where $A_k^l \in \mathcal{M}_{d\times d}$, and $\mathcal{U}$ a vector of functional. We define the new vector of functionals called *left multiplication of $\mathcal{U}$ by a matrix polynomial A_l*, and we denote it by $A_l\mathcal{U}$, to the map of $\mathbb{P}^d$ into $\mathcal{M}_{d\times d}$, described by

$$\left(A_l\mathcal{U}\right)(\mathcal{P}) = \sum_{k=0}^{l} \left(x^k\mathcal{U}\right)(\mathcal{P})\,(A_k^l)^T. \tag{4}$$

From the above definition we introduce the notion of *moments of order* $j \in \mathbb{N}$, associated with the vector of functionals $x^k\mathcal{U}$, which will be in general the following

$d \times d$ matrices

$$\mathcal{U}_j^k = \left(x^k \mathcal{U}\right)(\mathcal{X}_j) = \begin{bmatrix} u^1(x^{jd+k}) & \cdots & u^d(x^{jd+k}) \\ \vdots & \ddots & \vdots \\ u^1(x^{(j+1)d-1+k}) & \cdots & u^d(x^{(j+1)d-1+k}) \end{bmatrix},$$

with $j, k \in \mathbb{N}$, and from this moments, we construct the *block Hankel matrix of moments*

$$\mathcal{H}_n = \begin{bmatrix} \mathcal{U}_0^0 & \cdots & \mathcal{U}_0^n \\ \vdots & \ddots & \vdots \\ \mathcal{U}_n^0 & \cdots & \mathcal{U}_n^n \end{bmatrix}, \quad n \in \mathbb{N}.$$

We say that the vector of functionals $\mathcal{U}$ is *regular*, if the determinants of the principal minors of the above matrix are non-zero for every $n \in \mathbb{N}$. Notice that the two notions of regularity, that for the set of entries of $\{u_j\}_{j=1}^d$ and for $\mathcal{U}$ itself, are suitably compatibilized in [11, Th. 4]. Having in mind (2) it is obvious that $\mathcal{X}_j = (x^d)^j \mathcal{X}_0, \quad j \in \mathbb{N}$. Thus, from (3) we can express $\mathcal{P}_n(x)$ in the alternative way

$$\mathcal{P}_n(x) = \sum_{j=0}^{n} P_j^n \mathcal{X}_j\,, \quad P_j^n \in \mathcal{M}_{d\times d}\,, \tag{5}$$

where the matrix coefficients $P_j^n, \ j = 0, 1, \ \ldots, \ n$ are uniquely determined. Thus, it also occurs

$$\mathcal{P}_n(x) = W_n(x^d)\mathcal{X}_0\,, \tag{6}$$

where W_n is a matrix polynomial (i.e., W_n is a $d \times d$ matrix whose entries are polynomials) of degree n and dimension d, given by

$$W_n(x) = \sum_{j=0}^{n} P_j^n x^j\,, \quad P_j^n \in \mathcal{M}_{d\times d}. \tag{7}$$

Notice that the matrices $P_j^n \in \mathcal{M}_{d\times d}$ in (7) are the same as in (5). Within this context, we can describe the *matrix interpretation of multiple orthogonality for diagonal multi-indices*. Let $\{\mathcal{P}_n\}$ be a sequence of vector polynomials with polynomial entries as in (3), and a vector of functionals $\mathcal{U}$ as described above. $\{\mathcal{P}_n\}$ is said to be a *type II vector multiple orthogonal polynomial sequence* with respect to the vector of functionals $\mathcal{U}$, and a set of diagonal multi-indices, if

$$\left.\begin{array}{l} i)\ (x^k\mathcal{U})(\mathcal{P}_n) = 0_{d\times d}\,,\ k = 0, 1, \ldots, n-1\,, \\ ii)\ (x^n\mathcal{U})(\mathcal{P}_n) = \Delta_n\,, \end{array}\right\} \tag{8}$$

where Δ_n is a regular upper triangular $d \times d$ matrix (see [11, Th. 3] considering diagonal multi-indices).

Next, we introduce a few aspects of the duality theory, which will be useful in the sequel. We denote by $\mathbb{P}^*$ the dual space of $\mathbb{P}$, i.e. the linear space of linear functionals defined on $\mathbb{P}$ over $\mathbb{C}$. Let $\{P_n\}$ be a sequence of monic polynomials. We call $\{\ell_n\}$, $\ell_n \in \mathbb{P}^*$, the *dual sequence* of $\{P_n\}$ if $\ell_i(P_j) = \delta_{i,j}$, $i, j \in \mathbb{N}$ holds. Given a sequence of linear functionals $\{\ell_n\} \in \mathbb{P}^*$, by means of the shift $n \to nd$, the vector sequence of linear functionals $\{\mathcal{L}_n\}$, with

$$\mathcal{L}_n = \left[\ell_{nd} \cdots \ell_{(n+1)d-1}\right]^T, \quad n \in \mathbb{N},$$

is said to be the *vector sequence of linear functionals* associated with $\{\ell_n\}$.

It is very well known (see [17]) that a given sequence of type II polynomials $\{P_n\}$, simultaneously orthogonal with respect to a d linear functionals, or simply *d-orthogonal* polynomials, satisfy the following $(d+2)$-term recurrence relation

$$xP_{n+d}(x) = P_{n+d+1}(x) + \beta_{n+d}P_{n+d}(x) + \sum_{\nu=0}^{d-1} \gamma_{n+d-\nu}^{d-1-\nu} P_{n+d-1-\nu}(x), \tag{9}$$

$\gamma_{n+1}^0 \neq 0$ for $n \geq 0$, with the initial conditions $P_0(x) = 1$, $P_1(x) = x - \beta_0$, and

$$P_n(x) = (x - \beta_{n-1})P_{n-1}(x) - \sum_{\nu=0}^{n-2} \gamma_{n-1-\nu}^{d-1-\nu} P_{n-2-\nu}(x), \quad 2 \leq n \leq d.$$

E.g., if $d = 2$, the sequence of monic type II multiple orthogonal polynomials $\{P_n\}$ with respect to the regular system of functionals $\{u^1, u^2\}$ and normal multi-index satisfy, for every $n \geq 0$, the following four term recurrence relation (see [11, 25, Lemma 1-a])

$$xP_{n+2}(x) = P_{n+3}(x) + \beta_{n+2}P_{n+2}(x) + \gamma_{n+2}^1 P_{n+1}(x) + \gamma_{n+1}^0 P_n(x), \tag{10}$$

where $\beta_{n+2}, \gamma_{n+2}^1, \gamma_{n+1}^0 \in \mathbb{C}$, $\gamma_{n+1}^0 \neq 0$, $P_0(x) = 1$, $P_1(x) = x - \beta_0$ and $P_2(x) = (x - \beta_1)P_1(x) - \gamma_1^1 P_0(x)$.

Definition 1 ([17, Def. 4.1.]) A monic system of polynomials $\{S_n\}$ is said to be *d-symmetric* when it verifies

$$S_n(\xi_k x) = \xi_k^n S_n(x), \quad n \geq 0, \tag{11}$$

where $\xi_k = \exp(2k\pi i/(d+1))$, $k = 1, \ldots, d$, and $\xi_k^{d+1} = 1$. Notice that, if $d = 1$, then $\xi_k = -1$ and therefore $S_n(-x) = (-1)^n S_n(x)$ (see [13]).

We also assume (see [17, Def. 4.2.]) that the vector of linear functionals $\mathcal{L}_0 = \left[\ell_0 \cdots \ell_{d-1}\right]^T$ is said to be *d-symmetric* when the moments of its entries satisfy, for every $n \geq 0$,

$$\ell_\nu(x^{(d+1)n+\mu}) = 0\,, \quad \nu = 0, 1, \ldots, d-1\,, \quad \mu = 0, 1, \ldots, d\,, \quad \nu \neq \mu\,. \tag{12}$$

Observe that if $d = 1$, this condition leads to the well known fact $\ell_0(x^{2n+1}) = 0$, i.e., all the odd moments of a symmetric moment functional are zero (see [13, Def. 4.1, p.20]).

Under the above assumptions, we have the following

Theorem 1 (Cf. [17, Th. 4.1]) *For every sequence of monic polynomials $\{S_n\}$, d-orthogonal with respect to the vector of linear functionals $\mathcal{L}_0 = \left[\ell_0 \cdots \ell_{d-1}\right]^T$, the following statements are equivalent:*

(a) The vector of linear functionals $\mathcal{L}_0$ is d-symmetric.
(b) The sequence $\{S_n\}$ is d-symmetric.
(c) The sequence $\{S_n\}$ satisfies

$$x S_{n+d}(x) = S_{n+d+1}(x) + \gamma_{n+1} S_n(x)\,, \quad n \geq 0, \tag{13}$$

with $S_n(x) = x^n$ for $0 \leq n \leq d$.

Notice that (13) is a particular case of the $(d+2)$-term recurrence relation (9). Continuing the same trivial example above for $d = 2$, it directly implies that the sequence of polynomials $\{S_n\}$, satisfy the particular case of (9), i.e. $S_0(x) = 1$, $S_1(x) = 1$, $S_2(x) = x^2$ and

$$x S_{n+2}(x) = S_{n+3}(x) + \gamma_{n+1} S_n(x)\,, \quad n \geq 0.$$

Notice that the coefficients β_{n+2} and γ^1_{n+2} of polynomials S_{n+2} and S_{n+1} respectively, on the right hand side of (10) are zero.

On the other hand, the $(d+2)$-term recurrence relation (13) can be rewritten in terms of vector polynomials (3), and then we obtain what will be referred to as the *symmetric type II vector multiple orthogonal polynomial sequence* $\mathcal{S}_n = \left[S_{nd} \cdots S_{(n+1)d-1}\right]^T$. For $n \to dn + j$, $j = 0, 1, \ldots d-1$ and $n \in \mathbb{N}$, we have the following matrix three term recurrence relation

$$x\mathcal{S}_n = A\mathcal{S}_{n+1} + B\mathcal{S}_n + C_n\mathcal{S}_{n-1}\,, \quad n = 0, 1, \ldots \tag{14}$$

with $\mathcal{S}_{-1} = \left[0 \cdots 0\right]^T$, $\mathcal{S}_0 = \left[S_0 \cdots S_{d-1}\right]^T$, and matrix coefficients A, B, $C_n \in \mathcal{M}_{d\times d}$ given

$$A = \begin{bmatrix} 0 & 0 & \cdots & 0 \\ \vdots & \vdots & \ddots & \vdots \\ 0 & 0 & \cdots & 0 \\ 1 & 0 & \cdots & 0 \end{bmatrix}, \quad B = \begin{bmatrix} 0 & 1 & & \\ & \ddots & \ddots & \\ & & 0 & 1 \\ & & & 0 \end{bmatrix}, \quad \text{and} \tag{15}$$

$$C_n = \operatorname{diag}\left[\gamma_1 \cdots \gamma_{nd}\right].$$

Note that, in this case, one has

$$\mathcal{S}_0 = \mathcal{X}_0 = \left[1\ x \cdots x^{d-1}\right]^T.$$

Since $\{S_n\}$ satisfies (13), it is clear that this d-symmetric type II multiple polynomial sequence will be orthogonal with respect to certain system of d linear functionals, say $\{v^1, \ldots, v^d\}$. Hence, according to the matrix interpretation of multiple orthogonality, the corresponding type II vector multiple polynomial sequence $\{\mathcal{S}_n\}$ will be orthogonal with respect to a *symmetric vector of functionals* $\mathcal{V} = \left[v^1 \cdots v^d\right]^T$. The corresponding matrix orthogonality conditions for $\{\mathcal{S}_n\}$ and $\mathcal{V}$ are described in (8).

One of the main goals of this manuscript is to analyze symmetric sequences of type II vector multiple polynomials, orthogonal with respect to a symmetric vector of functionals. The remainder of this section will be devoted to the proof of one of our main results concerning the moments of the d functional entries of such symmetric vector of functionals $\mathcal{V}$. The following theorem states that, under certain conditions, the moment of each functional entry in $\mathcal{V}$ can be given in terms of the moments of other functional entry in the same $\mathcal{V}$.

Theorem 2 *If* $\mathcal{V} = \left[v^1 \cdots v^d\right]^T$ *is a symmetric vector of functionals, the moments of each functional entry* v^j, $j = 1, 2, \ldots, d$ *in* $\mathcal{V}$, *can be expressed for all* $n \geq 0$ *as*

(i) If $\mu = 0, 1, \ldots, j-2$

$$v^j(x^{(d+1)n+\mu}) = \frac{v_{j,\mu}}{v_{\mu+1,\mu}} v^{\mu+1}(x^{(d+1)n+\mu}),$$

where $v_{k,l} = v^k(S_l)$.

(ii) If $\mu = j-1$, *the value* $v^j(x^{(d+1)n+\mu})$ *depends on* v^j, *and it is different from zero.*

(iii) If $\mu = j, j+1, \ldots, d$

$$v^j(x^{(d+1)n+\mu}) = 0.$$

Proof In the matrix framework of multiple orthogonality, the type II vector polynomials $\mathcal{S}_n$ are multiple orthogonal with respect to the symmetric vector moment of functionals $\mathcal{V} : \mathbb{P}^d \to \mathcal{M}_{d\times d}$, with $\mathcal{V} = \left[v^1 \cdots v^d\right]^T$. If we multiply (14) by x^{n-1}, together with the linearity of $\mathcal{V}$, we get, for $n = 0, 1, \ldots$

$$\mathcal{V}\left(x^n \mathcal{S}_n\right) = A\mathcal{V}\left(x^{n-1}\mathcal{S}_{n+1}\right) + B\mathcal{V}\left(x^{n-1}\mathcal{S}_n\right) + C_n\mathcal{V}\left(x^{n-1}\mathcal{S}_{n-1}\right), \quad n = 0, 1, \ldots.$$

By the orthogonality conditions (8) for $\mathcal{V}$, we have

$$A\mathcal{V}\left(x^{n-1}\mathcal{S}_{n+1}\right) = B\mathcal{V}\left(x^{n-1}\mathcal{S}_n\right) = 0_{d\times d}, \quad n = 0, 1, \ldots,$$

and iterating the remain expression

$$\mathcal{V}\left(x^n \mathcal{S}_n\right) = C_n\mathcal{V}\left(x^{n-1}\mathcal{S}_{n-1}\right), \quad n = 0, 1, \ldots$$

we obtain

$$\mathcal{V}\left(x^n \mathcal{S}_n\right) = C_n C_{n-1}\cdots C_1\mathcal{V}\left(\mathcal{S}_0\right), \quad n = 0, 1, \ldots.$$

The above matrix $\mathcal{V}\left(\mathcal{S}_0\right)$ is given by

$$\mathcal{V}\left(\mathcal{S}_0\right) = \mathcal{V}_0^0 = \begin{bmatrix} v^1\left(\mathcal{S}_0\right) & \cdots & v^d\left(\mathcal{S}_0\right) \\ \vdots & \ddots & \vdots \\ v^1\left(\mathcal{S}_{d-1}\right) & \cdots & v^d\left(\mathcal{S}_{d-1}\right) \end{bmatrix}.$$

To simplify the notation, in the sequel $v_{i,j-1}$ denotes $v^i\left(\mathcal{S}_{j-1}\right)$. Notice that (8) leads to the fact that the above matrix is an upper triangular matrix, which in turn means that $v_{i,j-1} = 0$, for every $i, j = 1, \ldots, d-1$, that is

$$\mathcal{V}_0^0 = \begin{bmatrix} v_{1,0} & v_{2,0} & \cdots & v_{d,0} \\ & v_{2,1} & \cdots & v_{d,1} \\ & & \ddots & \vdots \\ & & & v_{d,d-1} \end{bmatrix}. \tag{16}$$

Let $\mathcal{L}_0$ be a d-symmetric vector of linear functionals as in Theorem 1. We can express $\mathcal{V}$ in terms of $\mathcal{L}_0$ as $\mathcal{V} = G_0\mathcal{L}_0$. Thus, we have

$$(G_0^{-1}\mathcal{V})(\mathcal{S}_0) = \mathcal{L}_0(\mathcal{S}_0) = I_d.$$

From (4) we have

$$\left(G_0^{-1}\mathcal{V}\right)(\mathcal{S}_0) = \mathcal{V}\left(\mathcal{S}_0\right)(G_0^{-1})^T = \mathcal{V}_0^0(G_0^{-1})^T = I_d.$$

Therefore, taking into account (16), we conclude

$$G_0 = (\mathcal{V}_0^0)^T = \begin{bmatrix} v_{1,0} & & & \\ v_{2,0} & v_{2,1} & & \\ \vdots & \vdots & \ddots & \\ v_{d,0} & v_{d,1} & \cdots & v_{d,d-1} \end{bmatrix}.$$

Observe that the matrix $(\mathcal{V}_0^0)^T$ is lower triangular, and every entry in their main diagonal is different from zero, so G_0 always exists and is a lower triangular matrix. Since $\mathcal{V} = G_0\mathcal{L}_0$, we finally obtain the expressions

$$\begin{aligned} v^1 &= v_{1,0}\ell_0, \\ v^2 &= v_{2,0}\ell_0 + v_{2,1}\ell_1, \\ v^3 &= v_{3,0}\ell_0 + v_{3,1}\ell_1 + v_{3,2}\ell_2, \\ \cdots & \\ v^d &= v_{d,0}\ell_0 + v_{d,1}\ell_1 + \cdots + v_{d,d-1}\ell_{d-1}, \end{aligned} \tag{17}$$

between the entries of $\mathcal{L}_0$ and $\mathcal{V}$.

Next, from (17), (12), together with the crucial fact that every value in the main diagonal of $\mathcal{V}_0^0$ is different from zero, it is a simple matter to check that the three statements of the lemma follow. We conclude the proof only for the functionals v^2 and v^1. The other cases can be deduced in a similar way.

From (17) we get $v^1(x^{(d+1)n+\mu}) = v_{1,0} \cdot \ell_0(x^{(d+1)n+\mu}) \neq 0$. Then, from (12) we see that for every $\mu \neq 0$ we have $v^1(x^{(d+1)n+\mu}) = 0$ (statement (iii)). If $\mu = 0$, we have $v^1(x^{(d+1)n}) = v_{1,0} \cdot \ell_0(x^{(d+1)n}) \neq 0$ (statement (ii)). Next, from (17) we get $v^2(x^{(d+1)n+\mu}) = v_{2,0} \cdot \ell_0(x^{(d+1)n+\mu}) + v_{2,1} \cdot \ell_1(x^{(d+1)n+\mu})$. Then, from (12) we see that for every $\mu \neq 1$, we have $v^2(x^{(d+1)n+\mu}) = 0$ (statement (iii)). If $\mu = 0$ we have

$$v^2(x^{(d+1)n}) = \frac{v_{2,0}}{v_{1,0}} v^1(x^{(d+1)n}) \text{ (statement } (i)).$$

If $\mu = 1$ then $v^2(x^{(d+1)n+1}) = v_{2,1} \cdot \ell_1(x^{(d+1)n+1}) \neq 0$ (statement (ii)).

Thus, the theorem follows. □

3 Representation of the Stieltjes and Weyl Functions

Let $\mathcal{U}$ a vector of functionals $\mathcal{U} = \begin{bmatrix} u^1 & \cdots & u^d \end{bmatrix}^T$. We define the *Stieltjes matrix function* associated to $\mathcal{U}$ (or *matrix generating function* associated to $\mathcal{U}$), $\mathcal{F}$ by (see [11])

$$\mathcal{F}(z) = \sum_{n=0}^{\infty} \frac{(x^n\mathcal{U})\,(\mathcal{X}_0(x))}{z^{n+1}}.$$

We define the above $\mathcal{F}(z)$ when the sequence of matrix moments $(x^n\mathcal{U})$ is bounded by a constant, let us say m, independent of n. Thus, $\mathcal{F}(z)$ is well defined for values $|z| > m$.

In this section we find the relation between the Stieltjes matrix function $\mathcal{F}$, associated to a certain d-symmetric vector of functionals $\mathcal{V}$, and certain interesting function associated to the corresponding block Jacobi matrix J. Here we deal with d-symmetric sequences of type II multiple orthogonal polynomials $\{S_n\}$, and hence J is a $(d+2)$-banded matrix with only two extreme non-zero diagonals, which is the block-matrix representation of the three-term recurrence relation with $d \times d$ matrix coefficients, satisfied by the vector sequence of polynomials $\{\mathcal{S}_n\}$ (associated to $\{S_n\}$), orthogonal with respect to $\mathcal{V}$.

Thus, the shape of a Jacobi matrix J, associated with the $(d+2)$-term recurrence relation (13) satisfied by a d-symmetric sequence of type II multiple orthogonal polynomials $\{S_n\}$ is

$$J = \begin{bmatrix} 0 & 1 & & & & & \\ 0 & 0 & 1 & & & & \\ \vdots & & \ddots & \ddots & & & \\ 0 & & & 0 & 1 & & \\ \gamma_1 & 0 & & & 0 & 1 & \\ & \gamma_2 & 0 & & & 0 & \ddots \\ & & \ddots & \ddots & & & \ddots \end{bmatrix}.$$

We can rewrite J as the block tridiagonal matrix

$$J = \begin{bmatrix} B & A & & & \\ C_1 & B & A & & \\ & C_2 & B & A & \\ & & \ddots & \ddots & \ddots \end{bmatrix}$$

associated to a three term recurrence relation with matrix coefficients, satisfied by the sequence of type II vector multiple orthogonal polynomials $\{\mathcal{S}_n\}$ associated to $\{S_n\}$. Here, every block matrix A, B and C_n has $d \times d$ size, and they are given in (15). When J is a bounded operator, it is possible to define the *resolvent operator* by

$$(zI - J)^{-1} = \sum_{n=0}^{\infty} \frac{J^n}{z^{n+1}}, \quad |z| > ||J||,$$

(see [10]) and we can put in correspondence the following block tridiagonal analytic function, known as the *Weyl function associated with J*

$$\mathcal{R}_J(z) = \sum_{n=0}^{\infty} \frac{\mathbf{e}_0^T \, J^n \, \mathbf{e}_0}{z^{n+1}}, \quad |z| > ||J||,$$

where $\mathbf{e}_0 = \begin{bmatrix} I_d & 0_{d\times d} & \cdots \end{bmatrix}^T$. If we denote by M_{ij} the $d \times d$ block matrices of a semi-infinite matrix M, formed by the entries of rows $d(i-1)+1, d(i-1)+2, \ldots, di$, and columns $d(j-1)+1, d(j-1)+2, \ldots, dj$, the matrix J^n can be written as the semi-infinite block matrix

$$J^n = \begin{bmatrix} J_{11}^n & J_{12}^n & \cdots \\ J_{21}^n & J_{22}^n & \cdots \\ \vdots & \vdots & \ddots \end{bmatrix}.$$

We can now formulate our first important result in this section. For more details we refer the reader to [7, Sec. 1.2] and [3].

Let $\{\mathcal{S}_n\}$ be a symmetric type II vector multiple polynomial sequence orthogonal with respect to the d-symmetric vector of functionals $\mathcal{V}$. Following [11, Th. 7], the matrix generating function associated to $\mathcal{V}$, and the Weyl function associated with J, the block Jacobi matrix corresponding to $\{\mathcal{S}_n\}$, can be put in correspondence by means of the matrix expression

$$\mathcal{F}(z) = R_J(z)\mathcal{V}(\mathcal{X}_0),$$

where, for the d-symmetric case, $\mathcal{V}(\mathcal{X}_0) = \mathcal{V}(\mathcal{S}_0) = \mathcal{V}_0^0$, which is explicitly given in (16).

First we study the case $d = 2$, and next we consider more general situations for $d > 2$ functional entries in $\mathcal{V}$. From Theorem 2, we obtain the entries for the representation of the Stieltjes matrix function $\mathcal{F}(z)$, associated with $\mathcal{V} = \begin{bmatrix} v^1 & v^2 \end{bmatrix}$, as

$$\mathcal{F}(z) = \sum_{n=0}^{\infty} \begin{bmatrix} v^1(x^{3n}) & \frac{v_{2,0}\cdot v^1(x^{3n})}{v_{1,0}} \\ 0 & v^2(x^{3n+1}) \end{bmatrix} / z^{3n+1} + \sum_{n=0}^{\infty} \begin{bmatrix} 0 & v^2(x^{3n+1}) \\ 0 & 0 \end{bmatrix} / z^{3n+2}$$
$$+ \sum_{n=0}^{\infty} \begin{bmatrix} 0 & 0 \\ v^1(x^{3n+3}) & \frac{v_{2,0}\cdot v^1(x^{3n+3})}{v_{1,0}} \end{bmatrix} / z^{3n+3}. \tag{18}$$

Notice that we have $\mathcal{F}(z) = \mathcal{F}_1(z) + \mathcal{F}_2(z) + \mathcal{F}_3(z)$. The following theorem shows that we can obtain $\mathcal{F}(z)$ in our particular case, analyzing two different situations.

Theorem 3 *Let $\mathcal{V} = \begin{bmatrix} v^1 & v^2 \end{bmatrix}^T$ be a symmetric vector of functionals, with $d = 2$. Then the Weyl function is given by*

$$\mathcal{R}_J(z) = \sum_{n=0}^{\infty} \frac{\begin{bmatrix} \frac{v^1(x^{3n})}{v_{1,0}} & 0 \\ 0 & \frac{v^2(x^{3n+1})}{v_{2,1}} \end{bmatrix}}{z^{3n+1}} + \sum_{n=0}^{\infty} \frac{\begin{bmatrix} 0 & \frac{v^2(x^{3n+1})}{v_{2,1}} \\ 0 & 0 \end{bmatrix}}{z^{3n+2}} + \sum_{n=0}^{\infty} \frac{\begin{bmatrix} 0 & 0 \\ \frac{v^1(x^{3n+3})}{v_{1,0}} & 0 \end{bmatrix}}{z^{3n+3}}$$

Proof It is enough to multiply (18) by $(\mathcal{V}_0^0)^{-1}$. The explicit expression for $\mathcal{V}_0^0$ is given in (16). □

Computations considering $d > 2$ functionals, can be cumbersome but doable. The matrix generating function $\mathcal{F}$ (as well as the Weyl function), will be the sum of $d+1$ matrix terms, i.e. $\mathcal{F}(z) = \mathcal{F}_1(z) + \cdots + \mathcal{F}_{d+1}(z)$, each of them of size $d \times d$.

Let us now outline the very interesting structure of $\mathcal{R}_J(z)$ for the general case of d functionals. We shall describe the structure of $\mathcal{R}_J(z)$ for $d = 3$, comparing the situation with the general case. Let $*$ denote every non-zero entry in a given matrix. Thus, there will be four J_{11} matrices of size 3×3. Here, and in the general case, the first matrix $[J_{11}^{(d+1)n}]_{d\times d}$ will always be diagonal, as follows

$$[J_{11}^{4n}]_{3\times 3} = \begin{bmatrix} * & 0 & 0 \\ 0 & * & 0 \\ 0 & 0 & * \end{bmatrix}.$$

Indeed, observe that for $n = 0$, $[J_{11}^{(d+1)n}]_{d\times d} = I_d$. Next, we have

$$[J_{11}^{4n+1}]_{3\times 3} = \begin{bmatrix} 0 & * & 0 \\ 0 & 0 & * \\ 0 & 0 & 0 \end{bmatrix}.$$

From (13) we know that the "distance" between the two extreme non-zero diagonals of J will always consist of d zero diagonals. It directly implies that, also in the general case, every entry in the last row of $[J_{11}^{(d+1)n}]_{d\times d}$ will always be zero, and therefore the unique non-zero entries in $[J_{11}^{(d+1)n}]_{d\times d}$ will be one step over the main diagonal.

Next we have

$$[J_{11}^{4n+2}]_{3\times 3} = \begin{bmatrix} 0 & 0 & * \\ 0 & 0 & 0 \\ * & 0 & 0 \end{bmatrix}.$$

Notice that for every step, the main diagonal in $[J_{11}^{(d+1)n}]_{d\times d}$ goes "one diagonal" up, but the other extreme diagonal of J is also moving upwards, with exactly d

zero diagonals between them. It directly implies that, no matter the number of functionals, only the lowest-left element of $[J_{11}^{(d+1)n+2}]_{d\times d}$ will be different from zero. Finally, we have

$$[J_{11}^{4n+3}]_{3\times 3} = \begin{bmatrix} 0 & 0 & 0 \\ * & 0 & 0 \\ 0 & * & 0 \end{bmatrix}.$$

Here, the last non-zero entry of the main diagonal in $[J_{11}^{4n}]_{3\times 3}$ vanishes. In the general case, it will occur exactly at step $[J_{11}^{(d+1)n+(d-1)}]_{d\times d}$, in which only the uppermost right entry is different from zero. Meanwhile, in matrices $[J_{11}^{(d+1)n+3}]_{d\times d}$ up to $[J_{11}^{(d+1)n+(d-2)}]_{d\times d}$ will be non-zero entries of the two extreme diagonals. In this last situation, the non-zero entries of $[J_{11}^{(d+1)n+(d-1)}]_{d\times d}$, will always be exactly one step under the main diagonal.

4 The Symmetrization Problem for Multiple OP

Throughout this section, let $\{A_n^1\}$ be an arbitrary and not necessarily symmetric sequence of type II multiple orthogonal polynomials, satisfying a $(d+2)$-term recurrence relation with known recurrence coefficients, and let J_1 be the corresponding $(d+2)$-banded matrix. We will consider next a generalization to the case of multiple orthogonality of the so-called *Darboux transformations with shift* (for a description of these transformations in the scalar case, see for example [12, Prop. 3.6]), which in a first step consist of finding factorizations of Jacobi matrices as described below. Let J_1 be such that the following LU factorization

$$J_1 - CI = LU = L_1 L_2 \cdots L_d U \tag{19}$$

is unique, where I is the identity matrix, U is an upper two-banded, semi-infinite, and invertible matrix, L is a $(d+1)$-lower triangular semi-infinite with ones in the main diagonal, and every L_i, $i = 1, \ldots, d$ is a lower two-banded, semi-infinite, and invertible matrix with ones in the main diagonal. The parameter C, which is called *the shift of the transformation,* is certain constant such that $\det(J_n - CI_n) \neq 0$ for all $n \in \mathbb{N}$. This number C is chosen such that the factorization of the matrix $L = L_1 L_2 \cdots L_d$ is unique, with all the matrices $L_1, L_2, \ldots L_d$ being lower two-banded, semi-infinite, and having ones in the main diagonal. See [9] for more details on this factorization in the case of d-diagonal Jacobi matrices.

We follow [9, Def. 3], where the authors generalize the concept of Darboux transformation to general Hessenberg banded matrices, in assuming that any circular permutation of $L_1 L_2 \cdots L_d U$ is a Darboux transformation of $J_1 - CI$. Thus we have d possible Darboux transformations of $J_1 - CI$, say $J_j - CI$, $j = 2, \ldots, d+1$, with $J_2 - CI = L_2 \cdots L_d U L_1$, $J_3 - CI = L_3 \cdots L_d U L_1 L_2, \ldots, J_{d+1} - CI = U L_1 L_2 \cdots L_d$.

Next, we solve the so called *symmetrization problem* in the theory of multiple orthogonal polynomials, i.e., starting with $\{A_n^1\}$, we find a total $d+1$ type II multiple orthogonal polynomial sequences $\{A_n^j\}$, $j = 1, \ldots, d+1$, satisfying $(d+2)$-term recurrence relation with known recurrence coefficients, which can be used to construct a new d-symmetric type II multiple orthogonal polynomial sequence $\{S_n\}$. It is worth pointing out that all the aforesaid sequences $\{A_n^j\}$, $j = 1, \ldots, d+1$ are of the same kind, with the same number of elements in their respective $(d+2)$-term recurrence relations, and multiple orthogonal with respect to the same number of functionals d.

Theorem 4 *Let $\{A_n^1\}$ be an arbitrary and not necessarily symmetric sequence of type II multiple orthogonal polynomials as stated above. Let $J_j - CI$, $j = 2, \ldots, d+1$, be the Darboux transformations of $J_1 - CI$ given by the d cyclic permutations of the matrices in the right hand side of (19). Let $\{A_n^j\}$, $j = 2, \ldots, d+1$, d new families of type II multiple orthogonal polynomials satisfying $(d+2)$-term recurrence relations given by the matrices J_j, $j = 2, \ldots, d+1$. Then, the sequence $\{S_n\}$ defined by*

$$\begin{cases} S_{(d+1)n}(x) = A_n^1(x^{d+1} + C), \\ S_{(d+1)n+1}(x) = x A_n^2(x^{d+1} + C), \\ \cdots \\ S_{(d+1)n+d}(x) = x^d A_n^{d+1}(x^{d+1} + C), \end{cases} \tag{20}$$

is a d-symmetric sequence of type II multiple orthogonal polynomials.

Proof Let $\{A_n^1\}$ satisfy the $(d+2)$-term recurrence relation given by (9)

$$x A_{n+d}^1(x) = A_{n+d+1}^1(x) + b_{n+d}^{[1]} A_{n+d}^1(x) + \sum_{\nu=0}^{d-1} c_{n+d-\nu}^{d-1-\nu,[1]} A_{n+d-1-\nu}^1(x),$$

$c_{n+1}^{0,[1]} \neq 0$ for $n \geq 0$, with the initial conditions $A_0^1(x) = 1$, $A_1^1(x) = x - b_0^{[1]}$, and

$$A_n^1(x) = (x - b_{n-1}^{[1]}) A_{n-1}^1(x) - \sum_{\nu=0}^{n-2} c_{n-1-\nu}^{d-1-\nu,[1]} A_{n-2-\nu}^1(x), \quad 2 \leq n \leq d,$$

with known recurrence coefficients. Set

$$\underline{\mathbf{A}}^j = \left[A_0^j(x) \; A_1^j(x) \; A_2^j(x) \cdots \right]^T.$$

Thus, in a matrix notation, we have

$$x\underline{\mathbf{A}}^1 = J_1\, \underline{\mathbf{A}}^1, \tag{21}$$

where

$$J_1 = \begin{bmatrix} b_d^{[1]} & 1 & & & & & \\ c_{d+1}^{d-1,[1]} & b_{d+1}^{[1]} & 1 & & & & \\ c_{d+1}^{d-2,[1]} & c_{d+2}^{d-1,[1]} & b_{d+2}^{[1]} & 1 & & & \\ \vdots & c_{d+2}^{d-2,[1]} & c_{d+3}^{d-1,[1]} & b_{d+3}^{[1]} & 1 & & \\ c_{d+1}^{0,[1]} & \cdots & c_{d+3}^{d-2,[1]} & c_{d+4}^{d-1,[1]} & b_{d+4}^{[1]} & 1 & \\ & \ddots & \ddots & \ddots & \ddots & \ddots & \ddots \end{bmatrix}.$$

Following [9] (see also [23]), the unique LU factorization for the square $(d+2)$-banded semi-infinite Hessenberg matrix $J_1 - CI$ is such that

$$U = \begin{bmatrix} \gamma_1 & 1 & & & \\ & \gamma_{d+2} & 1 & & \\ & & \gamma_{2d+3} & 1 & \\ & & & \gamma_{3d+4} & \ddots \\ & & & & \ddots \end{bmatrix} \tag{22}$$

is an upper two-banded, semi-infinite, and invertible matrix, and L is a $(d+1)$-lower triangular, semi-infinite, and invertible matrix with ones in the main diagonal. It is clear that the entries in L and U depend entirely on the known entries of J_1 and the constant parameter C. Thus, we rewrite (21) as

$$(x - C)\underline{\mathbf{A}}^1 = LU\,\underline{\mathbf{A}}^1. \tag{23}$$

Next, we define a new sequence of polynomials $\{A_n^{d+1}\}$ by $(x - C)\underline{\mathbf{A}}^{d+1} = U\,\underline{\mathbf{A}}^1$. Multiplying both sides of (23) by the matrix U, we have

$$x\left(U\,\underline{\mathbf{A}}^1\right) - C(U\,\underline{\mathbf{A}}^1) = UL\left(U\,\underline{\mathbf{A}}^1\right),$$

$$x(x - C)\underline{\mathbf{A}}^{d+1} - C(x - C)\underline{\mathbf{A}}^{d+1} = UL(x - C)\underline{\mathbf{A}}^{d+1},$$

and pulling out $(x - C)$ we get

$$(x - C)\underline{\mathbf{A}}^{d+1} = UL\,\underline{\mathbf{A}}^{d+1}, \tag{24}$$

which is the matrix form of the $(d+2)$-term recurrence relation satisfied by the new type II multiple polynomial sequence $\{A_n^{d+1}\}$.

Since L is given by (see [9])

$$L = L_1 L_2 \cdots L_d,$$

where every L_j is the lower two-banded, semi-infinite, and invertible matrix

$$L_j = \begin{bmatrix} 1 & & & & \\ \gamma_{d-j} & 1 & & & \\ & \gamma_{2d+1-j} & 1 & & \\ & & \gamma_{3d+2-j} & 1 & \\ & & & \ddots & \ddots \end{bmatrix}$$

with ones in the main diagonal, it is also clear that the entries in L_j will depend on the known entries in J_1. Under the same hypotheses, we can define new $d-1$ polynomial sequences starting with $\underline{\mathbf{A}}^1$ as follows: $\underline{\mathbf{A}}^2 = L_1^{-1}\underline{\mathbf{A}}^1$, $\underline{\mathbf{A}}^3 = L_2^{-1}\underline{\mathbf{A}}^2, \ldots,$ $\underline{\mathbf{A}}^d = L_{d-1}^{-1}\underline{\mathbf{A}}^{d-1}$ up to $\underline{\mathbf{A}}^{d+1} = L_d^{-1}\underline{\mathbf{A}}^d$. That is, $\underline{\mathbf{A}}^{j+1} = L_j^{-1}\underline{\mathbf{A}}^j$, $j = 1, \ldots d-1$. Combining this fact with (24) we deduce

$$(x - C)\underline{\mathbf{A}}^d = L_d U L_1 L_2 \cdots L_{d-1}\,\underline{\mathbf{A}}^d,$$
$$(x - C)\underline{\mathbf{A}}^{d-1} = L_{d-1} L_d U L_1 L_2 \cdots L_{d-2}\,\underline{\mathbf{A}}^{d-1},$$

up to the known expression (21)

$$(x - C)\underline{\mathbf{A}}^1 = L_1 L_2 \cdots L_d U\,\underline{\mathbf{A}}^1.$$

The above formulas mean that all these $d+1$ sequences $\{A^j\}$, $j = 1, \ldots, d+1$ are in turn type II multiple orthogonal polynomials. Finally, from these $d+1$ sequences, we construct the type II polynomials in the sequence $\{S_n\}$ as (20). Note that, according to (11), it directly follows that $\{S_n\}$ is a d-symmetric type II multiple orthogonal polynomial sequence, which proves our assertion. □

A particular example of (20) for $C = 0$ is given in [15, (2.16), (2.17) and (2.18)]. There, the author calls the sequence $\{S_n\}$ as Hermite-type d-OPS, because, up to some point, these polynomials have similar symmetric properties to the classical Hermite polynomials.

Next, we state the converse of the above theorem. That is, given a sequence of type II d-symmetric multiple orthogonal polynomials $\{S_n\}$ satisfying the high-term recurrence relation (13), we find a set of $d+1$ polynomial families $\{A_n^j\}$, $j = 1, \ldots, d+1$ of not necessarily symmetric type II multiple orthogonal polynomials, satisfying in turn $(d+2)$-term recurrence relations, so o they are themselves sequences of type II multiple orthogonal polynomials. When $d = 2$, this construction goes back to the work of Douak and Maroni (see [16]).

Theorem 5 *Let $\{S_n\}$ be a d-symmetric sequence of type II multiple orthogonal polynomials satisfying the corresponding high-order recurrence relation (13), and $\{A_n^j\}$, $j = 1, \ldots, d+1$, the sequences of polynomials given by (20). Then, each sequence $\{A_n^j\}$, $j = 1, \ldots, d+1$, satisfies the $(d+2)$-term recurrence relation*

$$xA_{n+d}^j(x) = A_{n+d+1}^j(x) + b_{n+d}^{[j]} A_{n+d}^j(x) + \sum_{\nu=0}^{d-1} c_{n+d-\nu}^{d-1-\nu,[j]} A_{n+d-1-\nu}^j(x),$$

$c_{n+1}^{0,[j]} \neq 0$ *for* $n \geq 0$, *with initial conditions* $A_0^j(x) = 1$, $A_1^j(x) = x - b_0^{[j]}$,

$$A_n^j(x) = (x - b_{n-1}^{[j]})A_{n-1}^j(x) - \sum_{\nu=0}^{n-2} c_{n-1-\nu}^{d-1-\nu,[j]} A_{n-2-\nu}^j(x), \quad 2 \leq n \leq d,$$

and therefore they are type II multiple orthogonal polynomial sequences.

Proof Since $\{S_n\}$ is a d-symmetric multiple orthogonal sequence, it satisfies (13) with $S_n(x) = x^n$ for $0 \leq n \leq d$. Shifting, for convenience, the multi-index in (13) as $n \to (d+1)n - d + j$, $\quad j = 0, 1, 2, \ldots .d$, we obtain the equivalent system of $(d+1)$ equations

$$\begin{cases} xS_{(d+1)n}(x) = S_{(d+1)n+1}(x) + \gamma_{(d+1)n-d+1} S_{(d+1)n-d}(x), & j = 0, \\ xS_{(d+1)n+1}(x) = S_{(d+1)n+2}(x) + \gamma_{(d+1)n-d+2} S_{(d+1)n-d+1}(x), & j = 1, \\ \vdots & \vdots \\ xS_{(d+1)n+d-1}(x) = S_{(d+1)n+d}(x) + \gamma_{(d+1)n} S_{(d+1)n-1}(x), & j = d-1, \\ xS_{(d+1)n+d}(x) = S_{(d+1)n+(d+1)}(x) + \gamma_{(d+1)n+1} S_{(d+1)n}(x), & j = d. \end{cases}$$

Substituting (20), having $C = 0$, into the above expressions yields

$$\begin{cases} 1) \ A_n^1(x^3) = A_n^2(x^3) + \gamma_{(d+1)n-(d-1)} A_{n-1}^2(x^3), \\ 2) \ A_n^2(x^3) = A_n^3(x^3) + \gamma_{(d+1)n-(d-2)} A_{n-1}^3(x^3), \\ \vdots \ \vdots \\ d) \ A_n^d(x^3) = A_n^{d+1}(x^3) + \gamma_{(d+1)n} A_{n-1}^{d+1}(x^3), \\ d+1) \ xA_n^{d+1}(x^3) = A_{n+1}^1(x^3) + \gamma_{(d+1)n+1} A_n^1(x^3). \end{cases}$$

Next, replacing $x^{d+1} \to x$, we get the following system of $(d+1)$ equations

$$\begin{cases} 1) \ A_n^1(x) = A_n^2(x) + \gamma_{(d+1)n-(d-1)} A_{n-1}^2(x), \\ 2) \ A_n^2(x) = A_n^3(x) + \gamma_{(d+1)n-(d-2)} A_{n-1}^3(x), \\ \vdots \ \vdots \\ d) \ A_n^d(x) = A_n^{d+1}(x) + \gamma_{(d+1)n} A_{n-1}^{d+1}(x), \\ d+1) \ xA_n^{d+1}(x) = A_{n+1}^1(x) + \gamma_{(d+1)n+1} A_n^1(x). \end{cases} \tag{25}$$

Having $x = 0$ in the last of the above equations, we can define the γ as

$$\gamma_{(d+1)n+1} = \frac{-A^1_{n+1}(0)}{A^1_n(0)}.$$

From now on we deal with the matrix representation of each equation in (25). Notice that the first d equations of (25), namely $A^j_n = A^{j+1}_n + \gamma_{(d+1)n+(j-d)} A^{j+1}_{n-1}$, $j = 1, \ldots, d$, can be written in the matrix way

$$\underline{\mathbf{A}}^j = L_j \, \underline{\mathbf{A}}^{j+1}, \tag{26}$$

where L_j is the lower two-banded, semi-infinite, and invertible matrix

$$L_j = \begin{bmatrix} 1 & & & & \\ \gamma_{d-j} & 1 & & & \\ & \gamma_{2d+1-j} & 1 & & \\ & & \gamma_{3d+2-j} & 1 & \\ & & & \ddots & \ddots \end{bmatrix}.$$

Similarly, the $(d+1)$-th equation in (25) can be expressed as

$$x \underline{\mathbf{A}}^{d+1} = U \, \underline{\mathbf{A}}^1, \tag{27}$$

where U is the upper two-banded, semi-infinite, and invertible matrix

$$U = \begin{bmatrix} \gamma_1 & 1 & & & \\ & \gamma_{d+2} & 1 & & \\ & & \gamma_{2d+3} & 1 & \\ & & & \gamma_{3d+4} & \ddots \\ & & & & \ddots \end{bmatrix}.$$

It is clear that the entries in the above matrices L_j and U are given in terms of the recurrence coefficients γ_{n+1} for $\{S_n\}$ satisfying (13). From (26) we have $\underline{\mathbf{A}}^1$ defined in terms of $\underline{\mathbf{A}}^2$, as $\underline{\mathbf{A}}^1 = L_1 \underline{\mathbf{A}}^2$. Likewise $\underline{\mathbf{A}}^2$ in terms of $\underline{\mathbf{A}}^3$ as $\underline{\mathbf{A}}^2 = L_2 \underline{\mathbf{A}}^3$, and so on up to $j = d$. Thus, it is easy to see that,

$$\underline{\mathbf{A}}^1 = L_1 L_2 \cdots L_d \underline{\mathbf{A}}^{d+1}.$$

Next, we multiply by x both sides of the above expression, and we apply (27) to obtain

$$x \underline{\mathbf{A}}^1 = L_1 L_2 \cdots L_d U \, \underline{\mathbf{A}}^1.$$

Since each L_j and U are lower and upper two-banded semi-infinite matrices, it follows easily that $L_1L_2\cdots L_d$ is a lower triangular $(d+1)$-banded matrix with ones in the main diagonal, so the above decomposition is indeed a LU decomposition of certain $(d+2)$-banded Hessenberg matrix $J_1 = L_1L_2\cdots L_dU$ (see for instance [9] and [23, Sec. 3.2 and 3.3]). The values of the entries in J_1 come from the usual definition for matrix multiplication, matching every entry in $J_1 = L_1L_2\cdots L_dU$, with L_j and U provided in (26) and (27) respectively.

On the other hand, starting with $\underline{\mathbf{A}}^2 = L_2\underline{\mathbf{A}}^3$ instead of $\underline{\mathbf{A}}^1$, and proceeding in the same fashion as above, we can reach

$$\begin{aligned}
x\underline{\mathbf{A}}^2 &= L_2\cdots L_dUL_1\,\underline{\mathbf{A}}^2,\\
x\underline{\mathbf{A}}^3 &= L_3\cdots L_dUL_1L_2\,\underline{\mathbf{A}}^3,\\
&\ \ \vdots\\
x\underline{\mathbf{A}}^{d+1} &= UL_1L_2\cdots L_d\,\underline{\mathbf{A}}^{d+1}.
\end{aligned}$$

Observe that J_j denotes a particular circular permutation of the matrix product $L_1L_2\cdots L_dU$. Thus, we have $J_1 = L_1L_2\cdots L_dU$, $J_2 = L_2\cdots L_dUL_1$, ..., $J_{d+1} = UL_1L_2\cdots L_d$. Using this notation, J_j is the matrix representation of the operator of multiplication by x

$$x\underline{\mathbf{A}}^j = J_j\,\underline{\mathbf{A}}^j,$$

which from (9) directly implies that each polynomial sequence $\{A^j\}$, $j = 1,\ldots d+1$, satisfies a $(d+2)$-term recurrence relation as in the statement of the theorem, with coefficients given in terms of the recurrence coefficients γ_{n+1} , from the high-term recurrence relation (13) satisfied by the symmetric sequence $\{S_n\}$.

This completes the proof. □

5 Matrix Multiple Orthogonality

For an arbitrary system of type II vector multiple polynomials $\{P_n\}$ orthogonal with respect to certain vector of functionals $\mathcal{U} = \left[u^1\ u^2\right]^T$, with

$$\mathcal{P}_n = \left[P_{3n}(x)\ P_{3n+1}(x)\ P_{3n+2}(x)\right]^T, \tag{28}$$

there exists a matrix decomposition

$$\mathcal{P}_n = W_n(x^3)\mathcal{X}_n \to \begin{bmatrix} P_{3n}(x)\\ P_{3n+1}(x)\\ P_{3n+2}(x)\end{bmatrix} = W_n(x^3)\begin{bmatrix}1\\ x\\ x^2\end{bmatrix}, \tag{29}$$

with W_n being the matrix polynomial (see [27])

$$W_n(x) = \begin{bmatrix} A_n^1(x) & A_{n-1}^2(x) & A_{n-1}^3(x) \\ B_n^1(x) & B_n^2(x) & B_{n-1}^3(x) \\ C_n^1(x) & C_n^2(x) & C_n^3(x) \end{bmatrix}. \tag{30}$$

Throughout this section, for simplicity of computations, we assume $d = 2$ for the vector of functionals $\mathcal{U}$, but the same results can be easily extended for an arbitrary number of functionals. We first show that, if a sequence of type II multiple 2-orthogonal polynomials $\{P_n\}$ satisfy a recurrence relation like (10), for example

$$x P_n(x) = P_{n+1}(x) + b_n P_n(x) + c_n P_{n-1}(x) + d_n P_{n-2}(x), \tag{31}$$

then there exists a sequence of matrix polynomials $\{W_n\}$, $W_n(x) \in \mathbb{P}^{3\times 3}$ associated to $\{P_n\}$ by (28) and (29), satisfying a matrix four term recurrence relation with matrix coefficients.

Theorem 6 *Let $\{P_n\}$ be a sequence of type II multiple polynomials, 2-orthogonal with respect to the system of functionals $\{u^1, u^2\}$, i.e., satisfying the four-term type recurrence relation (31). Let $\{W_n\}$, $W_n(x) \in \mathbb{P}^{3\times 3}$ associated to $\{P_n\}$ by (28) and (29). Then, the matrix polynomials W_n satisfy a matrix four term recurrence relation with matrix coefficients.*

Proof We first prove that the sequence of vector polynomials $\{W_n\}$ satisfy a four term recurrence relation with matrix coefficients, starting from the fact that $\{P_n\}$ satisfy (31). In order to get this result, we use the matrix interpretation of multiple orthogonality described in Sect. 2. From (28), using the matrix interpretation for multiple orthogonality, (31) can be rewritten as the matrix three term recurrence relation

$$x\mathcal{P}_n = A\mathcal{P}_{n+1} + B_n\mathcal{P}_n + C_n\mathcal{P}_{n-1}$$

or, equivalently

$$x \begin{bmatrix} P_{3n} \\ P_{3n+1} \\ P_{3n+2} \end{bmatrix} = \begin{bmatrix} 0&0&0 \\ 0&0&0 \\ 1&0&0 \end{bmatrix} \begin{bmatrix} P_{3n+3} \\ P_{3n+4} \\ P_{3n+5} \end{bmatrix}$$

$$+ \begin{bmatrix} b_{3n} & 1 & 0 \\ c_{3n+1} & b_{3n+1} & 1 \\ d_{3n+2} & c_{3n+2} & b_{3n+2} \end{bmatrix} \begin{bmatrix} P_{3n} \\ P_{3n+1} \\ P_{3n+2} \end{bmatrix} + \begin{bmatrix} 0 & d_{3n} & c_{3n} \\ 0 & 0 & d_{3n+1} \\ 0 & 0 & 0 \end{bmatrix} \begin{bmatrix} P_{3n-3} \\ P_{3n-2} \\ P_{3n-1} \end{bmatrix}$$

Multiplying the above expression by x we get

$$\begin{aligned} x^2\mathcal{P}_n &= Ax\mathcal{P}_{n+1} + B_n x\mathcal{P}_n + C_n x\mathcal{P}_{n-1} \\ &= A\left[x\mathcal{P}_{n+1}\right] + B_n\left[x\mathcal{P}_n\right] + C_n\left[x\mathcal{P}_{n-1}\right] \end{aligned}$$

$$
\begin{aligned}
&= AA\mathcal{P}_{n+2} + [AB_{n+1} + B_nA]\mathcal{P}_{n+1} \\
&\quad + [AC_{n+1} + B_nB_n + C_nA]\mathcal{P}_n \\
&\quad + [B_nC_n + C_nB_{n-1}]\mathcal{P}_{n-1} + C_nC_{n-1}\mathcal{P}_{n-2}
\end{aligned} \tag{32}
$$

The matrix A is nilpotent, so AA is the zero matrix of size 3×3. Having

$$
\begin{aligned}
A_n^{\langle 1\rangle} &= AB_{n+1} + B_nA, \quad B_n^{\langle 1\rangle} = AC_{n+1} + B_nB_n + C_nA, \\
C_n^{\langle 1\rangle} &= B_nC_n + C_nB_{n-1}, \; D_n^{\langle 1\rangle} = C_nC_{n-1},
\end{aligned}
$$

where the entries of $A_n^{\langle 1\rangle}$, $B_n^{\langle 1\rangle}$, and $C_n^{\langle 1\rangle}$ can be easily obtained using a computational software as Mathematica® or Maple®, from the entries of A_n, B_n, and C_n. Thus, we can rewrite (32) as

$$
x^2\mathcal{P}_n = A_n^{\langle 1\rangle}\mathcal{P}_{n+1} + B_n^{\langle 1\rangle}\mathcal{P}_n + C_n^{\langle 1\rangle}\mathcal{P}_{n-1} + D_n^{\langle 1\rangle}\mathcal{P}_{n-2}.
$$

We now continue in this fashion,multiplying again by x

$$
\begin{aligned}
x^3\mathcal{P}_n &= A_n^{\langle 1\rangle}x\mathcal{P}_{n+1} + B_n^{\langle 1\rangle}x\mathcal{P}_n + C_n^{\langle 1\rangle}x\mathcal{P}_{n-1} + D_n^{\langle 1\rangle}x\mathcal{P}_{n-2} \\
&= A_n^{\langle 1\rangle}A\mathcal{P}_{n+2} + \left[A_n^{\langle 1\rangle}B_{n+1} + B_n^{\langle 1\rangle}A\right]\mathcal{P}_{n+1} \\
&\quad + \left[A_n^{\langle 1\rangle}C_{n+1} + B_n^{\langle 1\rangle}B_n + C_n^{\langle 1\rangle}A\right]\mathcal{P}_n \\
&\quad + \left[B_n^{\langle 1\rangle}C_n + C_n^{\langle 1\rangle}B_{n-1} + D_n^{\langle 1\rangle}A\right]\mathcal{P}_{n-1} \\
&\quad + \left[C_n^{\langle 1\rangle}C_{n-1} + D_n^{\langle 1\rangle}B_{n-2}\right]\mathcal{P}_{n-2} + D_n^{\langle 1\rangle}C_{n-2}\mathcal{P}_{n-3}
\end{aligned}
$$

The matrix products $A_n^{\langle 1\rangle}A$ and $D_n^{\langle 1\rangle}C_{n-2}$ both give the zero matrix of size 3×3, and the remaining matrix coefficients are

$$
\begin{aligned}
A_n^{\langle 2\rangle} &= A_n^{\langle 1\rangle}B_{n+1} + B_n^{\langle 1\rangle}A, \qquad\qquad B_n^{\langle 2\rangle} = A_n^{\langle 1\rangle}C_{n+1} + B_n^{\langle 1\rangle}B_n + C_n^{\langle 1\rangle}A, \\
C_n^{\langle 2\rangle} &= B_n^{\langle 1\rangle}C_n + C_n^{\langle 1\rangle}B_{n-1} + D_n^{\langle 1\rangle}A, \; D_n^{\langle 2\rangle} = C_n^{\langle 1\rangle}C_{n-1} + D_n^{\langle 1\rangle}B_{n-2}.
\end{aligned} \tag{33}
$$

Using the expressions stated above, the matrix coefficients (33) can be easily obtained as well. Beyond the explicit expression of their respective entries, the key point is that they are structured matrices, namely $A_n^{\langle 2\rangle}$ is lower-triangular with one's in the main diagonal, $D_n^{\langle 2\rangle}$ is upper-triangular, and $B_n^{\langle 2\rangle}$, $C_n^{\langle 2\rangle}$ are full matrices. Therefore, the sequence of type II vector multiple orthogonal polynomials satisfy the following matrix four term recurrence relation

$$
x^3\mathcal{P}_n = A_n^{\langle 2\rangle}\mathcal{P}_{n+1} + B_n^{\langle 2\rangle}\mathcal{P}_n + C_n^{\langle 2\rangle}\mathcal{P}_{n-1} + D_n^{\langle 2\rangle}\mathcal{P}_{n-2}, \; n = 2, 3, \dots, \tag{34}
$$

Next, combining (29) with (34), we can assert that

$$x^3 W_n(x^3)\begin{bmatrix}1\\x\\x^2\end{bmatrix} = A_n^{\langle 2\rangle} W_{n+1}(x^3)\begin{bmatrix}1\\x\\x^2\end{bmatrix} + B_n^{\langle 2\rangle} W_n(x^3)\begin{bmatrix}1\\x\\x^2\end{bmatrix}$$
$$+ C_n^{\langle 2\rangle} W_{n-1}(x^3)\begin{bmatrix}1\\x\\x^2\end{bmatrix} + D_n^{\langle 2\rangle} W_{n-2}(x^3)\begin{bmatrix}1\\x\\x^2\end{bmatrix}, \quad n=2,3,\ldots.$$

Next, we take rid of the vector $\left[1\ x\ x^2\right]^T$ in the above equation. It could be argued that a matrix can have a nontrivial kernel and therefore this removal can not be performed, but it is the special way to construct the matrix polynomials W_n and the matrix coefficients $A_n^{\langle 2\rangle}, \ldots, D_n^{\langle 2\rangle}$ which allows us to do so (see for example [20] where similar techniques are employed). After the shift $x^3 \to x$ the above expression may be simplified as

$$xW_n(x) = A_n^{\langle 2\rangle} W_{n+1}(x) + B_n^{\langle 2\rangle} W_n(x) + C_n^{\langle 2\rangle} W_{n-1}(x) + D_n^{\langle 2\rangle} W_{n-2}(x), \tag{35}$$

where $W_{-1} = 0_{3\times 3}$ and W_0 is certain constant matrix, for every $n = 1, 2, \ldots$, which is the desired matrix four term recurrence relation for $W_n(x)$. □

This kind of matrix high-term recurrence relation completely characterizes certain type of orthogonality. Hence, we are going prove a Favard type Theorem which states that, under the assumptions of Theorem 6, the matrix polynomials $\{W_n\}$ are *type II matrix multiple orthogonal* with respect to a system of two matrices of measures $\{d\mathbf{M}^1, d\mathbf{M}^2\}$.

Next, we briefly review some of the standard facts on the theory of *matrix orthogonality*, or *orthogonality with respect to a matrix of measures* (see [18, 19] and the references therein). Let $W, V \in \mathbb{P}^{3\times 3}$ be two matrix polynomials, and let

$$\mathbf{M}(x) = \begin{bmatrix}\mu_{11}(x) & \mu_{12}(x) & \mu_{13}(x)\\ \mu_{21}(x) & \mu_{22}(x) & \mu_{23}(x)\\ \mu_{31}(x) & \mu_{32}(x) & \mu_{33}(x)\end{bmatrix},$$

be a matrix with positive Borel measures $\mu_{i,j}(x)$ in its entries. One says that $\mathbf{M}(E)$ is a positive definite weight matrix supported in the real line if $\mathbf{M}(E)$ is positive semidefinite for any Borel set $E \subset \mathbb{R}$, it has finite moments

$$\bar{\mu}_k = \int_E d\mathbf{M}(x)x^k, \quad k = 0, 1, 2, \ldots$$

of every order, and it satisfies that

$$\int_E V(x)d\mathbf{M}(x)V^*(x),$$

where $V^* \in \mathbb{P}^{3\times 3}$ is the adjoint matrix of $V \in \mathbb{P}^{3\times 3}$, is non-singular if the matrix leading coefficient of the matrix polynomial V is non-singular. Under these conditions, it is possible to associate to a weight matrix $\mathbf{M}$ the Hermitian sesquilinear form

$$\langle W, V\rangle = \int_E W(x)d\mathbf{M}(x)V^*(x).$$

We then say that a sequence of matrix polynomials $\{W_n\}$, $W_n \in \mathbb{P}^{3\times 3}$ with degree n and nonsingular leading coefficient, is orthogonal with respect to $\mathbf{M}$ if

$$\langle W_m, W_n\rangle = \Delta_n \delta_{m,n}\,, \tag{36}$$

where $\Delta_n \in \mathcal{M}_{3\times 3}$ is a positive definite upper triangular matrix, for $n \geq 0$.

Let $\boldsymbol{m} = (m_1, m_2) \in \mathbb{N}^2$ be a multi-index with length $|\mathbf{m}| := m_1 + \cdots + m_2$ and let $\{\mathbf{M}^1, \mathbf{M}^2\}$ be a set of matrix of measures as defined above. Let $\{W_{\mathbf{m}}\}$ be a sequence of matrix polynomials, with deg $W_{\mathbf{m}}$ is at most $|\mathbf{m}|$. $\{W_{\mathbf{m}}\}$ is said to be a *type II* multiple orthogonal with respect to the set of matrix measures $\{\mathbf{M}^1, \mathbf{M}^2\}$ and multi-index $\boldsymbol{m}$ if it satisfy the following orthogonality conditions

$$\int_E W(x)d\mathbf{M}^j(x)x^k = 0_{3\times 3}\,, \quad k = 0, 1, \ldots, n_j - 1\,, \quad j = 1, 2.$$

A multi-index $\boldsymbol{m}$ is said to be *normal* for the set of matrix measures $\{\mathbf{M}^1, \mathbf{M}^2\}$, if the degree of $W_{\mathbf{m}}$ is exactly $|\mathbf{m}| = m$. Thus, in what follows we will write $W_{\mathbf{m}} \equiv W_{|\mathbf{m}|} = W_m$. In this framework, let consider the sequence of *vector of matrix polynomials* $\{\mathcal{B}_n\}$ where

$$\mathcal{B}_n = \begin{bmatrix} W_{2n} \\ W_{2n+1} \end{bmatrix}. \tag{37}$$

We can define the vector of functionals $\mathfrak{M}(\mathcal{B}_n)$, associated to a set of positive definite matrix measures $\{\mathbf{M}^1, \mathbf{M}^2\}$ and with $\mathfrak{M} : \mathbb{P}^{6\times 3} \to \mathcal{M}_{6\times 6}$, by means of the action $\mathfrak{M}$ on $\mathcal{B}_n$, as follows

$$\mathfrak{M}(\mathcal{B}_n) := \left(\mathfrak{M}\cdot\mathcal{B}_n^T\right)^T =$$

$$\begin{bmatrix} \int_E W_{2n}(x)d\mathbf{M}^1(x) & \int_E W_{2n}(x)d\mathbf{M}^2(x) \\ \int_E W_{2n+1}(x)d\mathbf{M}^1(x) & \int_E W_{2n+1}(x)d\mathbf{M}^2(x) \end{bmatrix} \in \mathcal{M}_{6\times 6}\,. \tag{38}$$

Notice that

$$\int W_j(x)d\mathbf{M}^i = \Delta^i_j, \quad i = 1, 2, \text{ and } j = 0, 1, \tag{39}$$

with $\{\mathbf{M}^1, \mathbf{M}^2\}$ being a system of two matrix of measures as described above. Thus, we say that a sequence of vectors of matrix polynomials $\{\mathcal{B}_n\}$ is orthogonal with respect to $\mathfrak{M}$ if

$$\left.\begin{array}{l} i)\ \mathfrak{M}(x^k\mathcal{B}_n) = 0_{6\times 6}\,,\ k = 0, 1, \dots, n-1\,, \\ ii)\ \mathfrak{M}(x^n\mathcal{B}_n) = \Omega_n\,, \end{array}\right\} \tag{40}$$

where Ω_n is a regular block upper triangular 6×6 matrix, holds.

Now we are in a position to prove the following

Theorem 7 (Favard Type) *Let $\{\mathcal{B}_n\}$ a sequence of vectors of matrix polynomials of size 6×3, defined in (37), with W_n matrix polynomials satisfying the four term recurrence relation (35). The following statements are equivalent:*

(a) The sequence $\{\mathcal{B}_n\}_{n\geq 0}$ is orthogonal with respect to $\mathfrak{M}$, which is associated to a system of two matrix measures $\{\mathbf{M}^1, \mathbf{M}^2\}$ as described above.
(b) There are sequences of scalar 6×6 block matrices $\{A_n^{\langle 3\rangle}\}_{n\geq 0}$, $\{B_n^{\langle 3\rangle}\}_{n\geq 0}$, and $\{C_n^{\langle 3\rangle}\}_{n\geq 0}$, with $C_n^{\langle 3\rangle}$ block upper triangular non-singular matrix for $n \in \mathbb{N}$, such that the sequence $\{\mathcal{B}_n\}$ satisfy the matrix three term recurrence relation

$$x\mathcal{B}_n = A_n^{\langle 3\rangle}\mathcal{B}_{n+1} + B_n^{\langle 3\rangle}\mathcal{B}_n + C_n^{\langle 3\rangle}\mathcal{B}_{n-1}, \tag{41}$$

with $\mathcal{B}_{n-1} = \begin{bmatrix} 0_{3\times 3} & 0_{3\times 3}\end{bmatrix}^T$, $\mathcal{B}_0$ given, and $C_n^{\langle 3\rangle}$ non-singular.

Proof First we prove that (a) implies (b). Since the sequence of vector of matrix polynomials $\{\mathcal{B}_n\}$ is a basis in the linear space $\mathbb{P}^{6\times 3}$, we can write

$$x\mathcal{B}_n = \sum_{k=0}^{n+1} \widetilde{A}_k^n \mathcal{B}_k, \quad \widetilde{A}_k^n \in \mathcal{M}_{6\times 6}\,.$$

Then, from the orthogonality conditions (40), we get

$$\mathfrak{M}\,(x\mathcal{B}_n) = \mathfrak{M}(\widetilde{A}_k^n\mathcal{B}_k) = 0_{6\times 6}, \quad k = 0, 1, \dots, n-2.$$

Thus,

$$x\mathcal{B}_n = \widetilde{A}_{n+1}^n\mathcal{B}_{n+1} + \widetilde{A}_n^n\mathcal{B}_n + \widetilde{A}_{n-1}^n\mathcal{B}_{n-1}\,.$$

Having $\widetilde{A}_{n+1}^n = A_n^{\langle 3\rangle}$, $\widetilde{A}_n^n = B_n^{\langle 3\rangle}$, and $\widetilde{A}_{n-1}^n = C_n^{\langle 3\rangle}$, the result follows.

To proof that (*b*) implies (*a*), we know from Theorem 6 that the sequence of vector polynomials $\{W_n\}$ satisfy a four term recurrence relation with matrix coefficients. We can associate this matrix four term recurrence relation (35) with the block matrix three term recurrence relation (41). Then, it is sufficient to show that $\mathfrak{M}$ is uniquely determined by its orthogonality conditions (40), in terms of the sequence $\{C_n^{\langle 3\rangle}\}_{n\geq 0}$ in that (41).

Next, from (37) we can rewrite the matrix four term recurrence relation (35) into a matrix three term recurrence relation

$$x\mathcal{B}_n = \begin{bmatrix} 0_{3\times 3} & 0_{3\times 3} \\ A_{2n+1}^{\langle 2\rangle} & 0_{3\times 3} \end{bmatrix} \mathcal{B}_{n+1} + \begin{bmatrix} B_{2n}^{\langle 2\rangle} & A_{2n}^{\langle 2\rangle} \\ C_{2n+1}^{\langle 2\rangle} & B_{2n+1}^{\langle 2\rangle} \end{bmatrix} \mathcal{B}_n + \begin{bmatrix} D_{2n}^{\langle 2\rangle} & C_{2n}^{\langle 2\rangle} \\ 0_{3\times 3} & D_{2n+1}^{\langle 2\rangle} \end{bmatrix} \mathcal{B}_{n-1},$$

where the size of $\mathcal{B}_n$ is 6×3. We give $\mathfrak{M}$ in terms of its block matrix moments, which in turn are given by the matrix coefficients in (41). There is a unique vector moment functional $\mathfrak{M}$ and hence two matrix measures $d\mathbf{M}^1$ and $d\mathbf{M}^2$, such that

$$\mathfrak{M}(\mathcal{B}_0) = \begin{bmatrix} \int_E W_0(x)d\mathbf{M}^1(x) & \int_E W_0(x)d\mathbf{M}^2(x) \\ \int_E W_1(x)d\mathbf{M}^1(x) & \int_E W_1(x)d\mathbf{M}^2(x) \end{bmatrix} = C_0^{\langle 3\rangle} \in \mathcal{M}_{6\times 6}.$$

For the first moment of $\mathfrak{M}_0$ we thus get

$$\mathfrak{M}_0 = \mathfrak{M}(\mathcal{B}_0) = \begin{bmatrix} \Delta_0^1 & \Delta_0^2 \\ \Delta_1^1 & \Delta_1^2 \end{bmatrix} = C_0^{\langle 3\rangle}.$$

Hence, we have $\mathfrak{M}(\mathcal{B}_1) = 0_{6\times 6}$, which from (41) also implies $0_{6\times 6} = x\mathfrak{M}(\mathcal{B}_0) - B_0^{\langle 3\rangle}\mathfrak{M}(\mathcal{B}_0) = \mathfrak{M}_1 - B_0^{\langle 3\rangle}\mathfrak{M}_0$. Therefore

$$\mathfrak{M}_1 = B_0^{\langle 3\rangle} C_0^{\langle 3\rangle}.$$

By a similar argument, we have

$$\begin{aligned} 0_{6\times 6} &= \mathfrak{M}\Big((A_0^{\langle 3\rangle})^{-1}x^2\mathcal{B}_0 - (A_0^{\langle 3\rangle})^{-1}B_0^{\langle 3\rangle}x\mathcal{B}_0 - B_1^{\langle 3\rangle}(A_0^{\langle 3\rangle})^{-1}x\mathcal{B}_0 \\ &\quad + B_1(A_0^{\langle 3\rangle})^{-1}B_0^{\langle 3\rangle}\mathcal{B}_0 - C_1^{\langle 3\rangle}\mathcal{B}_0\Big) \\ &= (A_0^{\langle 3\rangle})^{-1}\mathfrak{M}_2 - \Big((A_0^{\langle 3\rangle})^{-1}B_0^{\langle 3\rangle} + B_1^{\langle 3\rangle}(A_0^{\langle 3\rangle})^{-1}\Big)\mathfrak{M}_1 \\ &\quad + \Big(B_1^{\langle 3\rangle}(A_0^{\langle 3\rangle})^{-1}B_0^{\langle 3\rangle} - C_1^{\langle 3\rangle}\Big)\mathfrak{M}_0 \in \mathcal{M}_{6\times 6}, \end{aligned}$$

which in turn yields the second moment of $\mathfrak{M}$

$$\mathfrak{M}_2 = \Big(B_0^{\langle 3\rangle} + A_0^{\langle 3\rangle}B_1^{\langle 3\rangle}(A_0^{\langle 3\rangle})^{-1}\Big)B_0^{\langle 3\rangle}C_0^{\langle 3\rangle} - A_0^{\langle 3\rangle}\Big(B_1^{\langle 3\rangle}(A_0^{\langle 3\rangle})^{-1}B_0^{\langle 3\rangle} - C_1^{\langle 3\rangle}\Big)C_0^{\langle 3\rangle}.$$

Repeated application of this inductive process, enables us to determine $\mathfrak{M}$ in a unique way through its moments, only in terms of the sequences of matrix coefficients $\{A_n^{\langle 3\rangle}\}_{n\geq 0}$, $\{B_n^{\langle 3\rangle}\}_{n\geq 0}$ and $\{C_n^{\langle 3\rangle}\}_{n\geq 0}$.

On the other hand, because of (38) and (41) we have

$$\mathfrak{M}\left(x\mathcal{B}_n\right) = 0_{6\times 6}, \quad n \geq 2.$$

Multiplying by x both sides of (41), from the above result we get

$$\mathfrak{M}\left(x^2\mathcal{B}_n\right) = 0_{6\times 6}, \quad n \geq 3.$$

The same conclusion can be drawn for $0 < k < n$

$$\mathfrak{M}\left(x^k\mathcal{B}_n\right) = 0_{6\times 6}, \quad 0 < k < n, \tag{42}$$

and finally

$$\mathfrak{M}\left(x^n\mathcal{B}_n\right) = C_n^{\langle 3\rangle}\mathfrak{M}\left(x^{n-1}\mathcal{B}_{n-1}\right).$$

Notice that the repeated application of the above argument leads to

$$\mathfrak{M}\left(x^n\mathcal{B}_n\right) = C_n^{\langle 3\rangle}C_{n-1}^{\langle 3\rangle}C_{n-2}^{\langle 3\rangle}\cdots C_1^{\langle 3\rangle}C_0^{\langle 3\rangle}. \tag{43}$$

From (42), (43) we conclude

$$\begin{aligned}\mathfrak{M}\left(x^k\mathcal{B}_n\right) &= C_n^{\langle 3\rangle}C_{n-1}^{\langle 3\rangle}C_{n-2}^{\langle 3\rangle}\cdots C_1^{\langle 3\rangle}C_0^{\langle 3\rangle}\delta_{n,k} \\ &= \Omega_n\delta_{n,k}, \quad n,k = 0,1,\ldots, \; k \leq n,\end{aligned}$$

which are exactly the desired orthogonality conditions (40) for $\mathfrak{M}$ stated in the theorem. □

6 Example with Multiple Hermite Polynomials

The monic type II multiple orthogonal polynomials $H_{n_1,n_2}^{\varsigma_1,\varsigma_2}(x)$ are orthogonal with respect to the pair of Hermite-like weights $w_j(x) = e^{-x^2+\varsigma_j x}$, $j = 1, 2$, so the following orthogonality conditions hold (see [4], [24, Â§23.5] or [31])

$$\int_{\mathbb{R}} x^k H_{n_1,n_2}^{\varsigma_1,\varsigma_2}(x)e^{-x^2+\varsigma_j x}dx = 0, \quad k = 0,1,\ldots,n_j-1, \quad j = 1,2$$

and

$$\int_{\mathbb{R}} x^{n_j} H_{n_1,n_2}^{\varsigma_1,\varsigma_2}(x) e^{-x^2+\varsigma_j x}\,dx = 2^{-|\boldsymbol{n}|}\sqrt{\pi}\, n_j! \prod_{i\neq j}(\varsigma_j-\varsigma_i)^{n_i} e^{\varsigma_j^2/4}, \qquad j=1,2.$$

Throughout this example we will use the recurrence coefficients given in [31]. The monic polynomials $H_n^{\boldsymbol{\varsigma}}(x) \equiv H_{n_1,n_2}^{\varsigma_1,\varsigma_2}$, where $\boldsymbol{\varsigma} = (\varsigma_1, \varsigma_2)$ and $n = n_1 + n_2$ satisfy the four term recurrence relation (31)

$$\begin{aligned} xH_n^{\boldsymbol{\varsigma}}(x) &= H_{n+1}^{\boldsymbol{\varsigma}}(x) + b_n H_n^{\boldsymbol{\varsigma}}(x) + c_n H_{n-1}^{\boldsymbol{\varsigma}}(x) + d_n H_{n-2}^{\boldsymbol{\varsigma}}(x)\,, \\ H_{-2}^{\boldsymbol{\varsigma}}(x) &= H_{-1}^{\boldsymbol{\varsigma}}(x) = 0,\ H_0^{\boldsymbol{\varsigma}}(x) = 1,\ H_1^{\boldsymbol{\varsigma}}(x) = x - b_0, \end{aligned} \tag{44}$$

where, following [31], the recurrence coefficients are given by

$$\begin{aligned} b_n &= \begin{cases} b_{2n} &= \dfrac{\varsigma_1}{2}, \quad \text{if } n_1 = n_2 = n, \\ b_{2n+1} &= \dfrac{\varsigma_2}{2}, \quad \text{if } n_1 = n+1,\ n_2 = n, \end{cases} \\ c_n &= \begin{cases} c_{2n} &= n, \quad \text{if } n_1 = n_2 = n, \\ c_{2n+1} &= \dfrac{2n+1}{2}, \quad \text{if } n_1 = n+1,\ n_2 = n, \end{cases} \\ d_n &= \begin{cases} d_{2n} &= \dfrac{n(\varsigma_1-\varsigma_2)}{4}, \quad \text{if } n_1 = n_2 = n, \\ d_{2n+1} &= \dfrac{n(\varsigma_2-\varsigma_1)}{4}, \quad \text{if } n_1 = n+1,\ n_2 = n. \end{cases} \end{aligned} \tag{45}$$

Observe that this notation means that (44) can be read for polynomials $H_{2n}^{\boldsymbol{\varsigma}}(x) = H_{n,n}^{\boldsymbol{\varsigma}}(x)$ (when $n_1 = n_2 = n$) and $H_{2n+1}^{\boldsymbol{\varsigma}}(x) = H_{n+1,n}^{\boldsymbol{\varsigma}}(x)$ (when $n_1 = n+1,\ n_2 = n$). It also implies that we take the diagonal multi-index $\boldsymbol{n} = (n_1, n_2) \in \mathcal{I}$, where

$$\mathcal{I} = \{(0,0),\ (1,0),\ (1,1),\ (2,1),\ (2,2),\ (3,2),\ (3,3),\ldots\}.$$

The associated four banded matrix J_1 is

$$J_1 = \begin{bmatrix} \varsigma_1/2 & 1 & & & & & \\ 1/2 & \varsigma_2/2 & 1 & & & & \\ \frac{(\varsigma_1-\varsigma_2)}{4} & 1 & \varsigma_1/2 & 1 & & & \\ & \frac{-(\varsigma_1-\varsigma_2)}{4} & 3/2 & \varsigma_2/2 & 1 & & \\ & & \frac{(\varsigma_1-\varsigma_2)}{2} & 2 & \varsigma_1/2 & 1 & \\ & & & \frac{-(\varsigma_1-\varsigma_2)}{2} & 5/2 & \varsigma_2/2 & \ddots \\ & & & & \ddots & \ddots & \ddots \end{bmatrix} \tag{46}$$

For the sake of simplicity, we henceforth assume a particular example in the sequel, having $\varsigma_1 = 1$ and $\varsigma_2 = 2$. Thus

$$J_1 = \begin{bmatrix} 1/2 & 1 & & & & & \\ 1/2 & 1 & 1 & & & & \\ -1/4 & 1 & 1/2 & 1 & & & \\ & 1/4 & 3/2 & 1 & 1 & & \\ & & -1/2 & 2 & 1/2 & 1 & \\ & & & 1/2 & 5/2 & 1 & \ddots \\ & & & & \ddots & \ddots & \ddots \end{bmatrix}, \tag{47}$$

and using Mathematica® we can obtain the explicit expression for the first polynomials of the sequence

$$\begin{aligned} H_0^{1,2}(x) &= 1 & H_3^{1,2}(x) &= x^3 - 2x^2 - \tfrac{x}{4} + \tfrac{3}{4}, \\ H_1^{1,2}(x) &= x - \tfrac{1}{2}, & H_4^{1,2}(x) &= x^4 - 3x^3 + \tfrac{1}{4}x^2 + 3x - \tfrac{5}{8}, \\ H_2^{1,2}(x) &= x^2 - \tfrac{3}{2}x, & H_5^{1,2}(x) &= x^5 - \tfrac{7}{2}x^4 - \tfrac{1}{4}x^3 + \tfrac{59}{8}x^2 - \tfrac{19}{8}x - \tfrac{19}{16}, \end{aligned}$$

$$\begin{aligned} H_6^{1,2}(x) &= x^6 - \tfrac{9}{2}x^5 + \tfrac{3}{4}x^4 + \tfrac{117}{8}x^3 - \tfrac{75}{8}x^2 - \tfrac{99}{16}x + \tfrac{19}{8}, \\ H_7^{1,2}(x) &= x^7 - 5x^6 + \tfrac{51}{2}x^4 - \tfrac{291}{16}x^3 - \tfrac{375}{16}x^2 + \tfrac{475}{32}x + \tfrac{61}{32}, \\ H_8^{1,2}(x) &= x^8 - 6x^7 + \tfrac{3}{2}x^6 + \tfrac{81}{2}x^5 - \tfrac{699}{16}x^4 - \tfrac{225}{4}x^3 + \tfrac{1049}{16}x^2 + \tfrac{21}{2}x - \tfrac{597}{64}, \end{aligned}$$

$$\begin{aligned} H_9^{1,2}(x) &= x^9 - \tfrac{13}{2}x^8 + \tfrac{1}{2}x^7 + \tfrac{243}{4}x^6 - \tfrac{1095}{16}x^5 \\ &\quad - \tfrac{4341}{32}x^4 + \tfrac{2897}{16}x^3 + \tfrac{1987}{32}x^2 - \tfrac{5129}{64}x - \tfrac{75}{128}, \\ H_{10}^{1,2}(x) &= x^{10} - \tfrac{15}{2}x^9 + \tfrac{5}{2}x^8 + \tfrac{345}{4}x^7 - \tfrac{2095}{16}x^6 - \tfrac{7983}{32}x^5 \\ &\quad + \tfrac{7805}{16}x^4 + \tfrac{4875}{32}x^3 - \tfrac{26485}{64}x^2 + \tfrac{2235}{128}x + \tfrac{1301}{32} \\ H_{11}^{1,2}(x) &= x^{11} - 8x^{10} + \tfrac{5}{4}x^9 + \tfrac{475}{4}x^8 - \tfrac{2945}{16}x^7 - \tfrac{3887}{8}x^6 + \tfrac{64343}{64}x^5 \\ &\quad + \tfrac{34055}{64}x^4 - \tfrac{11725}{8}x^3 - \tfrac{265}{64}x^2 + \tfrac{114113}{256}x - \tfrac{7439}{256} \end{aligned}$$

$$\cdots$$

6.1 *The Matrix Type II Multiple Orthogonal Polynomials*

From the above explicit polynomial expressions, and using the method detailed at the beginning of Sect. 5, it is not dificult to obtain the matrix cubic decomposition (30) of the vector polynomials

$$\mathcal{H}_n^{1,2} = W_n(x^3)\mathcal{X}_n \to \begin{bmatrix} H_{3n}^{1,2}(x) \\ H_{3n+1}^{1,2}(x) \\ H_{3n+2}^{1,2}(x) \end{bmatrix} = W_n(x^3) \begin{bmatrix} 1 \\ x \\ x^2 \end{bmatrix}.$$

The first cubic decompositions, which are in turn matrix polynomials, are

$$W_0(x) = \begin{bmatrix} 1 & 0 & 0 \\ -1/2 & 1 & 0 \\ 0 & -3/2 & 1 \end{bmatrix},$$

$$W_1(x) = \begin{bmatrix} 1 & 0 & 0 \\ -3 & 1 & 0 \\ -1/4 & -7/2 & 1 \end{bmatrix} x + \begin{bmatrix} 3/4 & -1/4 & -2 \\ -5/8 & 3 & 1/4 \\ -19/16 & -19/8 & 59/8 \end{bmatrix},$$

$$W_2(x) = \begin{bmatrix} 1 & 0 & 0 \\ -5 & 1 & 0 \\ 3/2 & -6 & 1 \end{bmatrix} x^2 + \begin{bmatrix} 117/8 & 3/4 & -9/2 \\ -291/16 & 51/2 & 0 \\ -225/4 & -699/16 & 81/2 \end{bmatrix} x + \begin{bmatrix} 19/8 & -99/16 & -75/8 \\ 61/32 & 475/32 & -375/16 \\ -597/64 & 21/2 & 1049/16 \end{bmatrix}, \tag{48}$$

$$W_3(x) = \begin{bmatrix} 1 & 0 & 0 \\ -15/2 & 1 & 0 \\ 5/4 & -8 & 1 \end{bmatrix} x^3 + \begin{bmatrix} 243/4 & 1/2 & -13/2 \\ -2095/16 & 345/4 & 5/2 \\ -3887/8 & -2945/16 & 475/4 \end{bmatrix} x^2$$

$$+ \begin{bmatrix} 2897/16 & -4341/32 & -1095/16 \\ 4875/32 & 7805/16 & -7983/32 \\ -11725/8 & 34055/64 & 64343/64 \end{bmatrix} x + \begin{bmatrix} -75/128 & -5129/64 & 1987/32 \\ 1301/32 & 2235/128 & -26485/64 \\ -7439/256 & 114113/256 & -265/64 \end{bmatrix}$$

$$\cdots$$

Next, we show that Theorem 6, and more specifically expression (35), hold in our numeric example for the above set of matrix polynomials. Combining (45) with (33), the first matrix coefficients in (35) are given by

$$n = 0 \begin{cases} A_0^{\langle 2\rangle} = \begin{bmatrix} 1 & 0 & 0 \\ 5/2 & 1 & 0 \\ 25/4 & 2 & 1 \end{bmatrix}, & B_0^{\langle 2\rangle} = \begin{bmatrix} 7/8 & 13/4 & 2 \\ 9/8 & 19/4 & 19/4 \\ 3/16 & 39/8 & 37/8 \end{bmatrix}, \\ C_0^{\langle 2\rangle} = \begin{bmatrix} 0 & 0 & 0 \\ 0 & 0 & 0 \\ 0 & 0 & 0 \end{bmatrix}, & D_0^{\langle 2\rangle} = \begin{bmatrix} 0 & 0 & 0 \\ 0 & 0 & 0 \\ 0 & 0 & 0 \end{bmatrix}, \end{cases}$$

$$
n=1\left\{\begin{array}{ll}
A_1^{\langle 2\rangle}=\begin{bmatrix}1&0&0\\2&1&0\\43/4&5/2&1\end{bmatrix}, & B_1^{\langle 2\rangle}=\begin{bmatrix}10&31/4&5/2\\63/4&67/8&37/4\\63/4&183/8&61/4\end{bmatrix},\\
C_1^{\langle 2\rangle}=\begin{bmatrix}3/8&39/8&9\\-9/16&27/8&27/8\\3/16&9/8&57/8\end{bmatrix}, & D_1^{\langle 2\rangle}=\begin{bmatrix}0&0&0\\0&0&0\\0&0&0\end{bmatrix},
\end{array}\right.
$$

$$
n=2\left\{\begin{array}{ll}
A_2^{\langle 2\rangle}=\begin{bmatrix}1&0&0\\5/2&1&0\\61/4&2&1\end{bmatrix}, & B_2^{\langle 2\rangle}=\begin{bmatrix}97/8&49/4&2\\171/4&41/2&55/4\\75/4&111/2&127/8\end{bmatrix},\\
C_2^{\langle 2\rangle}=\begin{bmatrix}129/8&153/16&261/8\\39/8&105/4&261/8\\3/2&-75/8&177/4\end{bmatrix}, & D_2^{\langle 2\rangle}=\begin{bmatrix}-3/32&3/8&-51/16\\0&3/32&15/16\\0&0&-3/8\end{bmatrix},
\end{array}\right.
$$

$$\ldots$$

Using the above matrices, and those for the matrix polynomials $W_n(x)$ in (48), it is trivial to check that

$$
\begin{aligned}
&\text{for } n=0 \to xW_0(x)=A_0^{\langle 2\rangle}W_1(x)+B_0^{\langle 2\rangle}W_0(x)+C_0^{\langle 2\rangle}W_{-1}(x)+D_0^{\langle 2\rangle}W_{n-2}(x),\\
&\text{for } n=1 \to xW_1(x)=A_1^{\langle 2\rangle}W_2(x)+B_1^{\langle 2\rangle}W_1(x)+C_1^{\langle 2\rangle}W_0(x)+D_1^{\langle 2\rangle}W_{-1}(x),\\
&\text{for } n=2 \to xW_2(x)=A_2^{\langle 2\rangle}W_3(x)+B_2^{\langle 2\rangle}W_2(x)+C_2^{\langle 2\rangle}W_1(x)+D_2^{\langle 2\rangle}W_0(x),\\
&\qquad\ldots
\end{aligned}
$$

6.2 *The Symmetric Type II Multiple Orthogonal Polynomial Sequence*

In what follows, we give an example satisfying the theory developed in Sect. 4. We begin with the multiple orthogonal polynomial sequence $\{H_n^{1,2}\}_{n\geq 1}$ and we find the associated symmetric multiple orthogonal polynomial sequence $\{S_n\}_{n\geq 1}$. All the matrices in the sequel are two, three or four banded so, we abbreviate them just giving their main ($\mathbf{0}$), sub $(-1,-2,\ldots)$ and/or supra $(+1,+2,\ldots)$ diagonals. Thus, having $C=1$ we seek the LU factorization of the semi-infinite four banded matrix

$$
J_1-I=\left\{\begin{array}{rcccccccccccc}
+1: & 1 & 1 & 1 & 1 & 1 & 1 & 1 & 1 & 1 & 1 & 1 & \cdots\\
\mathbf{0}: & -1/2 & 0 & -1/2 & 0 & -1/2 & 0 & -1/2 & 0 & -1/2 & 0 & -1/2 & \cdots\\
-1: & 1/2 & 1 & 3/2 & 2 & 5/2 & 3 & 7/2 & 4 & 9/2 & 5 & 11/2 & \cdots\\
-2: & -1/4 & 1/4 & -1/2 & 1/2 & -3/4 & 3/4 & -1 & 1 & -5/4 & 5/4 & -3/2 & \cdots
\end{array}\right.
$$

Then we have

$$L = \begin{cases} \mathbf{0}: & 1 \quad 1 \quad 1 \quad 1 \quad 1 \quad 1 \quad 1 \quad 1 \quad 1 \quad 1 \quad 1 \quad \cdots \\ -1: & -1 \;\; 1/2 \;\; -5/4 \;\; 6/5 \;\; -21/17 \;\; 29/14 \;\; -9/8 \;\; \frac{260}{81} \;\; \frac{-585}{601} \;\; \frac{1121}{234} \;\; \frac{-1973}{2476} \;\; \cdots \\ -2: & 1/2 \;\; 1/4 \;\; 1/2 \;\; 2/5 \;\; 15/34 \;\; 17/28 \;\; 7/18 \;\; 8/9 \;\; \frac{405}{1202} \;\; \frac{601}{468} \;\; \frac{351}{1238} \;\; \cdots \end{cases},$$

$$U = \begin{cases} +1: & 1 \quad 1 \quad 1 \quad 1 \quad 1 \quad 1 \quad 1 \quad 1 \quad 1 \quad 1 \quad 1 \quad \cdots \\ \mathbf{0}: & -1/2 \;\; 1 \;\; -1 \;\; 5/4 \;\; -17/10 \;\; 21/17 \;\; -18/7 \;\; 9/8 \;\; \frac{-601}{162} \;\; \frac{585}{601} \;\; \frac{-619}{117} \;\; \cdots \end{cases}.$$

Next, using the algorithm developed in [9, Th. 3], we obtain the new factorization of $L = L_1 L_2$, with

$$L_1 = \begin{cases} \mathbf{0}: & 1 \quad 1 \quad 1 \quad 1 \quad 1 \quad 1 \quad 1 \quad 1 \quad 1 \quad 1 \quad 1 \quad \cdots \\ -1: & 1 \;\; -1/4 \;\; 1/3 \;\; -6/19 \;\; 19/72 \;\; \frac{-108}{367} \;\; \frac{367}{1430} \;\; \frac{-2860}{10161} \;\; \frac{10161}{39910} \;\; \frac{-99775}{363631} \;\; \frac{363631}{1434202} \;\; \cdots \end{cases},$$

$$L_2 = \begin{cases} \mathbf{0}: & 1 \quad 1 \quad 1 \quad 1 \quad 1 \quad 1 \quad 1 \quad 1 \quad 1 \quad 1 \quad 1 \quad \cdots \\ -1: & -2 \;\; \frac{3}{4} \;\; \frac{-19}{12} \;\; \frac{144}{95} \;\; \frac{-1835}{1224} \;\; \frac{12155}{5138} \;\; \frac{-7903}{5720} \;\; \frac{319280}{91449} \;\; \frac{-29454111}{23985910} \;\; \frac{430977701}{85089654} \;\; \frac{-1865015451}{1775542076} \;\; \cdots \end{cases}.$$

Finally, Theorem 4 now yields the associated symmetric sequence

$$\begin{aligned} S_0^{1,2}(x) &= 1 & S_3^{1,2}(x) &= x^3 + \tfrac{1}{2}, \\ S_1^{1,2}(x) &= x, & S_4^{1,2}(x) &= x^4 - \tfrac{1}{2}x, \\ S_2^{1,2}(x) &= x^2, & S_5^{1,2}(x) &= x^5 + \tfrac{3}{2}x^2, \end{aligned}$$

$$\begin{aligned} S_6^{1,2}(x) &= x^6 + \tfrac{1}{2}x^3 - \tfrac{1}{2}, & S_9^{1,2}(x) &= x^9 + x^6 - \tfrac{5}{4}x^3 - \tfrac{1}{2}, \\ S_7^{1,2}(x) &= x^7 + \tfrac{3}{4}x^4 - \tfrac{5}{8}x, & S_{10}^{1,2}(x) &= x^{10} + \tfrac{2}{3}x^7 - \tfrac{3}{2}x^4 - \tfrac{7}{24}x, \\ S_8^{1,2}(x) &= x^8 - \tfrac{7}{4}x^2, & S_{11}^{1,2}(x) &= x^{11} + \tfrac{9}{4}x^8 - \tfrac{3}{2}x^5 - \tfrac{49}{16}x^2, \end{aligned}$$

$$\begin{aligned} S_{12}^{1,2}(x) &= x^{12} + x^9 - \tfrac{11}{4}x^6 - \tfrac{3}{2}x^3 + \tfrac{5}{8}, \\ S_{13}^{1,2}(x) &= x^{13} + \tfrac{25}{19}x^{10} - \tfrac{193}{76}x^7 - \tfrac{75}{38}x^4 + \tfrac{81}{152}x, \\ S_{14}^{1,2}(x) &= x^{14} - \tfrac{1}{5}x^{11} - \tfrac{119}{20}x^8 + \tfrac{3}{10}x^5 + \tfrac{207}{40}x^2, \\ S_{15}^{1,2}(x) &= x^{15} + \tfrac{3}{2}x^{12} - \tfrac{17}{4}x^9 - \tfrac{35}{8}x^6 + \tfrac{21}{8}x^3 + \tfrac{17}{16}, \\ S_{16}^{1,2}(x) &= x^{16} + \tfrac{89}{72}x^{13} - \tfrac{331}{72}x^{10} - \tfrac{1067}{288}x^7 + \tfrac{151}{48}x^4 + \tfrac{59}{64}x, \\ S_{17}^{1,2}(x) &= x^{17} + \tfrac{93}{34}x^{14} - \tfrac{333}{68}x^{11} - \tfrac{101}{8}x^8 + \tfrac{489}{136}x^5 + \tfrac{2361}{272}x^2, \end{aligned}$$

$$\cdots$$

which satisfy the four term recurrence relation $xS_{n+2}^{1,2}(x) = S_{n+3}^{1,2}(x) + \gamma_{n+1}S_n^{1,2}(x)$, with recurrence coefficients

$$\gamma_1 = \frac{-1}{2},\ \gamma_2 = 1,\ \gamma_3 = -2,\ \gamma_4 = 1,\ \gamma_5 = \frac{-1}{4},\ \gamma_6 = \frac{3}{4},\ \gamma_7 = -1,\ \gamma_8 = \frac{1}{3},$$
$$\gamma_9 = \frac{-19}{12},\ \gamma_{10} = \frac{5}{4},\ \gamma_{11} = \frac{-6}{19},\ \gamma_{12} = \frac{144}{95},\ \gamma_{13} = \frac{-17}{10},\ \gamma_{14} = \frac{19}{72},$$
$$\gamma_{15} = \frac{-1835}{1224},\ \gamma_{16} = \frac{21}{17},\ \gamma_{17} = \frac{-108}{367},\ \gamma_{18} = \frac{12155}{5138},\ldots$$

Acknowledgments We would like to thank the referee for carefully reading the manuscript, and for giving constructive comments, which helped us to improve the quality of the paper.

The work of the first author (AB) was supported by the Centre for Mathematics of the University of Coimbra - UIDB/00324/2020, funded by the Portuguese Government through FCT/MCTES. The work of the second author (EJH) was supported by Fundação para a Ciência e a Tecnologia (FCT) of Portugal, ref. SFRH/BPD/91841/2012, and partially supported by Dirección General de Investigación e Innovación, Consejería de Educación e Investigación of the Comunidad de Madrid (Spain), and Universidad de Alcalá under grant CM/JIN/2019-010, Proyectos de I+D para Jóvenes Investigadores de la Universidad de Alcalá 2019.

References

1. Álvarez-Fernández, C., Fidalgo, U., Mañas, M.: Multiple orthogonal polynomials of mixed type: Gauss–Borel factorization and the multi-component 2D Toda hierarchy. Adv. Math. **227**, 1451–1525 (2011)
2. Aptekarev, A.I.: Multiple orthogonal polynomials. J. Comput. Appl. Math. **99**, 423–447 (1998)
3. Aptekarev, A.I., Kaliaguine, V., Iseghem, J.V.: Genetic sum's representation for the moments of a system of Stieltjes functions and its application. Constr. Approx. **16**, 487–524 (2000)
4. Aptekarev, A.I., Branquinho, A., Van Assche, W.: Multiple orthogonal polynomials for classical weights. Trans. Amer. Math. Soc. **335**, 3887–3914 (2003)
5. Aptekarev, A.I., Kalyagin, V.A., Saff, E.B.: Higher-order three-term recurrences and asymptotics of multiple orthogonal polynomials. Constr. Approx. **30** (2), 175–223 (2009)
6. Barrios, D., Branquinho, A.: Complex high order Toda and Volterra lattices. J. Diff. Eq. Appl. **15**(2), 197–213 (2009)
7. Barrios, D., Branquinho, A., Foulquié-Moreno, A.: Dynamics and interpretation of some integrable systems via multiple orthogonal polynomials. J. Math. Anal. Appl. **361**(2), 358–370 (2010)
8. Barrios, D., Branquinho, A., Foulquié-Moreno, A.: On the relation between the full Kostant-Toda lattice and multiple orthogonal polynomials. J. Math. Anal. Appl. **377**(1), 228–238 (2011)
9. Barrios, D., Branquinho, A., Foulquié-Moreno, A.: On the full Kostant-Toda system and the discrete Korteweg-de Vries equations. J. Math. Anal. Appl. **401**(2), 811–820 (2013)
10. Beckermann, B.: On the convergence of bounded J-fractions on the resolvent set of the corresponding second order difference operator. J. Approx. Theory **99**(2), 369–408 (1999)
11. Branquinho, A., Cotrim, L., Foulquié-Moreno, A.: Matrix interpretation of multiple orthogonality. Numer. Algor.**55**, 19–37 (2010)
12. Bueno, M.I., Marcellán, F.: Darboux transformation and perturbation of linear functionals. Linear Algeb. Appl., **384**, 215–242 (2004)

13. Chihara, T.S.: An Introduction to Orthogonal Polynomials. Gordon and Breach, New York (1978)
14. Delvaux, S., López–García, A.: High order three-term recursions, Riemann-Hilbert minors and Nikishin systems on star-like sets. Constr. Approx. **37**(3), 383–453 (2013)
15. Douak, K.: The relation of the d-orthogonal polynomials to the Appell polynomials, J. Comput. Appl. Math., **70** (2), (1996), 279–295.
16. Douak, K., Maroni, P.: Les polynômes orthogonaux "classiques" de dimension deux. Analysis **12**, 71–107 (1992)
17. Douak, K., Maroni, P.: Une caractérisation des polynômes d-orthogonaux classiques. J. Approx. Theory **82**, 177–204 (1995)
18. Durán, A.J., On orthogonal polynomials with respect to a positive definite matrix of measures. Canad. J. Math. **47**, 88–112 (1995)
19. Durán, A.J., Grünbaum, A.: A survey on orthogonal matrix polynomials satisfying second order differential equations. J. Comput. Appl. Math. **178**, 169–190 (2005)
20. Durán, A.J., Van Assche, W.: Orthogonal matrix polynomials and higher-order recurrence relations. Linear Algebra Appl. **219**, 261–280 (1995)
21. Gardner, C.S., Greene, J.M., Kruskal, M.D., Miura, R.M.: A method for solving the Korteweg–de Vries equation. Phys. Rev. Lett. **19**, 1095–1097 (1967)
22. Haneczok, M., Van Assche, W.: Interlacing properties of zeros of multiple orthogonal polynomials, J. Math. Anal. Appl. **389**(1), 429–438 (2012)
23. Isaacson, E., and Bishop–Keller, H.: Analysis of Numerical Methods. Wiley, New York (1994)
24. Ismail, M.E.H.: Classical and quantum orthogonal polynomials in one variable. Encyclopedia of Mathematics and its Applications, vol. 98, Cambridge University Press, Cambridge (2005)
25. Kaliaguine, V.: The operator moment problem, vector continued fractions and an explicit form of the Favard theorem for vector orthogonal polynomials. J. Comput. Appl. Math. **65**(1–3), 181–193 (1995)
26. Marcellán, F., Sansigre, G.: Quadratic decomposition of orthogonal polynomials: a matrix approach, Numer. Algor. **3**, 285–298 (1992)
27. Maroni, P., Mesquita, T.A.: Cubic decomposition of 2-orthogonal polynomial sequences. Mediterr. J. Math. **10**, 843–863 (2013)
28. Maroni, P., Mesquita, T.A., da Rocha, Z.: On the general cubic decomposition of polynomial sequences. J. Differ. Equ. Appl. **17**(9), 1303–1332 (2011)
29. Nikishin, E.M., Sorokin, V.N.: Rational approximants and orthogonality. Translations of Mathematical Monographs, vol. 92. American Mathematical Society, Providence (1991)
30. Peherstorfer, F.: On Toda lattices and orthogonal polynomials. Proceedings of the fifth international symposium on orthogonal polynomials, special functions and their applications (Patras, 1999). J. Comput. Appl. Math. **133**(1–2), 519–534 (2001)
31. Van Assche, W., Coussement, E.: Some classical multiple orthogonal polynomials. J. Comput. Appl. Math. **127**(1–2), 317–347 (2001)

Darboux Transformations for Orthogonal Polynomials on the Real Line and on the Unit Circle

María José Cantero, Francisco Marcellán, Leandro Moral, and Luis Velázquez

Abstract Jacobi matrices constitute a communicating vessel between orthogonal polynomials on the real line (OPRL) and the theory of self-adjoint operators. In this context, commutation methods in operator theory lead to Darboux transformations of Jacobi matrices, which are equivalent to Christoffel transformations of OPRL. Jacobi matrices have unitary analogues known as CMV matrices, which provide a similar link between orthogonal polynomials on the unit circle (OPUC) and the theory of unitary operators. It has been recently shown that Darboux transformations also make sense in this setting, although some differences arise with respect to the Jacobi case. This paper is a survey on Darboux transformations for Jacobi and CMV matrices which present them in a unified way, highlighting their similarities and differences and connecting them to OPRL and OPUC.

Keywords Darboux transformations · Jacobi and CMV matrices · Orthogonal polynomials

1 Introduction

Darboux transformations have remarkable applications in various areas of mathematics and physics such as differential equations, spectral theory, integrable systems, signal theory or quantum physics. In a simple way, Darboux transfor-

---

M. J. Cantero (✉) · L. Moral · L. Velázquez
Department of Applied Mathematics & IUMA, Universidad de Zaragoza, Zaragoza, Spain
e-mail: mjcante@unizar.es; lmoral@unizar.es; velazque@unizar.es

F. Marcellán
Departamento de Matemáticas, Universidad Carlos III de Madrid, Leganés, Spain
e-mail: pacomarc@ing.uc3m.es

F. Marcellán, E. J. Huertas (eds.), *Orthogonal Polynomials: Current Trends and Applications*, SEMA SIMAI Springer Series 22,
https://doi.org/10.1007/978-3-030-56190-1_3

mations consist in the factorization of a non-negative self-adjoint operator, the commutation of whose factors gives a new non-negative self-adjoint operator. The spectrum of the new operator is closely related to that of the initial one. This may be used for the spectral analysis of non-trivial operators starting from simple ones, but also to generate operators with prescribed spectral properties.

OPRL are connected to self-adjoint operators via the Jacobi matrix encoding the corresponding three term recurrence relation [4, 7]. When applied to Jacobi matrices, Darboux transformations turn out to be equivalent to degree one polynomial modifications of the underline orthogonality measure on the real line, which define the Christoffel transformations of OPRL. The inverse process corresponds to the so-called Geronimus transformations, which not only divide the measure by the polynomial, but may add an arbitrary mass point at its zero. This shows that inverse Darboux transformations of Jacobi matrices have a degree of freedom given by a real parameter, a fact that may be exploited for different purposes such as generating flows of integrable systems, searching for bispectral situations or developing tools for signal processing.

The relevance of OPUC for the theory of unitary operators relies on their close connection to the orthogonal Laurent polynomials on the unit circle (OLPUC), together with the simplicity of the OLPUC recurrence relation, summarized by a five-diagonal unitary matrix known as a CMV matrix [1, 8, 10]. Darboux transformations of CMV matrices are built out of Darboux transformations of a Hermitian matrix obtained from the CMV and its adjoint, i.e. its inverse. As a consequence, Darboux for CMV is also related to polynomial modifications of the orthogonality measure, but given by a Laurent polynomial instead of just a polynomial. This causes strong similarities as well as important differences between Darboux for Jacobi and CMV.

This work gives a unified approach to the theory of Darboux transformations for Jacobi and CMV matrices, providing an ideal setting for the comparison between both of them. Darboux for Jacobi matrices has quite a long story, and has been used for the study of Toda lattices [9], generalizations of Bochner's theorem on the real line (see [5] and references therein) and related time-and-band limiting problems (see [3] and references therein). However, the theory of Darboux transformations for CMV matrices is of much more recent vintage [2].

The paper is organized as follows: Sect. 2 presents the different families of orthogonal polynomials involved in our discussion—OPRL, OPUC and OLPUC—their recurrence relations and related matrix representations. Darboux transformations for self-adjoint and unitary operators are introduced in Sect. 3. Section 4 applies Darboux machinery to Jacobi matrices, while Darboux transformations of CMV matrices are developed in Sect. 5, emphasizing the differences with respect to the Jacobi case.

2 Orthogonal Polynomials and Matrix Representations

Any positive Borel measure μ on the complex plane $\mathbb{C}$ defines the Hilbert space $L^2(\mu)$ of μ-square integrable functions with inner product

$$\langle f, g\rangle = \int_{\mathbb{C}} \overline{f(z)}\, g(z)\, \mathrm{d}\mu(z).$$

In what follows we will always suppose that μ is non-trivial, i.e. it has a support $\operatorname{supp}(\mu)$ with infinitely many points, and $\{z^n\}_{n\geq 0} \subset L^2(\mu)$. Then $\{z^n\}_{n\geq 0}$ is a linearly independent subset of $L^2(\mu)$ which may be orthonormalized. This yields a unique sequence of orthonormal polynomials with positive leading coefficients, which is an orthonormal basis of $L^2(\mu)$ whenever the polynomials are dense in $L^2(\mu)$.

The orthonormalization of $\{z^n\}_{n\geq 0}$ is equivalent to the Cholesky factorization

$$G = LL^+, \quad L \text{ lower triangular with positive diagonal}, \quad L^+ := \text{adjoint of } L,$$

of the Gram matrix

$$G = \left(\langle z^m, z^n\rangle\right)_{m,n\geq 0},$$

which has special properties depending on the location of $\operatorname{supp}(\mu)$. If $\operatorname{supp}(\mu) \subset \mathbb{R}$, then $G_{m,n} = \int_{\mathbb{R}} z^{m+n} \mathrm{d}\mu(z)$ and G is a Hankel matrix. On the other hand, when $\operatorname{supp}(\mu)$ lies on the unit circle $\mathbb{T} := \{z \in \mathbb{C} : |z| = 1\}$, then $G_{m,n} = \int_{\mathbb{T}} z^{n-m} \mathrm{d}\mu(z)$ so that G is a Toeplitz matrix. These two situations give rise to the two paradigmatic cases studied in the literature, OPRL and OPUC, which we will describe succinctly.

2.1 OPRL and Jacobi Matrices

If $\operatorname{supp}(\mu) \subset \mathbb{R}$, the corresponding orthonormal polynomials $(p_n)_{n\geq 0}$ defined by

$$\langle p_m, p_n\rangle = \int_{\mathbb{R}} p_m(z)\, p_n(z)\, \mathrm{d}\mu(z) = \delta_{m,n}, \qquad p_n(z) = \kappa_n z^n + \cdots, \qquad \kappa_n > 0,$$

are characterized by a three term recurrence relation [4, 8]

$$zp_n(z) = a_{n+1}p_{n+1}(z) + b_{n+1}p_n(z) + a_n p_{n-1}(z), \quad n \geq 0, \quad p_{-1} = 0, \quad p_0 = \kappa_0,$$

where $b_n \in \mathbb{R}$ and $a_n = \kappa_{n-1}/\kappa_n > 0$. This recurrence may be expressed as

$$zp(z) = \mathcal{J}p(z), \qquad \mathcal{J} = \begin{pmatrix} b_1 & a_1 & & \\ a_1 & b_2 & a_2 & \\ & a_2 & b_3 & a_3 \\ & & \ddots & \ddots & \ddots \end{pmatrix}, \qquad p = \begin{pmatrix} p_0 \\ p_1 \\ p_2 \\ \vdots \end{pmatrix}, \tag{1}$$

in terms of the *Jacobi matrix* $\mathcal{J}$.

When $\operatorname{supp}(\mu)$ is bounded, Weierstrass theorem implies that the set $\mathbb{P} := \mathbb{C}[z]$ of polynomials is dense in $L^2(\mu)$, hence the OPRL sequence $(p_n)_{n\geq 0}$ constitutes an orthonormal basis of $L^2(\mu)$. Then, (1) identifies $\mathcal{J}$ as a matrix representation of the multiplication operator

$$\begin{aligned} \mathsf{T}_\mu \colon L^2(\mu) &\to L^2(\mu) \\ f(z) &\mapsto zf(z), \end{aligned} \tag{2}$$

which is bounded because $\|zf(z)\| \leq K\|f(z)\|$, $K = \sup\{|z| : z \in \operatorname{supp}(\mu)\}$, as well as self-adjoint since $\langle zf(z), g(z)\rangle = \langle f(z), zg(z)\rangle$. The boundedness of T_μ translates into that of $\mathcal{J}$, which means that the sequences $(a_n)_{n\geq 1}$ and $(b_n)_{n\geq 1}$ are bounded.

For an unbounded $\operatorname{supp}(\mu)$ the multiplication operator (2) remains self-adjoint but becomes unbounded. Since Weierstrass theorem does not work for unbounded intervals, in the unbounded case $\mathbb{P}$ may not be dense in $L^2(\mu)$, thus $\mathcal{J}$ is not necessarily a matrix representation of T_μ. The polynomials are known to be dense in $L^2(\mu)$ when the related moment problem is determinate [7], i.e. when μ is the unique measure giving the OPRL $(p_n)_{n\geq 0}$, a condition which includes the bounded case and guarantees that $\mathcal{J}$ represents the whole self-adjoint operator T_μ. Then, $\mathcal{J}$ defines a self-adjoint operator in the Hilbert space of square summable sequences which is unitarily equivalent to T_μ. Its spectrum is given by $\operatorname{spec}\mathcal{J} = \operatorname{spec}\mathsf{T}_\mu = \operatorname{supp}(\mu)$, the eigenvalues being the mass points of μ.

2.2 *OPUC and Hessenberg Matrices*

When $\operatorname{supp}(\mu) \subset \mathbb{T}$, the corresponding orthonormal polynomials $(\varphi_n)_{n\geq 0}$ given by

$$\langle \varphi_n, \varphi_m \rangle = \int_{\mathbb{T}} \overline{\varphi_n(z)}\, \varphi_m(z)\, d\mu(z) = \delta_{n,m}, \qquad \varphi_n(z) = \kappa_n z^n + \cdots, \qquad \kappa_n > 0,$$

satisfy the recurrence relation [8]

$$z\varphi_n(z) = \rho_{n+1}\varphi_{n+1}(z) - a_{n+1}\varphi_n^*(z), \qquad \varphi_n^*(z) := z^n\overline{\varphi_n(1/\overline{z})}, \qquad n \geq 0, \tag{3}$$

where $a_n = \varphi_n(0)/\kappa_n$—known as Schur parameters, Geronimus parameters or Verblunsky coefficients of μ—are complex parameters such that $|a_n| < 1$, while $\rho_n = \sqrt{1-|a_n|^2} = \kappa_{n-1}/\kappa_n > 0$.

The multiplication operator (2) makes sense also for $\mathrm{supp}(\mu) \subset \mathbb{T}$, but in this case T_μ is unitary since it is onto and $\langle f(z), zg(z)\rangle = \langle z^{-1}f(z), g(z)\rangle$. If the OPUC sequence $(\varphi_n)_{n\geq 0}$ is an orthonormal basis of $L^2(\mu)$, the recurrence (3) yields a matrix representation of T_μ by expanding φ_n^* as a linear combination of $(\varphi_n)_{n\geq 0}$. However, φ_n^* is in general a linear combination of all the polynomials $\varphi_0, \varphi_1, \dots, \varphi_n$, hence

$$z\varphi(z) = \mathcal{H}\varphi(z), \qquad \mathcal{H} = \begin{pmatrix} h_{00} & h_{01} & & & \\ h_{10} & h_{11} & h_{12} & & \\ h_{20} & h_{21} & h_{22} & h_{23} & \\ \dots & \dots & \dots & \dots & \dots \end{pmatrix}, \qquad \varphi = \begin{pmatrix} \varphi_0 \\ \varphi_1 \\ \varphi_2 \\ \vdots \end{pmatrix},$$

thus the corresponding matrix representation of T_μ is given by a Hessenberg matrix $\mathcal{H}$ instead of a band one. Besides, $\mathcal{H}$ only represents the whole unitary operator T_μ when $\mathbb{P}$ is dense in $L^2(\mu)$, which is not always guaranteed because Weierstrass theorem does not hold for the polynomials on $\mathbb{T}$. In other words, $\mathcal{H}$ represents the multiplication operator $f(z) \mapsto zf(z)$ on the closure $\overline{\mathbb{P}}$ of $\mathbb{P}$, which is an isometric operator but, in general, is not onto since 1 may not lie in its range because z^{-1} is not necessarily in $\overline{\mathbb{P}}$. Hence, the Hessenberg matrix $\mathcal{H}$ is not always unitary.

The fact that Weierstrass theorem holds for Laurent polynomials on $\mathbb{T}$ suggests that the drawback regarding the unitarity of the matrix representation may be overcome if we substitute the OPUC by an orthonormal basis constituted by Laurent polynomials. Actually, this also will change the Hessenberg representation by a band one.

2.3 *OLPUC and CMV Matrices*

When $\mathrm{supp}(\mu) \subset \mathbb{T}$, the set $\Lambda := \mathbb{C}[z, z^{-1}]$ of Laurent polynomials is dense in $L^2(\mu)$. The orthonormalization of $\{1, z, z^{-1}, z^2, z^{-2}, \dots\}$ yields a sequence of orthonormal Laurent polynomials $(\chi_n)_{n\geq 0}$ which serves as an orthonormal basis of $L^2(\mu)$. The OLPUC χ_n are related to the OPUC φ_n by very simple relations, namely,

$$\chi_{2n}(z) = z^{-n}\varphi_{2n}^*(z), \qquad \chi_{2n+1}(z) = z^{-n}\varphi_{2n+1}(z).$$

The recurrence relation (3) for OPUC leads to another one for OLPUC which may be expressed as [1, 8, 10]

$$z\chi(z) = \mathcal{C}\chi(z), \qquad \chi = (\chi_0, \chi_1, \chi_2, \dots)^T,$$

where C is an infinite matrix given by

$$C = \begin{pmatrix} -a_1 & \rho_1 & & & & & \\ -\rho_1 a_2 & -\overline{a}_1 a_2 & -\rho_2 a_3 & \rho_2\rho_3 & & & \\ \rho_1\rho_2 & \overline{a}_1\rho_2 & -\overline{a}_2 a_3 & \overline{a}_2\rho_3 & & & \\ & & -\rho_3 a_4 & -\overline{a}_3 a_4 & -\rho_4 a_5 & \rho_4\rho_5 & \\ & & \rho_3\rho_4 & \overline{a}_3\rho_4 & -\overline{a}_4 a_5 & \overline{a}_4\rho_5 & \\ & & & & \dots & \dots & \dots \end{pmatrix}. \tag{4}$$

The matrix C, known in the literature as a *CMV matrix*, represents the multiplication operator (2), which now is unitary. Hence, unlike the Hessenberg representation linked to OPUC, the five-diagonal CMV representation (4) related to OLPUC is not only banded, but is always unitary. The spectral relations between C and T_μ are similar to those of the Jacobi case, but now they hold for any measure μ on $\mathbb{T}$.

3 The Darboux Transformations

Darboux transformations of Jacobi and CMV matrices are particular cases of Darboux transformations of self-adjoint and unitary operators. In this section we will describe this more general Darboux setting. This will allow us to distinguish later on the Darboux features which are specific of the Jacobi and CMV case from those already present for general self-adjoint and unitary operators. The self-adjoint case has its roots in a work by Moutard from 1875, with subsequent contributions from Darboux, Schrödinger, Dirac, von Neumann and others (see [6] and references therein). However, the application of Darboux methods to unitary operators is not usual in the literature.

3.1 Darboux Transformations of Self-Adjoint Operators

Let H be a bounded self-adjoint operator everywhere defined on a Hilbert space. The inequality $|\langle v|Hv\rangle| \leq \|H\|\|v\|^2$ shows that $H - \xi$ is positive definite for some $\xi \in \mathbb{R}$. Then, any factorization

$$H - \xi = AA^+,$$

leads to a new bounded self-adjoint operator G defined by

$$G - \xi = A^+A. \tag{5}$$

The above identities show that $\ker(AA^+) = \ker A^+$ and $\ker(A^+A) = \ker A$ are the ξ-eigenspaces of H and G respectively. The remaining eigenspaces of H and G are connected by the operators A and A^+, as follows from the equalities

$$HA = AA^+A + \xi A = AG, \qquad A^+H = A^+AA^+ + \xi A^+ = GA^+, \tag{6}$$

which imply that, except for the eigenvalue ξ, the operator A^+/A maps eigenvectors of H/G into eigenvectors of G/H with the same eigenvalue. Actually, a classical result due to von Neumann [11, Thm. 5.39] states that the operators AA^+ and A^+A are unitarily equivalent when restricted to $(\ker A^+)^\perp$ and $(\ker A)^\perp$ respectively, hence H and G are unitarily equivalent when restricted to the orthogonal complement of their ξ-eigenspaces. Therefore, the transformation $H \mapsto G$ is isospectral up to the eigenvalue ξ, i.e. it can modify the spectrum only by creating or annihilating ξ-eigenvectors.

We will refer to the transformation $H \mapsto G$ as a *Darboux transformation of a self-adjoint operator* H. Darboux transformations allow to translate spectral properties of known operators into spectral information about unknown ones, but they are also tools to generate new operators by adding/removing eigenvalues from a given one.

If $\ker A = \{0\}$, the operator A has an inverse on its range $\operatorname{ran} A$. Then (6) yields

$$G = A^{-1}HA, \tag{7}$$

which is an alternative to (5) for the definition of the Darboux transform G. In this case ξ is not an eigenvalue of G, but it is an eigenvalue of H whenever $\ker A^+ \neq \{0\}$, in which case the Darboux transformation annihilates the ξ-eigenvectors. If, instead, $\ker A^+ = \{0\}$, the Darboux transformation becomes strictly isospectral.

The above discussion can be generalized by choosing a real polynomial $\wp$ of arbitrary degree such that $\wp(H)$ is positive definite.

Theorem 1 *Let H be a bounded self-adjoint operator on a Hilbert space and $\wp$ a real polynomial such that $\wp(H)$ is positive definite. If $\wp(H) = AA^+$ with A having a bounded inverse on* $\operatorname{ran} A$*, then $G = A^{-1}HA$ defines a self-adjoint operator such that $\wp(G) = A^+A$ and* $\operatorname{spec} G$ *follows from* $\operatorname{spec} H$ *by removing the eigenvalues at the zeros of $\wp$.*

Proof First we must show that $G = A^{-1}HA$ is well defined, which amounts to prove that $\operatorname{ran}(HA) \subset \operatorname{ran} A$. Since H commutes with $\wp(H) = AA^+$, we find that $AA^+H = HAA^+$. Therefore, $AA^+H \ker A^+ = \{0\}$, which means that $H \ker A^+ \subset \ker A^+$. Bearing in mind that $H^+ = H$, this implies that $H(\ker A^+)^\perp \subset (\ker A^+)^\perp$. Since A has a bounded inverse on its range, $\operatorname{ran} A = (\ker A^+)^\perp$, thus $H \operatorname{ran} A \subset \operatorname{ran} A$.

The relation $G = A^{-1}HA$ implies that $\wp(G) = A^{-1}\wp(H)A = A^+A$ and $\wp(G^+) = \wp(G)^+ = A^+A$. Thus, G and G^+ commute with A^+A, which helps to prove the self-adjointness of G. Using that $AG = HA$ and $G^+A^+ = A^+H$ we find that

$$A^+AG^+ = G^+A^+A = A^+HA = A^+AG.$$

We conclude that $G^+ = G$ because $\ker A = \{0\}$ for A having an inverse on its range.

Regarding the spectrum, the relation $(G-z)^{-1} = A^{-1}(H-z)^{-1}A$ between the resolvents shows that $\operatorname{spec} G \subset \operatorname{spec} H$. The sum of the eigenspaces of H whose eigenvalues are zeros of $\wp$ constitute $\ker \wp(H) = \ker A^+$. An analogous statement for G together with the fact that $\ker \wp(G) = \ker A = \{0\}$ implies that G has no eigenvalues among the zeros of $\wp$. The statement about the spectrum is proved if we show that any point $z \in \operatorname{spec} H$ which is not a zero of $\wp$, necessarily lies in $\operatorname{spec} G$. Any such a point is characterized by the existence of a sequence of approximated eigenvectors $v_n \in (\ker A^+)^\perp$ of H, i.e.

$$(H-z)v_n \to 0, \qquad \|v_n\| \geq \epsilon > 0.$$

Since $v_n \in \operatorname{ran} A$, we can define $v_n' = A^{-1}v_n$, which satisfies $\|v_n'\| \geq \epsilon' > 0$ with $\epsilon' = \epsilon/\|A\|$. Therefore, $z \in \operatorname{spec} G$ because, bearing in mind that A^{-1} is bounded,

$$(G-z)v_n' = A^{-1}(H-z)v_n \to 0, \qquad \|v_n'\| \geq \epsilon' > 0.$$

□

Self-adjoint Multiplication Operators

Let μ be a non-trivial positive Borel measure with a bounded support on $\mathbb{R}$. The corresponding multiplication operator T_μ given by (2) is bounded and self-adjoint. Given $\xi \leq \inf \operatorname{supp}(\mu)$, the operator $\mathsf{T}_\mu - \xi$ is positive definite since $\langle f, (\mathsf{T}_\mu - \xi) f\rangle = \int_{\mathbb{R}} (z-\xi)|f(z)|^2 \, d\mu(z) \geq 0$ for $f \in L^2(\mu)$. Actually, $\langle f, (\mathsf{T}_\mu - \xi) f\rangle = 0$ only when ξ is a mass point of μ and f is proportional to the characteristic function 1_ξ of $\{\xi\}$. Therefore, $\langle f, (\mathsf{T}_\mu - \xi) f\rangle > 0$ if $f \in \mathbb{P} \setminus \{0\}$. When translated to the related Jacobi matrix $\mathcal{J}$, this means that $X^+(\mathcal{J} - \xi)X > 0$ for any column vector $X \neq 0$ with finitely many non-null entries. This implies the existence of a unique Cholesky factorization $\mathcal{J} - \xi = AA^+$, i.e. such that A is lower triangular with a positive upper diagonal [2, App. A]. The tridiagonal shape of $\mathcal{J} - \xi$ implies that A is indeed 2-band, pictorially,

$$A = \begin{pmatrix} + & & & \\ * & + & & \\ & * & + & \\ & & & \ddots \end{pmatrix}, \tag{8}$$

where, here and in what follows, the symbol $+$ stands for a positive coefficient, while $*$ denotes a possible non-zero coefficient. Therefore, the Darboux transformation $\mathcal{J} \mapsto \mathcal{K}$ given by $\mathcal{K} - \xi = A^+A$ yields a new bounded Jacobi matrix $\mathcal{K}$, whose related measure ν defines a new bounded self-adjoint

(continued)

multiplication operator T_ν, the Darboux transform of T_μ. The structure (8) implies that $\ker(\mathcal{K}-\xi) = \ker A = \{0\}$, so that $\ker(\mathsf{T}_\nu - \xi) = \{0\}$. Besides, if ξ is a mass point of μ, then $\ker(\mathsf{T}_\mu - \xi) = \operatorname{span}\{1_\xi\}$. In this case, the Darboux transformation $\mathsf{T}_\mu \mapsto \mathsf{T}_\nu$ removes the eigenvalue ξ, which means that ξ is a mass point of μ but not of ν.

3.2 *Darboux Transformations of Unitary Operators*

Consider a unitary operator U on a Hilbert space. Using the results of the previous subsection, we can define a Darboux transformation of U resorting to a bounded self-adjoint operator defined by U, for instance $\operatorname{Re} U = (U + U^+)/2$. If $\xi \in \mathbb{R}$ makes $\operatorname{Re} U - \xi$ positive definite, the corresponding Darboux transformation $U \mapsto V$ should start with a factorization $\operatorname{Re} U - \xi = AA^+$, generating a new unitary operator V such that $\operatorname{Re} V - \xi = A^+A$. However, $\operatorname{Re} V$ does not determine the unitary V. Instead, we can think on an approach to Darboux for unitaries defining V by a conjugation with A, analogously to (7), i.e. $V = A^{-1}UA$.

Indeed, inspired by Theorem 1, with no more effort we can develop a general result in which we substitute $\operatorname{Re} U - \xi = (U + U^{-1})/2 - \xi$ by an arbitrary Laurent polynomial mapping unitary operators into self-adjoint ones. These are those $\ell \in \Lambda$ satisfying $\ell(z) = \overline{\ell(1/\overline{z})}$, which have the explicit form

$$\ell(z) = \sum_{j=-n}^{n} c_j z^j, \qquad c_{-j} = \overline{c_j}. \tag{9}$$

We will refer to (9) as a *Hermitian Laurent polynomial* of degree $2n$, since it has $2n$ zeros counting multiplicity. The absence of Hermitian Laurent polynomials of degree one will be the origin of important differences between Darboux for Jacobi and CMV matrices.

These ideas lead to the following analogue of Theorem 1, which defines a mapping $U \mapsto V$ which we call *Darboux transformation of a unitary operator* U.

Theorem 2 *Let U be a unitary operator on a Hilbert space and ℓ a Hermitian Laurent polynomial such that $\ell(U)$ is positive definite. If $\ell(U) = AA^+$ with A having a bounded inverse on* ran A, *then $V = A^{-1}UA$ defines a unitary operator such that $\ell(V) = A^+A$ and* spec V *follows from* spec U *by removing the eigenvalues at the zeros of ℓ.*

Proof To prove that $V = A^{-1}UA$ is well defined we take into account that U^+ commutes with $\ell(U) = AA^+$. This implies that $U^+ \ker A^+ \subset \ker A^+$, which is equivalent to $U(\ker A^+)^\perp \subset (\ker A^+)^\perp$. Hence, $U \operatorname{ran} A \subset \operatorname{ran} A$ for A with bounded inverse on its range.

As in Theorem 1, $\ell(V) = A^{-1}\ell(U)A = A^+A$. Hence V and V^+ commute with A^+A, which combined with the identities $AV = UA$, $V^+A^+ = A^+U^+$ and the unitarity of U gives

$$A^+AV^+V = V^+A^+AV = A^+U^+UA = A^+A.$$

Since $\ker A = \{0\}$, this implies that $V^+V = 1$, which proves the unitarity of V, bearing in mind that $V^{-1} = A^{-1}U^{-1}A$ is everywhere defined.

The proof of the spectral relations between U and V is just as in Theorem 1. □

Unitary Multiplication Operators

Let T_μ be the unitary multiplication operator (2) given by a non-trivial positive Borel measure on $\mathbb{T}$. Given a Hermitian Laurent polynomial $\ell(z) = (z+z^{-1})/2-\xi$ with $0 < \epsilon = \inf\{\operatorname{Re} z-\xi : z \in \operatorname{supp}(\mu)\}$, the operator $\ell(\mathsf{T}_\mu)$ is positive definite. Indeed, $\langle f, \ell(\mathsf{T}_\mu)f\rangle = \int_{\mathbb{T}}(\operatorname{Re} z - \xi)|f(z)|^2\, d\mu(z) \geq \epsilon\|f\|^2$, which implies that the related CMV matrix C satisfies $X^+\ell(C)X \geq \epsilon\|X\|^2$. Therefore, there exists a unique Cholesky factorization $\ell(C) = AA^+$ [2, App. A]. Due to the five-diagonal shape of $\ell(C)$, the factor A becomes 3-band, i.e.

$$A = \begin{pmatrix} + & & & & \\ * & + & & & \\ * & * & + & & \\ & * & * & + & \\ & & \ddots & \ddots & \ddots \end{pmatrix}. \tag{10}$$

From $\|A^+X\|^2 = X^+\ell(C)X \geq \epsilon\|X\|^2$ we conclude that A^+ has an everywhere defined bounded inverse, so the same holds for A. Hence, the Darboux transform $\mathcal{D} = A^{-1}CA$ is well defined. It turns out that $\mathcal{D}$ is again a CMV matrix (see Sect. 5), whose measure ν provides a unitary multiplication operator defining a Darboux transformation $\mathsf{T}_\mu \mapsto \mathsf{T}_\nu$. This transformation is strictly isospectral since $\ker A = \ker A^+ = \{0\}$, which means that $\operatorname{supp}(\mu) = \operatorname{supp}(\nu)$. This would change if μ had a mass point ζ with $\operatorname{Re}\zeta = \xi$, while $0 < \epsilon = \inf\{\operatorname{Re} z - \xi : z \in \operatorname{supp}(\mu) \setminus \{\zeta\}\}$. Then, $\ker A = \{0\}$ but $\ker \ell(\mathsf{T}_\mu) = \operatorname{span}\{1_\zeta\}$, so that $\ker A^+ = \ker \ell(C)$ is non-trivial. Using that A^+A is unitarily equivalent to AA^+ restricted to $(\ker A^+)^\perp$, we get

$$\inf_X \frac{\|AX\|^2}{\|X\|^2} = \inf_X \frac{X^+A^+AX}{\|X\|^2} = \inf_{X\perp\ker A^+} \frac{X^+AA^+X}{\|X\|^2} = \inf_{f\perp 1_\zeta} \frac{\langle f, \ell(\mathsf{T}_\mu)f\rangle}{\|f\|^2}$$

$$= \inf_f \frac{\int_{\mathbb{T}\setminus\{\zeta\}}(\operatorname{Re} z - \xi)|f|^2\, d\mu(z)}{\int_{\mathbb{T}\setminus\{\zeta\}}|f|^2\, d\mu(z)} \geq \epsilon.$$

(continued)

This proves that A has a bounded inverse on $\operatorname{ran} A$, thus $\mathcal{D} = A^{-1}CA$ is again a well defined Darboux transformation. Actually, $\mathcal{D}$ is a CMV matrix (see Sect. 5). The related measure has no mass point at ζ because $\ker \ell(\mathcal{D}) = \ker A = \{0\}$. Therefore, the Darboux transformation $\mathsf{T}_\mu \mapsto \mathsf{T}_\nu$ removes the eigenvalue ζ in this case.

The previous Hilbert approach to Darboux transformations has some drawbacks which we will try to overcome. For instance, it is not directly applicable to unbounded self-adjoint operators. Even in the bounded case, some difficulties arise in connection with the domains of the operators. As an example, consider the so called inverse Darboux transformations $H \leftarrow\!\shortmid G$ of self-adjoint operators, which take G as a starting point. The relation (7) suggests that such an inverse transformation should be given by $H = AGA^{-1}$. However, this relation cannot be valid in general because H is everywhere defined, while A^{-1} has domain $\operatorname{ran} A$. This becomes even more important in the unitary case, where we have no alternative to the definition of Darboux transformations by conjugation with the factor A.

We will solve these problems simultaneously by dealing directly with the algebra of infinite matrices, avoiding any interpretation in terms of operators in Hilbert spaces. This has the advantage of allowing for the action of certain infinite matrices –e.g. band or triangular matrices, like A or A^{-1}– on non-square-summable vectors, which enlarges the domains solving the related problems (see the comments after Lemma 1). The Darboux transformations of infinite matrices keep many of the features already discussed for the operator setting, but have some advantages since no attention is paid to norms and unbounded situations. Nevertheless, working with infinite matrices without the help of norms and inner products is a source of other difficulties arising from the eventual manipulation of infinite sums. We must specify the matrix operations that will be admitted and their properties.

We say that a product AB of complex matrices A, B is *admissible* if every entry $(AB)_{ik} = \sum_j A_{ij} B_{jk}$ involves only a finite number of non-zero summands. *Hessenberg type matrices*, i.e. those with the shape

$$\begin{pmatrix} * & * & \cdots & * & & & \\ * & * & \cdots & * & * & & \\ * & * & \cdots & * & * & * & \\ \cdots & \cdots & \cdots & \cdots & \cdots & \cdots & \cdots \end{pmatrix}, \qquad \begin{pmatrix} * & * & \cdots & * & & & \\ * & * & \cdots & * & * & & \\ * & * & \cdots & * & * & * & \\ \cdots & \cdots & \cdots & \cdots & \cdots & \cdots & \cdots \end{pmatrix}^T,$$

lower Hessenberg type   upper Hessenberg type

generate admissible products: AB is admissible if either A is lower Hessenberg type or B is upper Hessenberg type. Band matrices are lower and upper Hessenberg type.

Most of the properties of finite matrix multiplication hold for admissible products. Exceptions are the associativity and the uniqueness of the inverse matrix. The later one is related to the fact that invertible matrices may have left and right kernels,

$$\ker_L A = \{X = (X_0, X_1, \dots) : XA = 0\}, \quad \ker_R A = \{X = (X_0, X_1, \dots)^T : AX = 0\}.$$

These kernels must be understood as sets of (not necessarily finite norm) vectors giving an admissible null product with the matrix. Thus $\ker_R A$ differs form what we called previously $\ker A$, the set of vectors with finite norm giving a null standard right product with the matrix A (see the comments just before Sect. 4.1).

Regarding the above pathologies, the following result shows the way we will circumvent them in the development of Darboux transformations for Jacobi and CMV matrices.

Proposition 1 ([2, Sect. 2]) *The associativity $(AB)C = A(BC)$ holds if either A and B are lower Hessenberg type, or B and C are upper Hessenberg type, or A is lower Hessenberg type and C is upper Hessenberg type.*

If a lower (upper) Hessenberg type matrix A has a lower (upper) Hessenberg type inverse, no other inverse exists. In this case, $\ker_R A = \{0\}$ ($\ker_L A = \{0\}$).

Whenever the associativity holds, we will omit parenthesis for multiple products. A Hessenberg type matrix A is called *invertible* if it has an inverse of the same type.

4 Darboux Transformations of Jacobi Matrices

Let $p = (p_0, p_1, \dots)^T$ and $q = (q_0, q_1, \dots)^T$ where $p_n, q_n \in \mathbb{R}[x]$ have positive leading coefficients and $\deg(p_n) = \deg(q_n) = n$. The matrices $\mathcal{J}$ and $\mathcal{K}$ given by

$$zp(z) = \mathcal{J}p(z), \qquad zq(z) = \mathcal{K}q(z),$$

are lower Hessenberg with positive upper diagonal, i.e. they have the shape

$$\begin{pmatrix} * & + & & & \\ * & * & + & & \\ * & * & * & + & \\ \dots & \dots & \dots & \dots & \dots \end{pmatrix}. \tag{11}$$

Other matrices relating p and q come into play, namely, the matrices A, B given by

$$p = Aq, \qquad \wp q = Bp,$$

where $\wp(z) = z - \xi$ is a real polynomial. The matrix A is lower triangular, while B is lower Hessenberg, both with positive upper diagonal. Thus, $\ker_R(A) = \{0\}$ and A is invertible since it has a lower triangular inverse. The following relations are the starting point to develop the Darboux transformations of Jacobi matrices.

Lemma 1 $\wp(\mathcal{J}) = AB, \quad \wp(\mathcal{K}) = BA, \quad \mathcal{J}A = A\mathcal{K}, \quad \mathcal{K}B = B\mathcal{J},$
$\mathcal{K} = A^{-1}\mathcal{J}A, \quad \mathcal{J} = A\mathcal{K}A^{-1}.$

Proof The first line follows from the linear independence of the sets p, q and

$$(\wp(\mathcal{J}) - AB)p = \wp p - \wp Aq = 0, \qquad (\wp(\mathcal{K}) - BA)q = \wp q - Bp = 0,$$
$$(\mathcal{J}A - A\mathcal{K})q = \mathcal{J}p - zAq = 0, \qquad (\mathcal{K}B - B\mathcal{J})p = \wp\mathcal{K}q - zBp = 0.$$

The two last identities stem from the two previous ones and Proposition 1. □

The relation $\mathcal{J} = A\mathcal{K}A^{-1}$ announces an anticipated advantage of the matrix approach regarding inverse Darboux transformations. In contrast to the operator setting, as invertible lower triangular matrices, A and A^{-1} have as "domain" and "range" the whole set of infinite vectors in the context of admissible matrix products. Hence, $\mathcal{J} = A\mathcal{K}A^{-1}$ makes sense in this context since $\mathcal{J}$ and $\mathcal{K}$ also have such a "domain" because, as band matrices, their product with any infinite vector is admissible. Similar manipulations are not possible for $\mathcal{K}B = B\mathcal{J}$ due to the following result.

Proposition 2 *B is not invertible and* $\ker \wp(\mathcal{J}) = \ker_R(B) = \operatorname{span}\{p(\xi)\}$.

Proof According to the Hessenberg structure (11) of B, $\dim \ker_R(B) = 1$, hence B cannot be invertible by Proposition 1. The equality $\ker_R \wp(\mathcal{J}) = \ker_R B$ follows from $\wp(\mathcal{J}) = AB$ and $\ker_R A = \{0\}$. Therefore, $\wp(\mathcal{J})p(\xi) = \wp(\xi)p(\xi) = 0$ implies that $p(\xi) \in \ker_R B$, which means that $p(\xi)$ spans $\ker_R(B)$. □

Suppose now that $\mathcal{J}$ and $\mathcal{K}$ are Jacobi matrices, so that p and q are orthonormal with respect to some measures on $\mathbb{R}$. According to Lemma 1, Darboux transformations correspond to the choice $B = A^+$ associated with the Cholesky factorization $\wp(\mathcal{J}) = AA^+$, which is possible iff $\wp(\mathcal{J}) > 0$, i.e. iff all the leading principal minors of $\wp(\mathcal{J})$ are positive, which means that $X^+\wp(\mathcal{J})X > 0$ for every column vector $X \neq 0$ with finitely many non-zero entries [2, App. A]. This choice may be characterized in terms of the orthogonality measures.

Theorem 3 $B = A^+ \Leftrightarrow$ *If p is orthonormal with respect to the measure $d\mu$, then q is orthonormal with respect to the measure $d\nu = \wp\, d\mu$.*

Proof Expressing the μ-orthonormality of p as $\int pp^+ d\mu = I$, we get

$$B^+ = \int p(Bp)^+ d\mu = \int \wp pq^+ d\mu = A \int qq^+ \wp\, d\mu.$$

The result follows from this equality and the fact that $\ker_R A = \{0\}$. □

The condition $B = A^+$ implies that A has the 2-band structure (8) with non-zero lower diagonal entries. Theorem 3 is valid for unbounded Jacobi matrices, actually it holds in the indeterminate case. This result raises the question of the positivity of $\wp\, d\mu$. In the bounded case, this is guaranteed by the positive definiteness of $\wp(\mathcal{J})$.

Proposition 3 *Let $\mathcal{J}$ be a Jacobi matrix and μ a related measure on $\mathbb{R}$. If $\wp \in \mathbb{P}$ is non-negative on* $\operatorname{supp}(\mu)$*, then $\wp(\mathcal{J}) > 0$, being both equivalent conditions when* $\operatorname{supp}(\mu)$ *is bounded.*

Proof The identity $\wp(\mathcal{J}) = \int \wp(\mathcal{J}) pp^+ \mathrm{d}\mu = \int pp^+ \wp \, \mathrm{d}\mu$ implies that $\wp(\mathcal{J}) > 0$ iff $\int |f|^2 \wp \, \mathrm{d}\mu > 0$ for $f \in \mathbb{P} \setminus \{0\}$. If $\wp \geq 0$ on $\operatorname{supp}(\mu)$, then $\int |f|^2 \wp \, \mathrm{d}\mu > 0$ for every $f \in \mathbb{P} \setminus \{0\}$ because $\operatorname{supp}(\mu)$ has infinitely many points. On the other hand, when $\operatorname{supp}(\mu)$ is bounded, $\mathbb{P}$ is dense in $L^2(\mu)$ and T_μ is bounded, i.e. continuous. Thus, $\langle f, \wp(\mathsf{T}_\mu) f \rangle = \int |f|^2 \wp \, \mathrm{d}\mu > 0$ for $f \in \mathbb{P} \setminus \{0\}$ means that this actually holds for $f \in L^2(\mu) \setminus \{0\}$, which implies that $\wp \geq 0$ on $\operatorname{supp}(\mu)$. □

The previous results show that, whenever $\mathcal{J} - \xi > 0$, a Darboux transformation $\mathcal{J} \mapsto \mathcal{K}$ is generated by the Cholesky factorization $\mathcal{J} - \xi = AA^+$ and $\mathcal{K} - \xi = A^+A$. This is equivalent to the *Chistoffel transformation* $\mathrm{d}\mu(z) \mapsto \mathrm{d}\nu(z) = (z - \xi)\,\mathrm{d}\mu(z)$, which preserves the support of the measure μ up to any mass point at ξ. In the determinate case $\operatorname{supp}(\mu) = \operatorname{spec} \mathsf{T}_\mu = \operatorname{spec} \mathcal{J}$, the mass points of μ being the eigenvalues of $\mathcal{J}$, thus the Darboux transformation is isospectral up to any eigenvalue of $\mathcal{J}$ at ξ, which is absent in $\mathcal{K}$.

The above picture solves an apparent contradiction which arises when applying Proposition 2 to $\mathcal{K}$, concluding that $\ker_R \wp(\mathcal{K}) = \operatorname{span}\{q(\xi)\}$. This is in contrast with the equality $\ker \wp(\mathcal{K}) = \ker A = \{0\}$, that follows from the arguments for general self-adjoint operators in Sect. 3.1. The only way to make both identities compatible is assuming that $q(\xi)$ has an infinite norm, so that it does not lie on $\ker \wp(\mathcal{K})$. This is indeed the case because $\mathcal{K}$ has no eigenvalue at ξ.

4.1 Inverse Darboux Transformations of Jacobi Matrices

In this subsection we will consider the inverse problem of the previous one: given a Jacobi matrix $\mathcal{K}$ and a polynomial $\wp(z) = z - \xi$ such that $\wp(\mathcal{K}) > 0$, find the Jacobi matrices $\mathcal{J}$ such that $\mathcal{K}$ is its Darboux transform. This amounts to find all the reversed Cholesky factorizations $\wp(\mathcal{K}) = A^+A$, with A having the shape (8). The condition $\wp(\mathcal{K}) > 0$, necessary for a reversed Cholesky factorization, almost guarantees it for a bounded $\mathcal{K}$ as a consequence of the following result.

Proposition 4 ([2, App. A]) *If $M = M^+ > 0$ is a bounded $(2n+1)$-band matrix, then $M = A^+A$ where A is n-band lower triangular with non-negative diagonal entries. When the lower diagonal entries of M are non-zero, the lower diagonal entries of A must be non-zero, while all but at most its first n diagonal entries must be positive.*

The last statement of this proposition was incorrectly stated in [2, App. A], which claimed that all the diagonal entries of A should be positive. When applied to a tridiagonal matrix $\wp(\mathcal{K}) > 0$ arising from a bounded Jacobi matrix $\mathcal{K}$, the above proposition guarantees the existence of a factorization $\wp(\mathcal{K}) = A^+A$ with

$$A = \begin{pmatrix} r_0 & & & \\ s_0 & r_1 & & \\ & s_1 & r_2 & \\ & & \ddots & \ddots \end{pmatrix}, \qquad \begin{array}{l} r_0 \geq 0, \\ r_{n+1} > 0, \quad s_n \neq 0, \quad n \geq 0. \end{array} \tag{12}$$

When $r_0 = 0$ we will refer to a *singular* factorization. Any non-singular factor A yields an inverse transform because $\wp(\mathcal{J}) = AA^+$ defines a Jacobi matrix $\mathcal{J}$. Nevertheless, Lemma 1 implies that, unlike the operator setting, the matrix approach allows us to define the inverse transformations by the conjugation $\mathcal{J} = A\mathcal{K}A^{-1}$. Inverse Darboux transformations also apply to the unbounded case, but the existence of the related reversed Cholesky factorizations is not covered by Proposition 4.

Denoting $\wp(\mathcal{K}) = (m_{i,j})$ and using the notation (12), $\wp(\mathcal{K}) = A^+A$ reads as

$$r_n^2 + s_n^2 = m_{n,n}, \qquad r_{n+1}s_n = m_{n+1,n}. \tag{13}$$

The solutions of these equations depend in general on a single real parameter, which may be chosen as r_0. Not all the choices $r_0 > 0$ give a non-singular factor A, since they must be consistent with $r_n > 0$ and $s_n \in \mathbb{R}\backslash\{0\}$ for all n. Actually, Proposition 4 guarantees the existence of a factor (12) when $\wp(\mathcal{K}) > 0$, but it can happen that there is a single solution corresponding to $r_0 = 0$, which means that no inverse Darboux transform exists.

The identification between Darboux and Christoffel transformations sheds light on this discussion. If ν is a measure related to the Jacobi matrix $\mathcal{K}$, inverse Darboux is equivalent to searching for solutions μ of $\wp\, d\mu = d\nu$ which are non-trivial positive Borel measures such that $\mathbb{P} \subset L^2(\mu)$. This requires $\wp \geq 0$ on supp(ν), which in view of Proposition 3 brings us back to the condition $\wp(\mathcal{K}) > 0$. Assuming this, the positive solutions are given by the so-called *Geronimus transformations*,

$$d\mu = \frac{d\nu}{\wp} + m\,\delta_\xi, \qquad m \geq 0, \tag{14}$$

which satisfy the rest of the requirements if the moments $\int z^n d\nu/\wp$ are finite for $n \geq 0$. In the bounded case, this is equivalent to ask for the finiteness of $\int d\nu/\wp$. Hence, the absence of non-singular reversed factorizations of $\wp(\mathcal{K}) > 0$ indicates that $\mathbb{P} \not\subset L^2(\mu)$, which in the bounded case means that $\int d\nu/\wp$ is infinite.

The freedom in the parameter r_0 for inverse Darboux is related to the freedom in the mass m for the Geronimus transformations. This relation can be made more precise. The search for such an explicit relation will also help to understand the choices of r_0 giving non-singular factorizations $\wp(\mathcal{K}) = A^+A$.

Theorem 4 *Suppose that $\wp(z) = z - \xi \geq 0$ on* $\text{supp}(\nu)$ *and* $\int d\nu/\wp$ *is finite. Then, any non-singular factorization* $\wp(\mathcal{K}) = A^+A$ *satisfies*

$$0 < r_0^2 \leq \frac{\int d\nu}{\int \frac{d\nu}{\wp}}. \tag{15}$$

This condition also guarantees the existence of such a factorization when $\text{supp}(\nu)$ *is bounded. The parameter r_0 and the mass m at ξ of the measure μ associated with the inverse Darboux transform given by* $\wp(\mathcal{J}) = A^+A$ *are related by*

$$r_0^2 = \frac{\int d\nu}{\int \frac{d\nu}{\wp} + m}. \tag{16}$$

Proof A non-singular factorization $\wp(\mathcal{K}) = AA^+$ is equivalent to the existence of a positive measure μ with finite moments such that $\wp\, d\mu = \nu$. If p and q are the OPRL related to μ and ν respectively, the first equation of $p = Aq$ gives $p_0 = r_0 q_0$, where $\int q_0^2\, d\nu = 1$. Using this and (14) we find that

$$1 = \int p_0^2\, d\mu = \int p_0^2 \frac{d\nu}{\wp} + m p_0^2 = r_0^2 q_0^2 \left(\int \frac{d\nu}{\wp} + m \right) = r_0^2 \frac{\int \frac{d\nu}{\wp} + m}{\int d\nu}.$$

This implies (16) and the necessary condition (15). That this condition is also sufficient in the bounded case follows by noticing that, choosing r_0 according to (15), then (16) and (14) define a measure μ which solves $\wp\, d\mu = d\nu$. By hypothesis, μ satisfies the requirements to be a measure related to a Jacobi matrix $\mathcal{J}$. Also, $\wp \geq 0$ on $\text{supp}(\mu)$, which by Proposition 3 means that $\wp(\mathcal{J}) > 0$. Thus, there exists a Cholesky factorization $\wp(\mathcal{J}) = AA^+$ and, according to Theorem 3, $\wp(\mathcal{K}) = A^+A$. Since the first diagonal entry of A satisfies (16), it is the chosen value of r_0. □

Chebyshev Polynomials

Let $d\nu(z) = \sqrt{1 - z^2}\, dz$, $z \in [-1, 1]$, be the measure corresponding to the second kind Chebyshev polynomials. The associated Jacobi matrix is

$$\mathcal{K} = \begin{pmatrix} 0 & \frac{1}{2} & & & \\ \frac{1}{2} & 0 & \frac{1}{2} & & \\ & \frac{1}{2} & 0 & \frac{1}{2} & \\ & & \ddots & \ddots & \ddots \end{pmatrix}.$$

(continued)

Consider $\wp(z) = z + 1$, which is non-negative on $[-1, 1]$. Since $\int d\nu = \pi/2$ and $\int d\nu/\wp = \pi$, the condition (15) for a non-singular factorization $\mathcal{K}+I = A^+A$ is $0 < r_0^2 \le 1/2$. As can be checked by induction, for any such a value of r_0 the Eqs. (13) lead to solutions $r_n, s_n > 0$ given by

$$r_n^2 = \frac{1}{2}\frac{n - 2r_0^2(n-1)}{n+1-2r_0^2 n}, \qquad s_n = \frac{1}{2r_{n+1}}, \qquad n \ge 0.$$

For each $r_0 \in (0, 1/\sqrt{2}]$, they provide the non-singular factor A leading to the inverse Darboux transform $\mathcal{J} = AA^+ - I$. According to (14) and (16), the corresponding measure μ is given by

$$d\mu(z) = \sqrt{\tfrac{1-z}{1+z}}\, dz + m\, \delta_{-1}(z)\, dz, \qquad m = \pi\left(\tfrac{1}{2r_0^2} - 1\right).$$

The mass m only vanishes for $r_0 = 1/\sqrt{2}$, in which case

$$A = \begin{pmatrix} \frac{1}{\sqrt{2}} & & & \\ \frac{1}{\sqrt{2}} & \frac{1}{\sqrt{2}} & & \\ & \frac{1}{\sqrt{2}} & \frac{1}{\sqrt{2}} & \\ & & \ddots & \ddots \end{pmatrix}, \qquad \mathcal{J} = AA^+ - I = \begin{pmatrix} -\frac{1}{2} & \frac{1}{2} & & \\ \frac{1}{2} & 0 & \frac{1}{2} & \\ & \frac{1}{2} & 0 & \frac{1}{2} \\ & & \ddots & \ddots & \ddots \end{pmatrix}.$$

On the other hand, when $r_0 = 1/2$, there is a mass $m = \pi$ and $r_n^2 = \frac{1}{2}\frac{n+1}{n+2}$, so that $\mathcal{J}$ is given by (1) with

$$b_1 = r_0^2 - 1 = -3/4, \qquad b_n = r_{n-1}^2 + s_{n-2}^2 - 1 = \frac{1}{2}\frac{1}{n(n+1)}, \qquad n \ge 2,$$

$$a_n = r_{n-1}s_{n-1} = \frac{1}{2}\frac{\sqrt{n(n+2)}}{n+1}, \qquad n \ge 1.$$

If we consider now as a starting point the first kind Chebyshev polynomials, with measure $d\nu(z) = \frac{1}{\sqrt{1-z^2}}\, dz$, $z \in [-1, 1]$, and Jacobi matrix

$$\mathcal{K} = \begin{pmatrix} 0 & \frac{1}{\sqrt{2}} & & \\ \frac{1}{\sqrt{2}} & 0 & \frac{1}{2} & \\ & \frac{1}{2} & 0 & \frac{1}{2} \\ & & \ddots & \ddots & \ddots \end{pmatrix},$$

(continued)

the same polynomial $\wp(z) = z + 1$ leads to the following solutions of (13),

$$r_n^2 = \frac{1}{2}\frac{1 - r_0^2(n-1)}{1 - r_0^2 n}, \quad n \geq 1, \qquad s_0 = \frac{1}{\sqrt{2}r_1}, \qquad s_n = \frac{1}{2r_{n+1}}, \quad n \geq 1.$$

No choice of $r_0 > 0$ yields $r_n > 0$ for all n, which means that there is no non-singular factorization $\wp(\mathcal{K}) = A^+A$. The factorization ensured by Proposition 4 is that one with $r_0 = 0$, which is singular and provides no inverse Darboux transform. This could be predicted just looking at the measure because $\int d\nu/\wp$ is infinite.

5 Darboux Transformations of CMV Matrices

In this section we will follow Sect. 3.2 to define the Darboux transformations of CMV matrices [2]. Many of the proofs are similar to those for Jacobi matrices, so we will omit them, paying attention only to the new ingredients. An advantage of CMV with respect to Jacobi is the absence of the subtleties of unbounded situations, thus the results typically hold for any CMV matrix. Also, the only requirement for a non-trivial positive Borel measure μ on $\mathbb{T}$ to generate a CMV matrix is the finiteness of $\int d\mu$, since it is equivalent to $\Lambda \subset L^2(\mu)$.

Any sequence $\chi = (\chi_0, \chi_1, \chi_2, \dots)^t$ of Laurent polynomials such that

$$\chi = T\eta, \qquad \eta = (1, z, z^{-1}, z^2, z^{-2}, \dots)^t, \qquad T \; \begin{matrix}\text{lower triangular with} \\ \text{positive diagonal}\end{matrix},$$

will be called a *zig-zag* sequence in Λ. Given two zig-zag sequences, χ, ω and a Hermitian Laurent polynomial $\ell(z) = \alpha z + \beta + \overline{\alpha} z^{-1}$ of degree 2, we can define the matrices $A, B, C, \mathcal{D}$ by

$$z\chi = C\chi, \qquad z\omega = \mathcal{D}\omega, \qquad \chi = A\omega, \qquad \ell\omega = B\chi.$$

The matrix A is triangular with positive diagonal, while $B, C, \mathcal{D}$ are lower Hessenberg type with the structure represented below, B having non-zero upper diagonal entries,

$$C, \mathcal{D} \to \begin{pmatrix} * & + & & & \\ * & * & * & + & \\ * & * & * & * & \\ \dots & \dots & \dots & \dots & \dots \end{pmatrix}, \qquad B \to \begin{pmatrix} * & * & * & & \\ * & * & * & * & \\ * & * & * & * & * \\ \dots & \dots & \dots & \dots & \dots \end{pmatrix}. \tag{17}$$

A difference with respect to the Jacobi case is that C and $\mathcal{D}$ are invertible because $z^{-1}\chi = C^{-1}\chi$ and $z^{-1}\omega = \mathcal{D}^{-1}\omega$ define lower Hessenberg type inverses. The

Hessenberg type structure is key to use Proposition 1 to prove the following results using steps similar to those in the proofs of the analogous results in Sect. 4.

Lemma 2 ([2, Sect. 2]) $\ell(\mathcal{C}) = AB$, $\ell(\mathcal{D}) = BA$, $\mathcal{C}A = A\mathcal{D}$, $\mathcal{D}B = B\mathcal{C}$, $\mathcal{D} = A^{-1}\mathcal{C}A$, $\mathcal{C} = A\mathcal{D}A^{-1}$.

Proposition 5 ([2, Sect. 2]) *B is not invertible and, if ζ_1, ζ_2 are the roots of ℓ, then*

$$\ker_R(\ell(\mathcal{C})) = \ker_R(B) = \begin{cases} \operatorname{span}\{\chi(\zeta_1), \chi(\zeta_2)\} & \text{if } \zeta_1 \neq \zeta_2, \\ \operatorname{span}\{\chi(\zeta_1), \chi'(\zeta_1)\} & \text{if } \zeta_1 = \zeta_2. \end{cases}$$

Assume now that χ, ω are orthonormal with respect to some measures on $\mathbb{T}$. Then $\mathcal{C}$, $\mathcal{D}$ are the related CMV matrices, and we have the following analogues of Theorem 3 and Proposition 3.

Theorem 5 ([2, Sect. 3])

$B = A^+ \Leftrightarrow$ *If χ is orthonormal with respect to the measure $d\mu$, then ω is orthonormal with respect to the measure $d\nu = \ell\, d\mu$.*

Proposition 6 ([2, Sect. 3]) *Let $\mathcal{C}$ be a CMV matrix and μ a related measure on $\mathbb{T}$. Then, a Hermitian $\ell \in \Lambda$ is non-negative on $\operatorname{supp}(\mu)$ iff $\ell(\mathcal{C}) > 0$.*

The previous results lead to the definition of the Darboux transformations $\mathcal{C} \mapsto \mathcal{D}$ of CMV matrices. If $\ell(\mathcal{C}) > 0$ for a CMV matrix $\mathcal{C}$, there is a unique Cholesky factorization $\ell(\mathcal{C}) = AA^+$, which defines $\mathcal{D} = A^{-1}\mathcal{C}A$ satisfying $\ell(\mathcal{D}) = AA^+$. Due to the five-diagonal structure of $\ell(\mathcal{C})$ and the fact that its lower diagonal entries are non-zero, the Cholesky factor A has the 3-band shape (10) with non-zero lower diagonal entries. Theorem 5 shows that $\mathcal{D}$ is a CMV matrix, that one related to the measure $d\nu = \ell\, d\mu$, where μ is associated with $\mathcal{C}$. The positivity of ν follows from Proposition 6, while the finiteness of $\int d\nu$ is a consequence of the fact that $\Lambda \subset L^2(\mu)$. The transformation $d\mu \mapsto d\nu = \wp\, d\mu$ preserves the support of μ up to the mass points at the zeros of ℓ. This means that the Darboux transformation $\mathcal{C} \mapsto \mathcal{D}$ is isospectral up to the eigenvalues of $\mathcal{C}$ at the zeros of ℓ, which are removed.

5.1 Inverse Darboux Transformations of CMV Matrices

Inverse Darboux transformations may also be considered for CMV matrices. Starting with a CMV matrix $\mathcal{D}$ such that $\ell(\mathcal{D}) > 0$ for some Hermitian polynomial ℓ of degree 2, we search for the reversed Cholesky factorizations $\ell(\mathcal{D}) = A^+A$, with A as in (10), which yield the CMV matrices $\mathcal{C} = A\mathcal{D}A^{-1}$ whose Darboux transform is $\mathcal{D}$. If ν is a measure related to $\mathcal{D}$, this is equivalent to find the CMV matrices C with a related measure on $\mathbb{T}$ having the form

$$d\mu = \frac{d\nu}{\ell} + \sum_{\ell(\zeta)=0} m_\zeta\, \delta_\zeta.$$

According to Proposition 6, the condition $\ell(\mathcal{D}) > 0$ guarantees that $d\nu/\ell$ is positive, which will be a measure related to a CMV matrix provided that $\int d\nu/\ell$ is finite. There are three possibilities according to the location of the roots ζ_1, ζ_2 of ℓ:

1. $\zeta_1 \neq \zeta_2,\ \zeta_1, \zeta_2 \in \mathbb{T} \Rightarrow d\mu = \frac{d\nu}{\ell} + m_1\, \delta_{\zeta_1} + m_2\, \delta_{\zeta_2}$.
2. $\zeta_1 = \zeta_2 = \zeta \in \mathbb{T} \Rightarrow d\mu = \frac{d\nu}{\ell} + m\, \delta_\zeta$.
3. $\zeta_1 \overline{\zeta}_2 = 1,\ \zeta_1, \zeta_2 \notin \mathbb{T} \Rightarrow d\mu = \frac{d\nu}{\ell}$.

Therefore, the inverse Darboux transforms of a CMV matrix $\mathcal{D}$ such that $\ell(\mathcal{D}) > 0$ constitute a set of CMV matrices parametrized by at most two real parameters, which will be empty if $\int d\nu/\ell$ is infinite.

Whenever $\ell(\mathcal{D}) > 0$, Proposition 4 guarantees the existence of a factorization $\ell(\mathcal{D}) = A^+ A$ with

$$A = \begin{pmatrix} r_0 & & & & \\ s_0 & r_1 & & & \\ t_0 & s_1 & r_2 & & \\ & t_1 & s_2 & r_3 & \\ & & \ddots & \ddots & \ddots \end{pmatrix}, \qquad \begin{array}{l} r_0, r_1 \geq 0, \\ r_{n+2} > 0, \quad t_n \neq 0, \quad n \geq 0. \end{array} \tag{18}$$

As in the Jacobi case, inverse Darboux for CMV requires non-singular factorizations, i.e. $r_0, r_1 > 0$. Nevertheless, in contrast to the Jacobi case, non-singular factorizations may not lead to CMV inverse Darboux transforms. To see this, denoting $\ell(\mathcal{D}) = (m_{i,j})$, let us write explicitly the factorization $\ell(\mathcal{D}) = A^+ A$, which becomes

$$r_n^2 + |s_n|^2 + |t_n|^2 = m_{n,n}, \quad r_{n+1} s_n + \overline{s_{n+1}} t_n = m_{n+1,n}, \quad r_{n+2} t_n = m_{n+2,n}. \tag{19}$$

Any solution of these equations is determined by r_0, r_1 and s_0, which gives a freedom of four real parameters in the factorization $\ell(\mathcal{D}) = A^+ A$, two more than those expected from the previous discussion. This indicates that most of the non-singular factors A will lead to matrices $C = A\mathcal{D}A^{-1}$ which are not CMV, which is in striking contrast with the Jacobi case. Actually, bearing in mind that admissible products of infinite matrices are subject to the rules of Proposition 1—implicitly assumed in all the previous manipulations—, we can be in troubles to show the unitarity of C.

An attempt to prove that $C^+ = C^{-1}$ is by noting that $C^{-1} = A\mathcal{D}^+ A^{-1}$, while Lemma 2 for $B = A^+$ yields $A^+ C = \mathcal{D} A^+$, so that $(C^+ A)A^{-1} = (A\mathcal{D}^+)A^{-1} = A\mathcal{D}^+ A^{-1}$. The second equality is due to the associativity guaranteed by the band structure of A and $\mathcal{D}^+$, see Proposition 1. However, we cannot use the associativity to cancel the factor A in $(C^+ A)A^{-1}$ because A^{-1} is lower triangular and we only know that C^+ has the upper Hessenberg type structure inferred from (17) because we do not know yet if C is a CMV matrix.

An alternative is to prove that $CC^+ = I$. Using again Lemma 2, taking care of the associativity rules in Proposition 1, we find that $(CC^+)A = C(C^+A) = C(A\mathcal{D}^+) = (CA)\mathcal{D}^+ = (A\mathcal{D})\mathcal{D}^+ = A(\mathcal{D}\mathcal{D}^+) = A$. However, the only conclusion from this identity is that the rows of $CC^+ - I$ belong to $\ker_L A = (\ker_R A^+)^+$, which is non-trivial by Proposition 5.

All this points out to the existence of non-unitary matrices $C = A\mathcal{D}A^{-1}$ coming from non-singular factorizations $\ell(\mathcal{D}) = A^+A$. Actually, it is possible to prove that a matrix C related to a ziz-zag sequence χ in Λ by $z\chi = C\chi$ is unitary iff it is a CMV matrix [2, Sect. 2]. Therefore, the existence of reversed Cholesky factorizations leading to matrices C which are not CMV, shown by parameter counting, also uncovers that such *spurious solutions* are indeed non-unitary. Fortunately, it is possible to give explicit conditions which select the values of the free parameters r_0, r_1, s_0 characterizing the non-singular factorizations leading to non-spurious solutions, i.e. the true CMV inverse Darboux transforms.

Theorem 6 ([2, Sect. 4]) *Let $\mathcal{D}$ be a CMV matrix with first Schur parameter b and $\sigma = \sqrt{1-|b|^2}$. If $\ell(\mathcal{D}) > 0$ for a Hermitian polynomial $\ell(z) = \alpha z + \beta + \overline{\alpha} z^{-1}$, the factorizations $\ell(\mathcal{D}) = A^+A$ leading to a CMV matrix $C = A\mathcal{D}A^{-1}$ are those with factors* (18) *satisfying*

$$r_0^2 = \beta - 2\,\mathrm{Re}(\alpha a), \qquad r_1 = r_0\frac{\sigma}{\rho}, \qquad s_0 = r_1\frac{a-b}{\sigma}, \tag{20}$$

for some $a \in \mathbb{C}$ with $|a| < 1$ and $\rho = \sqrt{1-|a|^2}$. Moreover, the parameter a is the first Schur parameter of C.

This theorem implies that, given a Hermitian Laurent polynomial ℓ and a CMV matrix $\mathcal{D}$, the inverse Darboux transforms are determined by their first Schur parameter. However, not every choice of this parameter necessarily yields a reversed Cholesky factorization $\ell(\mathcal{D}) = A^+A$, something which depends on the consistency of the conditions $r_n > 0$, $t_n \neq 0$ for the corresponding solutions of (19). Therefore, the inverse Darboux transforms are parametrized by at most two real parameters, which coincides with the result previously argued via orthogonality measures.

Constant Schur Parameters

Consider a CMV matrix $\mathcal{D}$ with constant Schur parameters $(b, b, b, \dots)$, $0 < b < 1$, and the Hermitian Laurent polynomial $\ell(z) = 2 - z - z^{-1}$. A simple

(continued)

inspection shows that a couple of solutions of (19) is given by

$$\text{Solution 1}: \quad r_0 = 2\sqrt{\frac{2b}{1+b}}, \quad s_0 = \sqrt{\frac{1-b}{1+b}}(b-1), \quad \begin{cases} r_n = 1+b, & n \geq 1, \\ s_n = 0, & n \geq 1, \\ t_n = b-1, & n \geq 0. \end{cases}$$

$$\text{Solution 2}: \quad r_0 = 2\sqrt{b}, \quad s_0 = -\sqrt{1-b^2}, \quad \begin{cases} r_{2n+1} = -t_{2n} = 1-b, & n \geq 0, \\ s_n = 0, & n \geq 1, \\ r_{2n} = -t_{2n-1} = 1+b, & n \geq 1. \end{cases}$$

Both give non-singular factorizations $\ell(\mathcal{D}) = A^+A$, but the first solution satisfies (20) for $a = (3b-1)/(1+b) \in (-1, 1)$, while the second one does not because the third condition in (20) yields $a = -1$ in this case. Therefore, the first solution leads to an inverse Darboux transform $C = A\mathcal{D}A^{-1}$, which turns out to be the CMV matrix with Schur parameters $(a, b, b, \dots)$ [2, Sect. 4], but the second solution yields a spurious solution which we know that is not even unitary. This is explicitly shown by calculating its first row, which gives the non-unit vector $(1, 2\sqrt{b(1+b)/(1-b)}, 0, 0, \dots)$.

Acknowledgments The work of the second author has been partially supported by the research project MTM2015-65888-C4-2-P from Dirección General de Investigación Científica y Técnica, Ministerio de Economía y Competitividad of Spain. The work of the rest of the authors has been supported in part by the research project MTM2017-89941-P from Ministerio de Economía, Industria y Competitividad of Spain and the European Regional Development Fund (ERDF), and by Project E26_17R of Diputación General de Aragón (Spain) and the ERDF 2014-2020 "Construyendo Europa desde Aragón".

References

1. Cantero, M.J., Moral, L., Velázquez, L., Five-diagonal matrices and zeros of orthogonal polynomials on the unit circle. Linear Algeb. Appl. **362**, 29–56 (2003)
2. Cantero, M.J., Marcellán, F., Moral, L., Velázquez, L.: Darboux transformations for CMV matrices. Adv. Math. **298**, 122–206 (2016)
3. Castro, M., Grünbaum, F.A.: The Darboux process and time-and-band limiting for matrix orthogonal polynomials. Linear Alg. Appl. **487**, 328–341 (2015)
4. Chihara, T.S.: An introduction to orthogonal polynomials. In: Mathematics and its Applications, vol. 13. Gordon and Breach Science Publishers, New York (1978)
5. Grünbaum, F.A.: The Darboux process and a noncommutative bispectral problem: some explorations and challenges. In: Geometric Aspects of Analysis and Mechanics, Progress in Mathematics, vol. 292, pp.161–177. Birkhäuser/Springer, New York (2011)
6. Mielnik, B., Rosas-Ortiz, O.: Factorization: little or great algorithm? J. Phys. A **37**(43), 10007–10035 (2004)
7. Simon, B.: The classical moment problem as a self-adjoint finite difference operator. Adv. Math. **137**, 82–203 (1998)

8. Simon, B.: Orthogonal Polynomials on the Unit Circle. Part 1 and 2. American Mathematical Society Colloquium Publications, vol. 54, vol. 1 and 2. American Mathematical Society, Providence (2005)
9. Teschl, G.: Jacobi Operators and Completely Integrable Nonlinear Lattices. Mathematical Surveys and Monographs, vol. 72. American Mathematical Society, Providence (2000).
10. Watkins, D.S.: Some perspectives on the eigenvalue problem. SIAM Rev. **35**(3), 430–471 (1993)
11. Weidmann, J.: Linear Operators in Hilbert Spaces. Springer, New York (1980)

Special Function Solutions of Painlevé Equations: Theory, Asymptotics and Applications

Alfredo Deaño

Abstract In this paper we review the construction of special function solutions of the Painlevé differential equations. We motivate their study using the theory of orthogonal polynomials, in particular deformation of classical weight functions, as well as unitarily invariant ensembles in random matrix theory. The asymptotic behavior of these Painlevé functions can be studied in at least two different regimes, using the Riemann-Hilbert approach and the classical saddle point method for integrals.

Keywords Painlevé-type functions · Orthogonal polynomials · Random matrices · Asymptotic representations in the complex plane

1 Painlevé Equations

The six Painlevé differential equations, P_I–P_{VI}, are nonlinear second order differential equations, whose solutions have appeared in many different areas of Mathematics in the last decades, including random matrix theory, continuous and discrete orthogonal polynomials, integrable systems and combinatorics, as well as in models in Statistical Physics. We refer the reader to [22, 35, 37, 40], among others, for more information.

The standard form of the Painlevé differential equations, as given in [51, Chapter 32], is the following:

$$u'' = 6u^2 + z, \tag{1}$$

$$u'' = zu + 2u^3 + \alpha, \tag{2}$$

A. Deaño (✉)
School of Mathematics, Statistics and Actuarial Science, University of Kent, Canterbury, UK
e-mail: A.Deano-Cabrera@kent.ac.uk

F. Marcellán, E. J. Huertas (eds.), *Orthogonal Polynomials: Current Trends and Applications*, SEMA SIMAI Springer Series 22,
https://doi.org/10.1007/978-3-030-56190-1_4

$$u'' = \frac{(u')^2}{u} - \frac{u'}{z} + \frac{\alpha u^2 + \beta}{z} + \gamma u^3 + \frac{\delta}{u}, \tag{3}$$

$$u'' = \frac{(u')^2}{2u} + \frac{3u^3}{2} + 4zu^2 + 2(z^2 - \alpha)u + \frac{\beta}{u}, \tag{4}$$

$$u'' = \left(\frac{1}{2u} + \frac{1}{u-1}\right)(u')^2 - \frac{u'}{z} + \frac{(u-1)^2}{z^2}\left(\alpha u + \frac{\beta}{u}\right) + \frac{\gamma u}{z} + \frac{\delta u(u+1)}{u-1}, \tag{5}$$

$$u'' = \frac{(u')^2}{2}\left(\frac{1}{u} + \frac{1}{u-1} + \frac{1}{u-z}\right) - \left(\frac{1}{z} + \frac{1}{z-1} + \frac{1}{u-z}\right)u' + \frac{u(u-1)(u-z)}{z^2(z-1)^2}\left[\alpha + \frac{\beta z}{u^2} + \frac{\gamma(z-1)}{(u-1)^2} + \frac{\delta z(z-1)}{(u-z)^2}\right]. \tag{6}$$

where $\alpha, \beta, \gamma, \delta \in \mathbb{C}$, see for instance [51, Chapter 32]. They have the so-called *Painlevé property*, which states that any solution is free from movable branch points (in the sense that these depend on initial or boundary values). This does not exclude fixed singularities, such as $z = 0$ in P_{III}, see (3), or $z = 0$ and $z = 1$ in P_{VI}, see (6), or movable poles.

Generic solutions of P_I–P_{VI} are sometimes called Painlevé transcendents, or *nonlinear special functions*, highlighting the role that they play in the nonlinear setting, which is analogue to classical special functions (such as Airy, Bessel or hypergeometric) in the linear world. In general, Painlevé transcendents cannot be expressed in terms of elementary or even classical special functions, but for specific values of the parameters $\alpha, \beta, \gamma, \delta$ in P_{II}–P_{VI}, it is known that families of rational and special function solutions exist; these appear for instance in the theory of semiclassical orthogonal polynomials, see for example [55] and the recent monograph [56], and in random matrix theory, see [37–39].

It is important to remark that in many applications, the standard form of the Painlevé ODEs given before is not very illuminating. In this sense, the so-called σ-equations ($S_I - S_{VI}$) often show more relevant properties, such as symmetry relations between the parameters. Also, as explained in detail in [35], Painlevé equations appear in a quite natural and systematic way when studying compatibility relations between 2×2 systems of *linear* differential equations:

$$\frac{d}{d\lambda}\Psi(\lambda, x) = A(\lambda, x)\Psi(\lambda, x), \qquad \frac{d}{dx}\Psi(\lambda, x) = U(\lambda, x)\Psi(\lambda, x), \tag{7}$$

where $\Psi : \mathbb{C}^2 \mapsto \mathbb{C}^{2\times 2}$, and A and U are rational functions of λ in the extended complex plane (possibly including the point at ∞). If one imposes compatibility of the two equations in (7), that is $\Psi_{\lambda x} = \Psi_{x\lambda}$, the following condition (sometimes called the zero curvature condition) holds:

$$\frac{d}{dx}A + AU = \frac{d}{d\lambda}U + UA. \tag{8}$$

This relation produces nonlinear differential equations relating the entries of the matrices A and U, and these identities are of Painlevé type for some standard forms of (7). The singularity structure of the system (7) in terms of the variable λ indicates which Painlevé equation we should look for, and in this respect, the linear system probably offers a clearer classification than the nonlinear ODEs.

The goal of this paper is to study families of special function solutions of Painlevé equations. These are families of solutions that can be constructed in terms of classical special functions, using Wronskian determinants generated from an initial seed function. In order to understand such families of solutions and give a systematic way to construct them, we need two important properties of the solutions of Painlevé equations: the *Hamiltonian formulation* and *Bäcklund transformations*, that we describe next.

1.1 Hamiltonian Formulation and Tau Functions

The Hamiltonian formulation provides a systematic way to study families of solutions of Painlevé equations. We refer the reader to the works of Okamoto [45–48], Forrester and Witte [38, 39] and also the monograph of Forrester [37], for more detailed information.

It is known that any Painlevé equation can be written as a Hamiltonian system of differential equations:

$$\frac{dq}{dz} = \frac{\partial H}{\partial p}, \qquad \frac{dp}{dz} = -\frac{\partial H}{\partial q}, \tag{9}$$

where $H(q, p, z, \mathbf{v})$ is a function that depends on p, q, the variable z and (in general) a vector of parameters $\mathbf{v}$.

As a first example, the Hamiltonian function for $\mathrm{P_{II}}$ is

$$H_{\mathrm{II}} = \tfrac{1}{2}p^2 - \left(q^2 + \tfrac{1}{2}z\right)p - \left(\alpha + \tfrac{1}{2}\right)q, \tag{10}$$

and then (9) gives the following system:

$$\frac{dq}{dz} = p - q^2 - \tfrac{1}{2}z, \qquad \frac{dp}{dz} = 2qp + \alpha + \tfrac{1}{2}. \tag{11}$$

Eliminating p, we see that q satisfies $\mathrm{P_{II}}$, given by (2), and eliminating q we get

$$p\frac{d^2p}{dz^2} = \frac{1}{2}\left(\frac{dp}{dz}\right)^2 + 2p^3 - zp^2 - \frac{\left(\alpha + \frac{1}{2}\right)^2}{2}, \tag{12}$$

an equation that is not independent of P_{II} but appears often in applications and is known as P_{34}; finally, the function $\sigma(z) = H_{II}(q, p, z)$ satisfies the σ equation S_{II}:

$$\left(\frac{d^2\sigma}{dz^2}\right)^2 + 4\left(\frac{d\sigma}{dz}\right)^3 + 2\frac{d\sigma}{dz}\left(z\frac{d\sigma}{dz} - \sigma\right) = \frac{1}{4}\left(\alpha + \frac{1}{2}\right)^2. \tag{13}$$

As pointed out by Clarkson [23], if we have $\sigma(z)$, then we can recover $p(z)$ and $q(z)$ as follows:

$$p(z) = -2\sigma'(z), \qquad q(z) = \frac{4\sigma''(z) + 2\alpha + 1}{8\sigma'(z)}. \tag{14}$$

In this sense, it seems reasonable to focus on the Hamiltonian (or the σ function), since properties of the solutions of P_{II} and P_{34} can be obtained from it.

If we write the Hamiltonian as $H(z) = H(q(z), p(z), z)$, we define a *tau function* as follows:

$$H(z) = \frac{d}{dz} \log \tau(z). \tag{15}$$

1.2 Bäcklund Transformations and Toda Equation

The Painlevé equations P_{II}–P_{VI} admit transformations that relate two solutions (of the same or different equation) with different values of the parameters. These transformations can be thought of as nonlinear analogues of the relations between contiguous functions for classical special functions, such as hypergeometric ones.

We illustrate the main idea using P_{II} as an example again: if $q(z)$ satisfies P_{II} with parameter α, then $v(z) = -q(z)$ verifies P_{II} with parameter $-\alpha$, an easy fact that can be checked directly from the differential equation; therefore, we can write $S : \mathcal{F} \mapsto \mathcal{F}$, where $\mathcal{F} = \{q(z; \alpha), \alpha \in \mathbb{C}\}$ is the space of solutions of P_{II} with parameter α, and

$$S(q(z, \alpha)) = -q(z, -\alpha), \tag{16}$$

is a Bäcklund transformation, that satisfies $S \circ S = I$. A less trivial transformation is given by

$$q(z, \alpha \pm 1) = T_{\pm}(q(z, \alpha)) = \begin{cases} -q, & \alpha = \mp\frac{1}{2}, \\ -q - \dfrac{2\alpha \pm 1}{\pm 2q' + 2q^2 + z}, & \text{otherwise.} \end{cases} \tag{17}$$

Bäcklund transformations can be understood in a more general context, in the sense that they do not just transform the solutions of the differential equations, but the Hamiltonian and the tau function as well.

In the references [22, Section 4.1] and [40, §19] we can find other examples with other Panlevé equations. When there are more parameters present, the main idea of the Bäcklund transformations is similar, but the parameters can change independently in several directions in parameter space (instead of just $\alpha \mapsto \alpha \pm 1$ as in P_{II}). In order to choose these directions suitably, the general formulation uses the theory of affine Weyl groups, see for example [37, §8.2.2].

We take P_{II} as an example again, following the ideas exposed in [38, Section 3]. The system (11) can be written in a more symmetric way by defining the parameters

$$\alpha_0 = \tfrac{1}{2} - \alpha, \qquad \alpha_1 = \tfrac{1}{2} + \alpha, \tag{18}$$

which satisfy $\alpha_0 + \alpha_1 = 1$, as well as the functions

$$f_0 = 2q^2 - p + z, \qquad f_1 = p. \tag{19}$$

Thus, (11) becomes

$$f_0' = -2qf_0 + \alpha_0, \qquad f_1' = 2qf_1 + \alpha_1, \tag{20}$$

and the Hamiltonian is

$$H_{II} = -\tfrac{1}{2} f_0 f_1 - \alpha_1 q. \tag{21}$$

Given this construction, the group of transformations $W = \langle s_0, s_1, \pi \rangle$ acts on the roots α_0, α_1 and on the functions f_0, f_1, defined in (18) and (19), in the following way:

	α_0	α_1	f_0	f_1	q
s_0	$-\alpha_0$	$\alpha_1 + 2\alpha_0$	f_0	$f_1 - \dfrac{4\alpha_0 q}{f_0} + \dfrac{2\alpha_0^2}{f_0^2}$	$q - \dfrac{\alpha_0}{f_0}$
s_1	$\alpha_0 + 2\alpha_1$	$-\alpha_1$	$f_0 + \dfrac{4\alpha_1 q}{f_1} + \dfrac{2\alpha_1^2}{f_1^2}$	f_1	$q + \dfrac{\alpha_1}{f_1}$
π	α_1	α_0	f_1	f_0	$-q$

The composition $T_1 := \pi s_1$ shifts the parameters as follows:

$$T_1(\alpha_0, \alpha_1) = (\alpha_1 + 2\alpha_0, -\alpha_0),$$

since $T_1(\alpha_0) = \pi s_1(\alpha_0) = \pi(\alpha_0 + 2\alpha_1) = \alpha_1 + 2\alpha_0$. It follows that this transformation corresponds to $\alpha \mapsto \alpha - 1$, and its inverse, $T_2 = s_1\pi$, corresponds to $\alpha \mapsto \alpha + 1$.

When we apply this transformation to the Hamiltonian, we obtain

$$T_1(\mathrm{H_{II}}) = \pi s_1\left(-\tfrac{1}{2}f_0f_1 - \alpha_1 q\right) = \pi\left(-\tfrac{1}{2}f_0f_1 - \alpha_1 q\right) = \mathrm{H_{II}} + q. \tag{22}$$

It seems natural to iterate this calculation, in order to generate a sequence of Hamiltonian functions: $H_{\mathrm{II},n} = T_1^n(H_{\mathrm{II}})$. As a consequence of (22), we have

$$T_1^{n+1}(\mathrm{H_{II}}) - T_1^n(\mathrm{H_{II}}) = \mathrm{H}_{\mathrm{II},n+1} - \mathrm{H}_{\mathrm{II},n} = q_n, \tag{23}$$

using the notation $q_n = q(z, \alpha - n)$, given an initial value of α. A similar calculation yields

$$T_1^{n+1}(\mathrm{H_{II}}) - T_1^n(\mathrm{H_{II}}) - (T_1^n(\mathrm{H_{II}}) - T_1^{n-1}(\mathrm{H_{II}})) = q_n - T_1^{-1}(q_n), \tag{24}$$

and since

$$H_{\mathrm{II},n} = T_1^n H_{\mathrm{II}} = \frac{d}{dz}\tau_n(z),$$

we conclude that

$$\begin{aligned} T_1^{n+1}(\mathrm{H_{II}}) - T_1^n(\mathrm{H_{II}}) - (T_1^n(\mathrm{H_{II}}) - T_1^{n-1}(\mathrm{H_{II}})) &= \mathrm{H}_{\mathrm{II},n+1} - 2\mathrm{H}_{\mathrm{II},n} + \mathrm{H}_{\mathrm{II},n-1} \\ &= \frac{d}{dz}\ln\frac{\tau_{n+1}(z)\tau_{n-1}(z)}{\tau_n^2(z)}. \end{aligned}$$

Combining this with (24), we have

$$q_n - T_1^{-1}(q_n) = \frac{d}{dz}\ln\frac{\tau_{n+1}(z)\tau_{n-1}(z)}{\tau_n^2(z)}. \tag{25}$$

On the other hand, since $T_1T_2 = I$, where $T_2 = s_1\pi$, we have

$$q_n - T_1^{-1}(q_n) = q_n - T_2(q_n) = q_n - s_1\pi(q_n) = 2q_n + \frac{\alpha_1}{f_{1,n}} = \frac{d}{dz}\ln f_{1,n}, \tag{26}$$

where we write $f_{1,n} := f_1(z; \alpha - n)$ again. We differentiate (21), bearing in mind that $2q' = f_1 - f_0$, which results from combining the first equation in (11) with (19), then

$$-2\frac{d}{dz}\mathrm{H}_{\mathrm{II},n} = f_{1,n} \Rightarrow \frac{d}{dz}\ln f_{1,n} = \frac{d}{dz}\ln\left(\frac{d}{dz}\mathrm{H}_{\mathrm{II},n}\right) = \frac{d}{dz}\ln\left(\frac{d^2}{dz^2}\ln\tau_n(z)\right).$$

Combining this equation with (26) and (25), integrating and taking exponentials, we have

$$\frac{\tau_{n+1}(z)\tau_{n-1}(z)}{\tau_n^2(z)} = C\frac{d^2}{dz^2}\ln\tau_n(z), \tag{27}$$

where C is a constant, which is a *Toda equation*.

In the references [38] and [39], the authors present this and other examples with different Painlevé equations. In those cases, the calculations are more involved, since the number of parameters is larger and then there are several Bäcklund transformations, changing parameters in different directions. Also, in some cases changes of variable and modifications of the tau functions are needed as well, in order to arrive at a Toda equation.

1.3 Special Function Solutions of Painlevé Equations

The iterative process that we just described, involving transformations of the Hamiltonian and the tau function, can be used in general to construct sequences of solutions of Painlevé equations. Furthermore, the Toda equation (27) allows us to compute the tau functions in a simple way, starting from suitable initial values. In the case of special function solutions, the combination of (27) and $\tau_0(z) = 1$ allows us to deduce a remarkable structure in the form of Wronskian determinants.

If we take the example of P_{II} again, suppose that $\tau_0(z) = 1$, then $H_0 = 0$, so from (21) we have

$$-\tfrac{1}{2}f_{0,0}f_{1,0} - \alpha_1 q_0 = 0,$$

where once again we have written $f_{0,n} := f_0(z;\alpha - n)$, for $n \geq 0$, and similarly for f_1. This equality can be achieved if we set $\alpha_1 = 0$ (and then $\alpha = -\frac{1}{2}$ from (18)), and $f_{1,0} = 0$.

As a consequence,

$$H_1 - H_0 = q_0 \Rightarrow \frac{d}{dz}\log\tau_1 = q_0,$$

and if we differentiate again and use the second equation in (11), we obtain

$$\frac{d^2}{dz^2}\tau_1(z) + \frac{z}{2}\tau_1(z) = 0,$$

which is a linear differential equation that can be solved in terms of classical Airy functions:

$$\tau_1(z) = \varphi(z) = C_1\mathrm{Ai}(-2^{-1/3}z) + C_2\mathrm{Bi}(-2^{-1/3}z). \tag{28}$$

This function $\varphi(z)$ is usually called the *seed function*.

The key observation is that the Toda equation, together with the initial value $\tau_0(z) = 1$, allows us to write $\tau_n(z)$ as an $n \times n$ Wronskian determinant:

$$\tau_n(z) = \det\left(\frac{d^{j+k}\varphi(z)}{dz^{j+k}}\right)_{j,k=0,1,\ldots n-1}, \tag{29}$$

with initial values $\tau_0(z) = 1$ and $\tau_1(z) = \varphi(z)$. A proof of this result can be found for instance in [57, Theorem 6.3], as a consequence of the Jacobi identity for determinants and the structure of the Wronskian.

An alternative derivation of these special function solutions, explained in [22], is to assume that the solution that we seek solves, apart from the corresponding Painlevé equation, a Riccati equation:

$$u'(z) = p_2(z)u(z)^2 + p_1(z)u(z) + p_0(z), \tag{30}$$

for some coefficients $p_0(z)$, $p_1(z)$, $p_2(z)$ to be determined. The combination of both differential equations imposes restrictions on the parameters of the Painlevé equation and fixes the form of the coefficients $p_0(z)$, $p_1(z)$ and $p_2(z)$. Finally, the Riccati equation can be linearised by means of $u(z) = \pm\varphi'(z)/\varphi(z)$, resulting in a linear second order differential equation for $\varphi(z)$.

In the case of P_{II}, we obtain

$$p_2(z) = \pm 1, \qquad p_1(z) = 0, \qquad p_0(z) = \pm\tfrac{1}{2}z, \qquad \alpha = \pm\tfrac{1}{2},$$

and therefore

$$u'(z) = \pm u(z)^2 \pm \tfrac{1}{2}z.$$

If we write $u(z) = \mp\varphi'(z)/\varphi(z)$, we arrive at

$$\varphi''(z) + \frac{z}{2}\varphi(z) = 0,$$

which is the same Airy equation as before.

We observe that this approach is conceptually simpler, and it imposes specific values of α in P_{II} in a quite straightforward way. In exchange, it concerns solutions of P_{II} only, whereas the Hamiltonian formulation describes a bigger picture, including solutions of P_{34} and σ_{II}.

Formula (29) describes a whole sequence of tau functions in terms of Airy functions when $\alpha = \frac{n}{2}$, $n \in \mathbb{Z}$. Using the fact that

$$\sigma(z) = H_{\mathrm{II}}(z) = \frac{d}{dz}\log\tau(z), \tag{31}$$

Table 1 Classical special functions that correspond to special function solutions of P_{II}–P_{VI}.

P_{II}	Airy: $\mathrm{Ai}(z)$, $\mathrm{Bi}(z)$
P_{III}	Bessel: $J_\nu(z)$, $Y_\nu(z)$, $I_\nu(z)$, $K_\nu(z)$
P_{IV}	Parabolic cylinder (Weber): $U(a, z)$, $D_\nu(z)$
P_V	Kummer (Whittaker): ${}_1F_1(a; c; z)$
P_{VI}	Gauss: ${}_2F_1(a, b; c; z)$

and then (14), we can construct special function (Airy) solutions of P_{II}. We remark that proving the fact that Airy solutions happen *only* for these values of α requires extra work, we refer the reader to [40, Theorem 21.1] in particular for P_{II}.

Special function solutions of Painlevé equations $P_{III}-P_{VI}$ have been obtained in the literature, from the work of Okamoto [45], see also Forrester and Witte [38, 39], and Clarkson [22]. One obtains a similar determinant structure:

$$\tau_n(z) = \det\left(\frac{\mathrm{D}^{j+k}}{\mathrm{D}z^{j+k}}\varphi(z)\right)_{j,k=0,1,\ldots n-1}, \tag{32}$$

with initial values $\tau_0(z) = 1$ and $\tau_1(z) = \varphi(z)$ a suitable seed function. Here D is a differential operator that can be more general than the derivative, and sometimes further changes of variable or modifications of the original tau function are needed in the calculation of the Toda equation.

The correspondence between Painlevé equations and classical special functions is given in Table 1.

Finally, we mention that once we have properties of $\tau_n(z)$, we can translate them to solutions of Painlevé equations. For example, in P_{II} we have the formulas (14), that give special function solutions $p_n(z)$ and $q_n(z)$ to P_{34} and P_{II} respectively, in terms of $\sigma_n(z) = \frac{d}{dz}\log\tau_n(z)$. Similar formulas for P_{IV} can be found for instance in [24].

2 Motivation 1. Orthogonal Polynomials

2.1 General Theory

The classical theory of orthogonal polynomials can be found, with different points of view, in many standard references, such as the books by Chihara [20], Ismail [41] or Szegő [52], as well as in many chapters of books on special functions, as well as Chapter 18 of the DLMF [51]. For completeness, we recall below the facts that we need in order to make the connection with Painlevé functions.

Given a weight function $w(x)$, positive and integrable in a subset of the real line $J \subset \mathbb{R}$, and such that the moments

$$\mu_m = \int_J x^m w(x)dx, \qquad m \geq 0 \tag{33}$$

are defined, we can generate a sequence of orthogonal polynomials (OPs in the sequel) $P_n(x)$, for $n = 0, 1, 2, \ldots$, such that $\deg(P_n) = n$ and the polynomials satisfy the $L^2(J, w(x))$ orthogonality

$$\int_J P_n(x)P_m(x)w(x)dx = \varkappa_n\delta_{mn}, \tag{34}$$

where $\varkappa_n \neq 0$ and δ_{mn} is the classical Kronecker delta. Furthermore, the family of OPs is unique once we fix the normalization; two standard ones are monic polynomials, $P_n(x) = x^n + \ldots$ and orthonormal ones:

$$\pi_n(x) = \gamma_n x^n + \ldots, \qquad \int_J \pi_n^2(x)w(x)dx = 1.$$

An essential property of the OPs, that follows directly from orthogonality, is that they satisfy a three term recurrence relation:

$$x P_n(x) = P_{n+1}(x) + \alpha_n P_n(x) + \beta_n P_{n-1}(x), \tag{35}$$

with initial data $P_{-1}(x) = 0$ and $P_0(x) = 1$.

In the sequel it will be important to observe that the OPs, as well as the coefficients α_n and β_n in (35), can be constructed from Hankel determinants generated with the moments (33) of the weight function. Namely, if

$$H_n = \det(\mu_{j+k})_{j,k=0,1,\ldots,n-1}, \qquad n \geq 1, \tag{36}$$

then

$$P_n(x) = \frac{1}{H_n}\begin{vmatrix} \mu_0 & \mu_1 & \ldots & \mu_{n-1} & \mu_n \\ \mu_1 & \mu_2 & \ldots & \mu_n & \mu_{n+1} \\ \vdots & \vdots & \ddots & \vdots & \vdots \\ \mu_{n-1} & \mu_n & \ldots & \mu_{2n-2} & \mu_{2n-1} \\ 1 & x & \ldots & x^{n-1} & x^n \end{vmatrix} = x^n + \ldots \tag{37}$$

is the n-th OP with respect to $w(x)$. Additionally, the Hankel determinant H_n and $P_n(x)$ admit n-fold integral representations, see [41, Sec. 2.1]:

$$\begin{aligned} H_n &= \frac{1}{n!}\int_J \cdots \int_J \prod_{1\leq j<k\leq n} (x_k - x_j)^2 \prod_{j=1}^n w(x_j)\mathrm{d}x_j, \\ P_n(x) &= \frac{1}{n!H_n}\int_J \cdots \int_J \prod_{j=1}^n (x - x_j) \prod_{1\leq j<k\leq n} (x_k - x_j)^2 \prod_{j=1}^n w(x_j)\mathrm{d}x_j. \end{aligned} \tag{38}$$

These formulas, using determinants and iterated integrals, are not very convenient to describe properties of OPs or for numerical evaluation, but they are a key ingredient to connect OPs with the theory of random matrices, and they are very useful in the asymptotic analysis, both as $n \to \infty$ and as $z \to \infty$.

2.2 Connection with Painlevé Equations

The connection between orthogonal polynomials and Painlevé equations is well known, and one standard way to study it is via specific deformations of classical weights. In the literature, quite a few examples have been considered already: we mention the work of Magnus [43] and the following examples of deformations of classical weights on the real line:

- Basor et al. [2]:

$$w(x) = (1-x)^{\alpha}(1+x)^{\beta}e^{-tx}, \qquad \alpha, \beta > -1, \quad x \in (-1,1),$$

- Chen and Its [18]:

$$w(x;s) = x^{\alpha}e^{-x-t/x}, \qquad \alpha, t > 0, \quad x \in [0,\infty),$$

- Chen and Dai [17]:

$$w(x;s) = x^{\alpha}(1-x)^{\beta}e^{-t/x}, \qquad \alpha, \beta, t > 0, \quad x \in [0,1],$$

- Boelen and Van Assche [8], Filipuk et al. [33], Clarkson and Jordaan [24]:

$$w(x;t) = x^{\lambda}e^{-x^2+tx}, \qquad \lambda > -1, \quad x \in (0,\infty).$$

- Clarkson et al. [25]:

$$w(x) = |x|^{2\lambda+1}e^{-x^4+tx^2}, \qquad \lambda > -1, \quad x \in \mathbb{R}$$

- Chen and Zhang [19]:

$$w(x;t) = (x-t)^{\gamma}x^{\alpha}(1-x)^{\beta}, \qquad t < 0, \quad \alpha, \beta > 0, \quad \gamma \in \mathbb{R}, \quad x \in [0,1],$$

We refer the reader to the recent monograph of Van Assche [56] for more examples, involving discrete OPs and OPs on the unit circle as well.

In all these examples, the deformation of a classical weight (Jacobi, Laguerre or Hermite, respectively) is introduced by multiplication by an exponential or algebraic factor that contains a parameter $t \in \mathbb{R}$:

$$w(x;t) = w_0(x)W(x,t).$$

If this new weight function satisfies the conditions above, we can consider deformed OPs with respect to $w(x;t)$:

$$\int_J P_n(x;t)P_m(x;t)w(x;t)\mathrm{d}x = \varkappa_n(t)\delta_{mn}.$$

These new OPs will naturally depend on t, and they satisfy a recurrence relation very similar to (35), but with coefficients depending on t:

$$xP_n(x;t) = P_{n+1}(x;t) + \alpha_n(t)P_n(x;t) + \beta_n(t)P_{n-1}(x;t).$$

The connection with Painlevé equations can occur in at least two different situations:

1. For *finite n*, one can identify the recurrence coefficients $\alpha_n(t)$ and $\beta_n(t)$, as well as other quantities related to the OPs, as solutions of Painlevé differential equations in the deformation parameter t (in the notation of the previous examples). In this situation, one typically encounters special function solutions of Painlevé equations, because the recurrence coefficients can be written in terms of the Hankel determinants H_n, and the moments contained in these are, in many cases, classical special functions.
2. As $n \to \infty$, it can happen that there are critical transitions as other parameters tend to critical values (possibly coupled with n). In this setting, solutions of Painlevé equations are often needed in order to describe the asymptotic behaviour. In this case, one typically needs fully transcendental solutions of Painlevé equations. For instance, if we consider OPs on the real line with respect to the weight

$$w(x;t) = e^{-V(x;t)}, \qquad V(x;t) = \frac{x^4}{4} + \frac{t}{2}x^2,$$

where $t \in \mathbb{R}$, there is a critical transition when $t = t_c = -2$, when the support of the equilibrium measure, that determines the behaviour of the zeros of the OPs as $n \to \infty$, changes from one interval to two. The asymptotic analysis near this critical case makes use of a particular (transcendental) solution of $\mathrm{P_{II}}$, the Hastings–McLeod solution. It was studied by Bleher and Its [5], in the context of random matrix theory, and it is included as a particular case in the analysis by Claeys et al. [21].

In this paper we are interested in the first situation, because of the connection with special function solutions of Painlevé equations. In order to identify which Painlevé equation is relevant (for instance in the examples given above), there are several procedures; one possibility is to use ladder operators, as described for example by Ismail [41, Chapter 3]: if the weight function is given by $w(x;t) = \exp(-v(x;t))$, then the OPs with respect to $w(x;t)dx$ satisfy

$$\begin{aligned} P_n'(x;t) &= -B_n(x;t)P_n(x;t) + \beta_n(t)A_n(x;t)P_{n-1}(x;t) \\ P_{n-1}'(x;t) &= -A_{n-1}(x;t)P_n(x;t) + (B_n(x;t) + v'(x))P_{n-1}(x;t), \end{aligned} \tag{39}$$

with coefficients

$$\begin{aligned} A_n(x;t) &= \frac{1}{\varkappa_n(t)} \int \frac{v'(x;t) - v'(s;t)}{x-s} P_n^2(s;t)w(s;t)ds, \\ B_n(x;t) &= \frac{1}{\varkappa_{n-1}(t)} \int \frac{v'(x;t) - v'(s;t)}{x-s} P_n(s;t)P_{n-1}(s;t)w(s;t)ds, \end{aligned} \tag{40}$$

where as before $\varkappa_n(t) = \int P_n^2(x;t)w(x;t)dx$. Equations (39) are often called lowering and raising relations (respectively), since they decrease/increase the degree of the polynomials by means of a differential operator in x.

In many references, the relevant Painlevé equation is usually obtained from these ladder relations via a series of ad-hoc manipulations, differentiation with respect to t and identities for quantities related to the OPs, together with suitable changes of variable. An alternative is to consider the following vector:

$$\psi_n(x;t) = \begin{pmatrix} P_n(x;t) \\ P_{n-1}(x;t) \end{pmatrix}, \tag{41}$$

and write (39) in matrix form:

$$\psi_n'(x;t) = \begin{pmatrix} -B_n(x;t) & \beta_n(t)A_n(x;t) \\ -A_{n-1}(x;t) & B_n(x;t) + v'(x;t) \end{pmatrix} \psi_n(x;t). \tag{42}$$

This can be seen as (part of) the x equation of a Lax pair for $\psi_n(x;t)$, and we could try to identify it among the different Painlevé equations studied from this perspective in [35], in particular using the singularity structure in the spectral variable t. One could also carry out a similar calculation but differentiating with respect to t, in order to obtain the other equation of the Lax pair.

Another approach that we mention next is systematic but laborious.

2.3 The Riemann–Hilbert Formulation

The Riemann–Hilbert (RH) formulation of OPs is originally due to Fokas, Its and Kitaev [34], and it has been extensively used in the literature in the last decades, often in connection with large n asymptotic behaviour of OPs, using the Deift–Zhou method of steepest descent. We refer the reader to [28] or [29] for more details. The methodology, however, can also be used in order to obtain algebraic and differential identities for OPs for finite n, as explained in [41, Chapter 22].

We recall the basic ideas of the Riemann–Hilbert formulation, for simplicity assuming that we deal with OPs on the real line, with respect to a weight function $w(x)$: we seek a 2×2 matrix $Y_n : \mathbb{C} \mapsto \mathbb{C}^{2\times 2}$ with the following properties:

1. $Y_n(z)$ is analytic $\mathbb{C}\setminus\mathbb{R}$.
2. If $x\in\mathbb{R}$, the function $Y(x)$ admits boundary values $Y_{n\pm}(x) = \lim_{\varepsilon\to 0} Y_n(x\pm i\varepsilon)$, which are related by a jump condition

$$Y_{n+}(x) = Y_{n-}(x)\begin{pmatrix} 1 & w(x) \\ 0 & 1 \end{pmatrix} \tag{43}$$

3. As $z\to\infty$, the matrix $Y_n(z)$ satisfies

$$Y_n(z) = \left(I + \frac{Y_{1,n}}{z} + \frac{Y_{2,n}}{z^2} + O(z^{-3})\right)\begin{pmatrix} z^n & 0 \\ 0 & z^{-n} \end{pmatrix}. \tag{44}$$

If the support of the weight function is a finite or semi–infinite interval (as in the case of Laguerre or Jacobi), then one needs to specify the local behaviour of Y as z tends to these endpoints in order to have a unique solution of the Riemann–Hilbert problem.

The proof of existence and uniqueness of solution for this RH problem is very instructive and can be found in [28, Cap. 3]. In particular, it can be shown that $\det Y(z) = 1$, and therefore the matrix $Y(z)$ is invertible for all $z\in\mathbb{C}$. The solution is constructed as follows:

$$Y_n(z) = \begin{pmatrix} P_n(z) & (CP_n w)(z) \\ -2\pi i \varkappa_{n-1}^{-1}(z) P_{n-1}(z) & -2\pi i \varkappa_{n-1}^{-1}(CP_{n-1}w)(z) \end{pmatrix}, \tag{45}$$

where $P_n(z)$ is the n-th OP with respect to $w(z)$,

$$(Cf)(z) = \frac{1}{2\pi i}\int_\Gamma \frac{f(s)}{s-z}\mathrm{d}s \tag{46}$$

is the Cauchy transform of the function $f(z)$, and as before $\varkappa_n = \int_{\mathbb{R}} P_n^2(z) w(z) dz$.

We observe that the matrix $Y(z)$ does not only contain the OPs, but the functions of the second kind as well, and the L^2 norms. Furthermore, it is possible to identify the entries of the matrix $Y_{1,n}$ in (44) in terms of known quantities: if $P_n(z) = z^n + \sigma_{n,n-1} z^{n-1} + \ldots$, then

$$(Y_{1,n})_{11} = \sigma_{n,n-1}, \qquad (Y_{1,n})_{12} = -\frac{\varkappa_n}{2\pi i}, \qquad (Y_{1,n})_{21} = -\frac{2\pi i}{\varkappa_{n-1}}, \tag{47}$$

which in turn leads to formulas for the recurrence coefficients in terms of the entries of $Y_{1,n}$:

$$\alpha_n = (Y_{1,n})_{11} - (Y_{1,n+1})_{11}, \qquad \beta_n = (Y_{1,n})_{12}(Y_{1,n})_{21}. \tag{48}$$

The main idea in order to use this construction for the purposes of deriving algebraic and differential identities is to observe that if we define

$$U_n(z) = Y_n(z) \begin{pmatrix} w(z)^{1/2} & 0 \\ 0 & w(z)^{-1/2} \end{pmatrix}, \tag{49}$$

then the jump relation for this new matrix becomes

$$U_{n+}(x) = U_{n-}(x) \begin{pmatrix} 1 & 1 \\ 0 & 1 \end{pmatrix},$$

which is clearly constant with respect to x. We note that in this transformation new jumps can appear for example if the weight function has finite or semi–infinite support; consider the case $w(x) = x^{\alpha} e^{-x}$, with $\alpha > -1$ on $(0, \infty)$, then $w(z)^{\pm 1/2}$ in (49) will have an extra cut for instance on $\mathbb{R}^-$.

If the jumps of $U_n(z)$ are constant on all contours, it can be shown that the derivative $U_n'(z)$ has the same jumps as $U_n(z)$, and if we take the product $U_n'(z)U_n^{-1}(z)$, then this product does not have any jumps, only possible isolated singularities (poles) at the endpoints of the support of the weight function (if any). It follows that

$$V_n(z) = f(z)U_n'(z)U_n^{-1}(z), \tag{50}$$

where $f(z)$ has zeros that cancel the possible poles of $U_n'(z)U_n^{-1}(z)$, is an entire function, with known asymptotic behaviour as $z \to \infty$, from (44). By Liouville's theorem, we can determine the form of $V_n(z)$ precisely, and then we obtain a differential relation for $U_n(z)$, that we can translate back to $Y_n(z)$.

A similar idea can be applied, in a very systematic way, to different operations on $U_n(z)$: for example, a shift in n instead to the derivative with respect to z, in which case we consider $U_{n+1}(z)U_n^{-1}(z)$, and we recover (among other identities), the three term recurrence relation; also, if the weight depends on an extra parameter t, then we can differentiate with respect to it, consider $\dot{U}_n(z)U_n^{-1}(z)$, and try to obtain another differential relation in t.

This procedure can be quite laborious, in particular if the weight function is complicated, but in return it has the advantage that it offers a very systematic way to obtain identities for the OPs and related quantities. Also, if one can obtain two differential identities (in x and in t), it is usually not too complicated to identify it with a known Lax pair for one of the Painlevé equations.

3 Motivation 2: Random Matrix Theory

The standard way to define a unitarily invariant ensemble of Hermitian random matrices is to take a probability law of the form

$$dP_N(M) = \frac{1}{Z_N} \exp\left(-\mathrm{tr}(V(M))\right) dM, \tag{51}$$

where M is an $N \times N$ Hermitian matrix, Z_N is the partition function that normalises the measure to be a probability one, $V(M)$ is a smooth function (often called the potential) with sufficient growth at infinity, and dM is the standard Lebesgue measure on the entries of M. The most classical and well studied case corresponds to the quadratic potential $V(M) = \frac{M^2}{2}$, which leads to the *Gaussian Unitary Ensemble* (GUE).

It can be shown [28] that this measure is invariant under unitary transformations, that is $dP(UMU^*) = dP(M)$ for any unitary $N \times N$ matrix U. Therefore the usual first step is to rewrite the probability law in terms of eigenvalues and eigenvectors, which leads to a remarkable decoupling: in terms of the (real) eigenvalues λ_j, we have

$$dP(\lambda_1, \lambda_2, \ldots, \lambda_N) = \frac{1}{Z'_N} \prod_{1 \leq j < k \leq N} (\lambda_k - \lambda_j)^2 \prod_{j=1}^{N} e^{-V(\lambda_j)} d\lambda_j. \tag{52}$$

This formula allows for an interpretation of the eigenvalues as random variables on the real line, that interact with a repulsion given by the Vandermonde determinant

$$\Delta(\lambda)^2 = \prod_{1 \leq j < k \leq N} (\lambda_k - \lambda_j)^2 = \det \begin{pmatrix} 1 & \lambda_1 & \ldots & \lambda_1^{N-1} \\ 1 & \lambda_2 & \ldots & \lambda_2^{N-1} \\ \vdots & \vdots & \vdots & \vdots \\ 1 & \lambda_N & \ldots & \lambda_N^{N-1} \end{pmatrix}^2$$

and a confinement given by the potential term $\prod_{j=1}^{N} e^{-V(\lambda_j)}$. The partition function is a normalising factor to make (52) a probability measure:

$$Z'_N = \int_{\mathbb{R}} \cdots \int_{\mathbb{R}} \prod_{1 \leq j < k \leq N} (\lambda_k - \lambda_j)^2 \prod_{j=1}^{N} e^{-V(\lambda_j)} d\lambda_j. \tag{53}$$

Two related and important objects in random matrix theory are the following:

- *Gap probability*, which is the probability that a certain subset of the real line does not contain any eigenvalues, and can be studied both for finite N and in the large N limit. For example, the probability that there are no eigenvalues in (s, ∞), for a given $s \in \mathbb{R}$, is

$$E_N(0;(s,\infty)) = \frac{1}{Z_N}\int_{-\infty}^{s}\cdots\int_{-\infty}^{s}\prod_{1\le j<k\le N}(\lambda_k-\lambda_j)^2\prod_{j=1}^{N}\mathrm{e}^{-V(\lambda_j)}d\lambda_j. \tag{54}$$

- *Averages of characteristic polynomials* with respect to a random matrix ensemble:

$$\begin{aligned} E_N(s,\mu) &= \left\langle \prod_{k=1}^{N}(s-\lambda_k)^{\mu}\right\rangle \\ &= \int_{-\infty}^{s}\cdots\int_{-\infty}^{s}\prod_{k=1}^{N}(x-\lambda_k)^{\mu}\prod_{1\le j<k\le N}(\lambda_k-\lambda_j)^2\prod_{j=1}^{N}e^{-V(\lambda_j)}d\lambda_j. \end{aligned} \tag{55}$$

The connection between these formulas and the integral representation (38) for the Hankel determinants constructed with the moments of the weight function $w(\lambda) = e^{-V(\lambda)}$ is the essential element in order to study the partition function and associated quantities using orthogonal polynomials.

The connection with Painlevé equations in this context is given by the fact that in several ensembles, the previous integrals can be written in terms of classical special functions; then, Heine's formula (38) allows us to rewrite the multiple integral as a Hankel determinant, which is then related to a determinant for special function solutions of one of the Painlevé equations.

For example, if we consider GUE, the gap probability in (s,∞) is

$$E_N(0;(s,\infty)) = \frac{1}{Z_N^{\mathrm{GUE}}}\int_{-\infty}^{s}\cdots\int_{-\infty}^{s}\prod_{1\le j<k\le N}(\lambda_k-\lambda_j)^2\prod_{j=1}^{N}e^{-\frac{\lambda_j^2}{2}}d\lambda_j, \tag{56}$$

where Z_N^{GUE} is the partition function for GUE, which can be computed explicitly. Applying Heine's formula, we obtain

$$\begin{aligned} E_N(0;(s,\infty)) &= \frac{1}{Z_N^{\mathrm{GUE}}}\det\left(\int_{-\infty}^{s}\lambda^{j+k}e^{-\frac{\lambda^2}{2}}d\lambda\right)_{j,k=0,\dots,N-1} \\ &= \frac{1}{Z_N^{\mathrm{GUE}}}\det\left(\int_{-\infty}^{s}(s-\lambda)^{j+k}e^{-\frac{\lambda^2}{2}}d\lambda\right)_{j,k=0,\dots,N-1}. \end{aligned} \tag{57}$$

Using the integral representation

$$U(a,z) = \frac{e^{-\frac{1}{4}z^2}}{\Gamma\left(\frac{1}{2}+a\right)}\int_0^{\infty}t^{a-\frac{1}{2}}\mathrm{e}^{-\frac{1}{2}t^2-zt}\mathrm{d}t, \qquad \mathrm{Re}\, a > -\tfrac{1}{2}. \tag{58}$$

for parabolic cylinder functions (or Hermite–Weber functions) [51, 12.5.1], this determinant can be written as follows:

$$
\begin{aligned}
E_N(0; (s,\infty)) &= \frac{1}{Z_N^{\mathrm{GUE}}} \det\left(e^{-\frac{1}{4}s^2}\Gamma(j+k+1)U(j+k+\tfrac{1}{2},-s)\right)_{j,k=0,\dots,N-1} \\
&= \frac{e^{-\frac{Ns^2}{2}}}{Z_N^{\mathrm{GUE}}} \det\left(\frac{d^{j+k}}{ds^{j+k}}\left(e^{\frac{1}{4}s^2}U(\tfrac{1}{2},-s)\right)\right)_{j,k=0,\dots,N-1},
\end{aligned}
\tag{59}
$$

where in the last equality we have applied the formula

$$
\frac{d^m}{dz^m}\left(e^{\frac{1}{4}z^2}U(a,z)\right) = (-1)^m\left(\tfrac{1}{2}+a\right)_m \mathrm{e}^{\frac{1}{4}z^2}U(a+m,z), \tag{60}
$$

see [51, 12.8.9]. This last determinant has structure of a Wronskian, and corresponds to a special function solutions of P_{IV}, we refer the reader for instance to [24]. The seed function is not completely general in this example, since it only involves one parabolic cylinder function U, instead of a general solution of Weber's differential equation, as in the next section.

4 Asymptotic Analysis

The asymptotic analysis of special function solutions of Painlevé equations makes use of the determinant expressions given before, but the fact that $\tau_n(z)$ can be written in terms of classical special function does not mean that the analysis is trivial, compared to fully transcendental solutions. One can consider at least two different asymptotic regimes: as $n \to \infty$, with z in compact subsets of $\mathbb{C}$ (or $\mathbb{R}$), and when n is fixed and $z \to \infty$ in the complex plane, with possible uniformity in other parameters that appear in the determinant.

4.1 Asymptotics as $n \to \infty$

In this regime, one methodology that has been successfully applied to rational solutions of P_{II}–P_{IV} by Balogh et al. [1], Bertola and Bothner [3], Bothner et al. [11], Buckingham [12], Buckingham y Miller [13, 14], is based on identifying the entries of the determinant $\tau_n(z)$ as the moments of a certain weight function $w(s; z)$, possibly on a suitable contour in the complex plane. Once this identification is made, one can work with the sequence of OPs with respect to this weight, using for example the Deift–Zhou method of steepest descent applied to the corresponding Riemann–Hilbert problem, in order to obtain the large n asymptotic behaviour.

Next, we consider an example applied to special function solutions of P_{IV}: in [16], the authors consider Hankel determinants constructed with moments of the following weight function:

$$w(x; v, s, \alpha) = e^{-x^2}|x - v|^\alpha \begin{cases} s, & \text{if } x < v, \\ 1, & \text{if } x > v, \end{cases} \tag{61}$$

where $v \in \mathbb{R}$, $s \in [0, 1]$ and $\alpha \in (-1, \infty)$. This is a generalisation of the classical Hermite weight on the real line, $w(x) = w(x; v, 0, 1) = e^{-x^2}$, introducing the following modifications:

- The parameter v is a shift of the Gaussian; this parameter can be moved to the exponential term by the change of variable $u = x - v$.
- If $\alpha \neq 0$, the weight has a root–type singularity at $x = v$; a more general situation is studied by Charlier in [15].
- The parameter s introduces a jump singularity, if $s \neq 1$. Moreover, if $s = 0$, then the weight is effectively supported on the semi–infinite interval $[v, \infty)$.

Using the integral representation (58), the moments of the weight function (61) are given by

$$\mu_m = e^{-\frac{v^2}{2}} 2^{-\frac{\alpha+m+1}{2}} \Gamma(\alpha + m + 1) \left[s\, U(\alpha + m + \tfrac{1}{2}, -\sqrt{2}v) + U(\alpha + m + \tfrac{1}{2}, \sqrt{2}v) \right], \tag{62}$$

for $m \geq 0$. Applying (60), we can write the Hankel determinant as follows:

$$\begin{aligned} H_n(v, s, \alpha) &= \det \left(\mu_{j+k}\right)_{j,k=0,\ldots,n-1} \\ &= e^{-nv^2} 2^{-n^2 - \frac{n(\alpha-1)}{2}} \Gamma(\alpha + 1)^n \det \left(\frac{d^{j+k}}{dv^{j+k}} \varphi(v, s, \alpha) \right)_{j,k=0,\ldots,n-1}, \end{aligned}$$

where

$$\varphi(v, s, \alpha) = \left[s\, U(\alpha + \tfrac{1}{2}, -\sqrt{2}v) + U(\alpha + \tfrac{1}{2}, \sqrt{2}v) \right] e^{\frac{v^2}{2}}, \tag{63}$$

which is a seed function for special function solutions of P_{IV}, see for example [24, §3.3]. It is remarkable to observe how the different modifications of the weight function match the different parameters of this seed function.

The main asymptotic results obtained in [16] for the Hankel determinants can be summarised as follows:

- If $s = 0$, as $n \to \infty$ we have

$$\log \frac{H_n(\sqrt{2n}t, 0, \alpha)}{H_n(0, 0, \alpha)} = C_1(t)n^2 + C_2(t, \alpha)n + C_3(t, \alpha) + O(n^{-1}),$$

where $t \in (-1, 1)$, $C_1(t)$, $C_2(t, \alpha)$ and $C_3(t, \alpha)$ are explicit constants (in n) and the error term is uniform for t in compact subsets of $(-1, 1)$.

- If $s \in (0, 1]$, as $n \to \infty$ we have

$$\log \frac{H_n(\sqrt{2n}t, s, \alpha)}{H_n(\sqrt{2n}t, 1, \alpha)} = \left[2 \arcsin t + 2t\sqrt{1-t^2} + \pi\right] \frac{\log s}{2\pi} n + \frac{(\log s)^2}{4\pi^2} \log n + \tilde{c}_0 + O\left(\frac{\log n}{n}\right),$$

where the constant $\tilde{c}_0 = \tilde{c}_0(t, s, \alpha)$ is again explicit.

These are relative asymptotics, based on deformation (with respect to the parameter v in the first case and the parameter s in the second), and taking as initial values cases where the determinant (or at least its large n asymptotics) can be computed explicitly. The determinant $H_n(0, 0, \alpha)$ was studied in [27], related to the generalised GUE ensemble in random matrix theory, and $H_n(\sqrt{2n}t, 1, \alpha)$ was analysed by Krasovsky in [42].

We note that the parameter v is coupled with n by the formula $v = \sqrt{2n}t$, and $t \in (-1, 1)$, which gives uniformity for t in the bulk of the spectrum. The case when t tends to ± 1 in a double scaling regime could be tackled in principle using tools similar to the ones explained in [9, 58].

We also remark that the asymptotic result is clearly not uniform in s, since it is not possible to recover the case $s = 0$ just by taking a limit $s \to 0^+$ from the result when $s \in (0, 1]$, due to the $\log s$ terms. In [16], the authors propose the coupling of the two parameters as $s = e^{-n\lambda}$, with $\lambda \geq 0$; in this setting, it can be shown that there exists a critical value $\lambda_c(t)$ (which is explicit as a function of t), and the uniform asymptotics for $H_n(\sqrt{2n}t, e^{-n\lambda}, \alpha)$ can be determined using deformation, starting from $s = 1$ or from $s = 0$. The regime $0 \leq \lambda \leq \lambda_c(t)$ is particularly delicate, since the equilibrium measure associated to this problem is supported on two disjoint intervals (a so–called two-cut case), and the asymptotics are given in terms of elliptic functions.

4.2 Asymptotics as $z \to \infty$

If n is fixed and $z \to \infty$ in $\tau_n(z)$, the first step is analogous to the one for the large n asymptotics: we relate the entries of $\tau_n(z)$ with the moments of a suitable weight function, say $w(s; z)$, on a certain subset of the real line or more generally on a suitable curve Γ in the complex plane. This transforms the Wronskian determinant into a Hankel or Toeplitz determinant. It is worth noting that for many classical special functions, there is choice between different integral representations, with contour integrals usually giving more flexibility in terms of the parameters.

As an example, the large z asymptotic behaviour of special function (Airy) solutions of P_{II} are studied in [26]. The seed function is given by (28):

$$\varphi(z) = C_1 \,\mathrm{Ai}(-2^{-1/3}z) + C_2 \,\mathrm{Bi}(-2^{-1/3}z). \tag{64}$$

We use the classical contour integral representations of the Airy functions:

$$\begin{aligned} \mathrm{Ai}\,(z) &= \frac{1}{2\pi i}\int_{\infty e^{-\pi i/3}}^{\infty e^{\pi i/3}} \exp\left(\tfrac{1}{3}s^3 - zs\right) ds, \\ \mathrm{Bi}\,(z) &= \frac{1}{2\pi}\left(\int_{-\infty}^{\infty e^{-\pi i/3}} + \int_{-\infty}^{\infty e^{\pi i/3}}\right) \exp\left(\tfrac{1}{3}s^3 - zs\right) ds, \end{aligned} \tag{65}$$

for $z \in \mathbb{C}$, see [51, formulae 9.5.4 and 9.5.5]. Bearing this in mind, it is natural to take the weight function

$$w(s; z) = \exp\left(\tfrac{1}{3}s^3 + 2^{-1/3}zs\right), \tag{66}$$

with moments

$$\begin{aligned} \mu_m^{\mathrm{Ai}}(z) &= \frac{C_1}{2\pi i}\int_{\infty e^{-\pi i/3}}^{\infty e^{\pi i/3}} s^m w(s; z) ds, \\ \mu_m^{\mathrm{Bi}}(z) &= \frac{C_2}{2\pi}\left(\int_{-\infty}^{\infty e^{-\pi i/3}} + \int_{-\infty}^{\infty e^{\pi i/3}}\right) s^m w(s; z) ds. \end{aligned} \tag{67}$$

for $m \geq 0$.

Next, we apply the classical Heine's formula [41, Corollary 2.1.3], [52, Section 2.2], that gives the multiple integral representation (38) for the Hankel determinant obtained before:

$$\tau_n(z) = \det\left(\frac{d^{j+k}}{dz^{j+k}}\varphi(z)\right)_{j,k=0}^{n-1} = 2^{-\frac{n(n-1)}{3}} \det\left(\mu_{j+k}^{\mathrm{Ai}}(z) + \mu_{j+k}^{\mathrm{Bi}}(z)\right)_{j,k=0}^{n-1} \tag{68}$$

Finally, we apply the (classical) method of steepest descent to the n-fold resulting integral. In the one variable setting, this method is explained in many references, such as [7, 44, 49, 53]. The multivariate case is less explored, although we refer the reader to [6], [7, Chapter 8] or [32], as well as [30], in the context of β-ensembles in random matrix theory and multivariate hypergeometric functions.

In this example, additional technical details to take into account are the following:

1. The seed function $\varphi(z)$ satisfies the following rotational symmetry:

$$\varphi(ze^{\pm\frac{2\pi i}{3}}) = \widetilde{C}_{1\pm}\mathrm{Ai}(-2^{-1/3}z) + \widetilde{C}_{2\pm}\mathrm{Bi}(-2^{-1/3}z),$$

with new constants

$$\widetilde{C}_{1\pm} = \frac{C_1}{2}\mathrm{e}^{\pm\pi i/3} + \frac{3C_2}{2}\mathrm{e}^{\mp\pi i/6}, \quad \widetilde{C}_{2\pm} = \frac{C_1}{2}\mathrm{e}^{\mp\pi i/6} + \frac{C_2}{2}\mathrm{e}^{\pm\pi i/3},$$

where C_1 and C_2 are the original ones. The tau function then satisfies the rotational symmetry

$$\tau_n(e^{\pm\frac{2\pi i}{3}}z) = e^{\pm\frac{2\pi i}{3}n(n-1)}\tau_n(z), \qquad n \geq 1.$$

This allows us to restrict the asymptotic analysis to two different sectors of $\mathbb{C}$:

- The non-oscillatory sector $|\arg(-z)| < \frac{\pi}{3}$.
- The real axis, where oscillatory behavior occurs.

2. In (68), the determinant combines some integrals corresponding to Ai and others corresponding to Bi. In the saddle point method, the contributions from the stationary points will be different in each case, because the contours of integration are different, and this has to be taken into account both in the non-oscillatory and in the oscillatory regimes. We refer the reader to [26] for more details.

A by-product of this asymptotic analysis is the calculation of zero-free regions for $\tau_n(z)$ in the complex plane, at least asymptotically as $z \to \infty$. This type of result is of importance in the theory of Painlevé differential equations, since these zeros become poles of the Painlevé functions, using (14) again in the case of P_{II}, and similar formulas for other Painlevé equations.

5 Conclusions and Related Problems

Special function solutions of Painlevé equations are families of solutions that can be written as determinants involving classical special functions, and they are relevant in the theory of orthogonal polynomials and in random matrix theory. Despite the fact that they are fully identified for P_{II}–P_{VI}, in the sense of the correspondence shown in Table 1, many of their analytical, asymptotic and numerical aspects are not fully explored yet.

The asymptotic analysis (both for large n and large z) of these and similar examples is to a large extent open, with the added technicality (in the case of P_{III}, P_V and P_{VI}) of the branch points at the origin and/or at 1, which brings the analysis to a Riemann surface more general than $\mathbb{C}$. Related to this analysis, an important aspect is the (asymptotic) distribution of zeros of the tau functions, which become poles of the Painlevé solutions; determining pole-free regions in the complex plane has always been a major theme in the study of solutions of Painlevé equations.

Finally, the numerical evaluation of these special function solutions looks deceptively simple, given the expressions in terms of determinants; even with classical special functions, efficient evaluation of the Wronskian determinants for fixed (but perhaps large) values of n and $z \in \mathbb{C}$ does not seem to be an easy task, and the Toda equation, useful as it is for analytical and asymptotic properties, involves derivatives as well as contiguous tau functions.

In the literature, the following approaches have been proposed in order to compute solutions of Painlevé equations numerically:

- Integrate the differential equation numerically, either as an initial value problem, in the spirit of Fornberg and Weideman [36], see also [31], or as a boundary value problem.
- Solve the Riemann–Hilbert problem associated with the linear system (7), as proposed by Olver and Trogdon, see [54] and also [50]. In this approach, the monodromy data, rather than initial or boundary data, are considered to be the "natural" piece of information in order to determine the solution that we are looking for. Then the idea is combined with the Deift–Zhou method of steepest descent [28, 29], discretising the contours that result from the deformations by numerical collocation.
- A third option, proposed by Bornemann [10] for solutions that are relevant in random matrix theory, is based on writing gap probabilities as Fredholm determinants; for instance, the probability of having no eigenvalues in an interval $[-\theta, \theta]$ in GUE is given by

$$\begin{aligned} A(\theta) &= \sum_{m=1}^{N} \frac{(-1)^k}{k!} \int_{-\theta}^{\theta} \cdots \int_{-\theta}^{\theta} \det\left(K_N(\lambda_j, \lambda_k)\right)_{j,k=1,\ldots,m} d\lambda_1 \cdots d\lambda_m \\ &= \det(I - K_N), \end{aligned}$$

  cf. [28, pg. 109], where K_N is a Fredholm operator with a suitable kernel. This operator can be discretised for numerical evaluation. We refer the reader to [4] for an example related to P_{IV}, but this approach, though promising, is still under research in other cases.

Acknowledgments The author gratefully acknowledges financial support from the EPSRC grant "Painlevé equations: analytical properties and numerical computation", reference EP/P026532/1. The author wishes to thank the organisers of the VII Iberoamerican Workshop in Orthogonal Polynomials and Applications (EIBPOA2018), at Universidad Carlos III de Madrid (Spain), for their invitation to deliver a plenary talk on material related to the content of this paper in July 2018, and the referee for useful comments and corrections on the first version of this paper.

References

1. Balogh, F., Bertola, M., Bothner, T.: Hankel determinant approach to generalized Vorob'ev-Yablonski polynomials and their roots. Constr. Approx. **44**(3), 417–453 (2016)
2. Basor, E., Chen, Y., Ehrhardt, T.: Painlevé V and time-dependent Jacobi polynomials. J. Phys. A **43**(1), 015204, 25 (2010)
3. Bertola, M., Bothner, T.: Zeros of large degree Vorob'ev-Yablonski polynomials via a Hankel determinant identity. Int. Math. Res. Not. IMRN **2015**(19), 9330–9399 (2015)
4. Bertola, M., Elias Rebelo, J.G., Grava, T.: Painlevé IV critical asymptotics for orthogonal polynomials in the complex plane. SIGMA Symm. Integr. Geom. Method Appl. **14**, Paper No. 091, 34 (2018)
5. Bleher, P.M., Its, A.R.: Asymptotics of the partition function of a random matrix model. Ann. Inst. Four. **55**(6), 1943–2000 (2005)
6. Bleistein, N.: Saddle point contribution for an n-fold complex-valued integral (2012)
7. Bleistein, N., Handelsman, R.A.: Asymptotic Expansions of Integrals, 2nd edn. Dover Publications, New York (1986)
8. Boelen, L., Van Assche, W.: Discrete Painlevé equations for recurrence coefficients of semiclassical Laguerre polynomials. Proc. Amer. Math. Soc. **138**(4), 1317–1331 (2010)
9. Bogatskiy, A., Claeys, T., Its, A.: Hankel determinant and orthogonal polynomials for a Gaussian weight with a discontinuity at the edge. Comm. Math. Phys. **347**(1), 127–162 (2016)
10. Bornemann, F.: On the numerical evaluation of Fredholm determinants. Math. Comp. **79**(270), 871–915 (2010)
11. Bothner, T., Miller, P.D., Sheng, Y.: Rational solutions of the Painlevé-III equation. Stud. Appl. Math. **141**(4), 626–679 (2018)
12. Buckingham, R.: Large-degree asymptotics of rational Painlevé-IV functions associated to generalized hermite polynomials. Int. Math. Res. Not. IMRN **07**, rny172 (2018)
13. Buckingham, R.J., Miller, P.D.: Large-degree asymptotics of rational Painlevé-II functions: noncritical behaviour. Nonlinearity **27**(10), 2489–2578 (2014)
14. Buckingham, R.J., Miller, P.D.: Large-degree asymptotics of rational Painlevé-II functions: critical behaviour. Nonlinearity **28**(6), 1539–1596 (2015)
15. Charlier, C.: Asymptotics of Hankel determinants with a one-cut regular potential and Fisher–Hartwig singularities. Int. Math. Res. Not. **2018**, 62 (2018)
16. Charlier, C., Deaño, A.: Asymptotics for Hankel determinants associated to a Hermite weight with a varying discontinuity. SIGMA Symm. Integr. Geom. Method Appl. **14**, Paper No. 018, 43 (2018)
17. Chen, Y., Dai, D.: Painlevé V and a Pollaczek-Jacobi type orthogonal polynomials. J. Approx. Theory **162**(12), 2149–2167 (2010)
18. Chen, Y., Its, A.: Painlevé III and a singular linear statistics in Hermitian random matrix ensembles. I. J. Approx. Theory **162**(2), 270–297 (2010)
19. Chen, Y., Zhang, L.: Painlevé VI and the unitary Jacobi ensembles. Stud. Appl. Math. **125**(1), 91–112 (2010)
20. Chihara, T.S.: An introduction to Orthogonal Polynomials. Mathematics and its Applications, vol. 13. Gordon and Breach Science Publishers, New York (1978)
21. Claeys, T., Kuijlaars, A.B.J., Vanlessen, M.: Multi-critical unitary random matrix ensembles and the general Painlevé II equation. Ann. Math. **168**(2), 601–641 (2008)
22. Clarkson, P.A.: Painlevé equations—nonlinear special functions. In: Orthogonal Polynomials and Special Functions. Lecture Notes in Mathematics, vol. 1883, pp. 331–411. Springer, Berlin (2006)

23. Clarkson, P.A.: On Airy solutions of the second Painlevé equation. Stud. Appl. Math. **137**(1), 93–109 (2016)
24. Clarkson, P.A., Jordaan, K.: The relationship between semiclassical Laguerre polynomials and the fourth Painlevé equation. Constr. Approx. **39**(1), 223–254 (2014)
25. Clarkson, P.A., Jordaan, K., Kelil, A.: A generalized Freud weight. Stud. Appl. Math. **136**(3), 288–320 (2016)
26. Deaño, A.: Large z asymptotics for special function solutions of Painlevé II in the complex plane. SIGMA Symm. Integr. Geom. Method Appl. **14**, Paper No. 107, 19 (2018)
27. Deaño, A., Simm, N.J.: On the probability of positive-definiteness in the gGUE via semi-classical Laguerre polynomials. J. Approx. Theory **220**, 44–59 (2017)
28. Deift, P.A.: Orthogonal Polynomials and Random Matrices: A Riemann-Hilbert Approach. Courant Lecture Notes in Mathematics, vol. 3. New York University, Courant Institute of Mathematical Sciences, New York; American Mathematical Society, Providence (1999)
29. Deift, P., Kriecherbauer, T., McLaughlin, K.T.-R., Venakides, S., Zhou, X.: Strong asymptotics of orthogonal polynomials with respect to exponential weights. Comm. Pure Appl. Math. **52**(12), 1491–1552 (1999)
30. Desrosiers, P., Liu, D.-Z.: Asymptotics for products of characteristic polynomials in classical β-ensembles. Constr. Approx. **39**(2), 273–322 (2014)
31. Fasondini, M., Fornberg, B., Weideman, J.A.C.: Methods for the computation of the multivalued Painlevé transcendents on their Riemann surfaces. J. Comput. Phys. **344**, 36–50 (2017)
32. Fedoryuk, M.V.: Asymptotic methods in analysis. In: Analysis I, pp. 83–191. Springer, Berlin (1989)
33. Filipuk, G., Van Assche, W., Zhang, L.: The recurrence coefficients of semi-classical Laguerre polynomials and the fourth Painlevé equation. J. Phys. A **45**(20), 205201, 13 (2012)
34. Fokas, A.S., Its, A.R., Kitaev, A.V.: The isomonodromy approach to matrix models in 2D quantum gravity. Comm. Math. Phys. **147**(2), 395–430 (1992)
35. Fokas, A.S., Its, A.R., Kapaev, A.A., Yu, V.: Novokshenov. *Painlevé Transcendents. The Riemann-Hilbert Approach.* Mathematical Surveys and Monographs, vol. 128. American Mathematical Society, Providence (2006)
36. Fornberg, B., Weideman, J.A.C.: A computational exploration of the second Painlevé equation. Found. Comput. Math. **14**(5), 985–1016 (2014)
37. Forrester, P.J.: Log-Gases and Random Matrices. London Mathematical Society Monographs Series, vol. 34. Princeton University Press, Princeton (2010)
38. Forrester, P.J., Witte, N.S.: Application of the τ-function theory of Painlevé equations to random matrices: PIV, PII and the GUE. Comm. Math. Phys. **219**(2), 357–398 (2001)
39. Forrester, P.J., Witte, N.S.: Application of the τ-function theory of Painlevé equations to random matrices: P_V, P_{III}, the LUE, JUE, and CUE. Comm. Pure Appl. Math. **55**(6), 679–727 (2002)
40. Gromak, V.I., Laine, I., Shimomura, S.: Painlevé Differential Equations in the Complex Plane. De Gruyter Studies in Mathematics, vol. 28. Walter de Gruyter, Berlin (2002)
41. Ismail, M.E.H.: Classical and Quantum Orthogonal Polynomials in One Variable. Encyclopedia of Mathematics and its Applications, vol. 98. Cambridge University Press, Cambridge (2009). With two chapters by Walter Van Assche, With a foreword by Richard A. Askey, Reprint of the 2005 original
42. Krasovsky, I.V.: Correlations of the characteristic polynomials in the Gaussian unitary ensemble or a singular Hankel determinant. Duke Math. J. **139**(3), 581–619 (2007)
43. Magnus, A.P.: Painlevé-type differential equations for the recurrence coefficients of semi-classical orthogonal polynomials. In: Proceedings of the Fourth International Symposium on Orthogonal Polynomials and Their Applications (Evian-Les-Bains, 1992), vol.57, pp. 215–237 (1995)
44. Miller, P.D.: Applied Asymptotic Analysis. Graduate Studies in Mathematics, vol. 75. American Mathematical Society, Providence (2006)
45. Okamoto, K.: Studies on the Painlevé equations. III. Second and fourth Painlevé equations, P_{II} and P_{IV}. Math. Ann. **275**(2), 221–255 (1986)

46. Okamoto, K.: Studies on the Painlevé equations. I. Sixth Painlevé equation P_{VI}. Ann. Mat. Pura Appl. **146**(4), 337–381 (1987)
47. Okamoto, K.: Studies on the Painlevé equations. II. Fifth Painlevé equation P_V. Jpn. J. Math. **13**(1), 47–76 (1987)
48. Okamoto, K.: Studies on the Painlevé equations. IV. Third Painlevé equation P_{III}. Funkcial. Ekvac. **30**(2–3), 305–332 (1987)
49. Olver, F.W.J.: Asymptotics and Special Functions. AKP Classics. A K Peters, Wellesley (1997) Reprint of the 1974 original [Academic Press, New York; MR0435697 (55 #8655)].
50. Olver, S.: Numerical solution of Riemann-Hilbert problems: Painlevé II. Found. Comput. Math. **11**(2), 153–179 (2011)
51. Olver, F.W.J., Olde Daalhuis, A.B., Lozier, D.W., Schneider, B.I., Boisvert, R.F., Clark, C.W., Miller, B.R., Saunders, B.V. (eds.): NIST Digital Library of Mathematical Functions (2020). http://dlmf.nist.gov/, Release 1.0.16 of 18 Sep 2017
52. Szegő, G.: Orthogonal Polynomials, 4 edn. American Mathematical Society, Providence (1975). American Mathematical Society, Colloquium Publications, Vol. XXIII
53. Temme, N.M.: Asymptotic Methods for Integrals. Series in Analysis, vol. 6. World Scientific Publishing, Hackensack (2015)
54. Trogdon, T., Olver, S.: Riemann-Hilbert Problems, Their Numerical Solution, and the Computation of Nonlinear Special Functions. Society for Industrial and Applied Mathematics (SIAM), Philadelphia (2016)
55. Van Assche, W.: Discrete Painlevé equations for recurrence coefficients of orthogonal polynomials. In: Difference Equations, Special Functions and Orthogonal Polynomials, pp. 687–725. World Scientific Publishing, Hackensack (2007)
56. Van Assche, W.: Orthogonal Polynomials and Painlevé Equations. Number 27 in Australian Mathematical Society Lecture Series. Cambridge University Press, Cambridge (2018)
57. Vein, R., Dale, P.: Determinants and Their Applications in Mathematical Physics. Applied Mathematical Sciences, vol. 134. Springer, New York (1999)
58. Wu, X.-B., Xu, S.-X., Zhao, Y.-Q.: Gaussian unitary ensemble with boundary spectrum singularity and σ-form of the Painlevé II equation. Stud. Appl. Math. **140**(2), 221–251 (2018)

Discrete Semiclassical Orthogonal Polynomials of Class 2

Diego Dominici and Francisco Marcellán

Abstract In this contribution, discrete semiclassical orthogonal polynomials of class $s \leq 2$ are studied. By considering all possible solutions of the Pearson equation, we obtain the canonical families in each class. We also consider limit relations between these and other families of orthogonal polynomials.

Keywords Orthogonal polynomials · Discrete semiclassical orthogonal polynomials · Moments · Stieltjes transform

1 Introduction

Semiclassical linear functionals with respect to the derivative operator have received increasing attention in the literature of orthogonal polynomials taking into account their connections with many interesting problems in mathematical physics and numerical quadratures. From the so called Pearson equation associated with those linear functionals, ladder operators for the corresponding sequences of orthogonal polynomials can be deduced from the structure relations they satisfy (see [36, 37, 39]). As a direct consequence, a simple computation allows to deduce the corresponding holonomic equations, i.e. second order linear differential equations with polynomial coefficients of bounded degrees (depending on the degree of the polynomial) associated with the class of the linear functional (see [30, 39]).

D. Dominici (✉)
Johannes Kepler University Linz, Linz, Austria

State University of New York at New Paltz, New Paltz, NY, USA
e-mail: diego.dominici@dk-compmath.jku.at

F. Marcellán
Departamento de Matemáticas, Universidad Carlos III de Madrid, Leganés, Spain
e-mail: pacomarc@ing.uc3m.es

F. Marcellán, E. J. Huertas (eds.), *Orthogonal Polynomials: Current Trends and Applications*, SEMA SIMAI Springer Series 22,
https://doi.org/10.1007/978-3-030-56190-1_5

The case of semiclassical linear functionals with respect to the difference operator in a uniform lattice is related to discretizations of holonomic equations, a fact pointed out in the classical case by Nikiforov, Uvarov and Suslov in a stimulating monograph on classical discrete orthogonal polynomials [42].

In both situations, a key problem is the classification of semiclassical linear functionals by using a hierarchy according to their class. This represents an alternative method to the Askey tableau, where classical orthogonal polynomials appear as a hierarchy of hypergeometric polynomials (see [31]).

Examples of semiclassical linear functionals with respect to the derivative operator when the class is either $s = 1$ (see [9]) or $s = 2$ (see [35]) have been studied in the literature. Nevertheless, some of them are related to perturbations of classical linear functionals by the addition of Dirac functionals, or their derivatives, supported on convenient points [6, 33]. The important fact is that these orthogonal polynomials, the so called Krall-type orthogonal polynomials, are eigenfunctions of higher order differential operators. They are related to bispectral problems and have been intensively used in the generation of exceptional polynomials, an useful tool in the study of integrable systems.

In the discrete case, there is a limited development in these topics (see [34]). The case $s = 0$ has been deeply studied by many authors (see [1, 24]) and the approach to the case $s = 1$ has been done by Dominici and Marcellán in a recent paper (see [17]). Some examples of semiclassical discrete polynomials for $s = 1$ appeared in [47].

The aim of this contribution is twofold. First, we deal with three different perturbations of discrete linear functionals based on the so called linear spectral transformations (Uvarov, Christoffel and Geronimus) [50], truncations and symmetrization processes, respectively. Second, from the above transformations we are able to generate semiclassical linear functionals of class $s = 1$ and $s = 2$. Some of them appear in a natural way in the literature but others are new. We put attention only in the representation of the corresponding Stieltjes functions and, as a consequence, we deduce the corresponding class. Notice that an interesting and open problem is to analyze the coefficients of the three term recurrence relation they satisfy, in particular the so called Laguerre-Freud equations. For generalized Charlier polynomials such equations appeared in the monograph by W. Van Assche [49] as well as in [29]. They lead to limiting cases of the discrete Painlevé equation corresponding to D_4^c in the Sakai's classification. A generalization of the Krawtchouk polynomials and the fifth Painlevé equation are studied in [11]. Some interesting examples of semiclassical extensions of Meixner polynomials appeared in [10] and [21]. They yield a limit case of asymmetric discrete Painlevé IV equations. Discrete orthogonal polynomials with respect to the hypergeometric weight have been analyzed in [22] and [19].

The structure of the paper is the following. In Sect. 2, a basic background about discrete semiclassical linear functionals is given. In particular, the class is defined from the degrees of polynomials appearing in the discrete Pearson equation they satisfy. We emphasize the role of the Stieltjes function associated with every linear functional. In the semiclassical case, the corresponding Stieltjes function satisfies a first order linear difference equation with polynomial coefficients. They provide the key information about the class of the linear functional. Section 3 deals with the three canonical perturbations of linear functionals we will deal in the sequel. The corresponding Stieltjes functions are deduced. In Sect. 4, classical discrete linear functionals are revisited. In Sect. 5 we analyze semiclassical linear functionals of class $s = 1$ in such a way that the results given in [17] are completed. Finally, in Sect. 6, we study examples of discrete semiclassical linear functionals of class $s = 2$ generated by the three canonical perturbations introduced in Sect. 3.

2 Basic Background on Discrete Semiclassical Linear Functionals

Let $\mathbb{N}_0$ denote the set $\mathbb{N} \cup \{0\} = 0, 1, 2, \ldots$. Throughout the paper we will use the notation

$$\mathbf{a} = \left(a_1, \ldots, a_p\right), \quad \mathbf{b} = \left(b_1, \ldots, b_q\right), \quad p, q \in \mathbb{N}_0,$$

$$(x + \mathbf{a}) = \prod_{i=1}^{p} (x + a_i), \quad (x + \mathbf{b}) = \prod_{i=1}^{q} (x + b_i), \tag{1}$$

and

$$(\mathbf{a})_n = \prod_{i=1}^{p} (a_i)_n, \quad (\mathbf{b})_n = \prod_{i=1}^{q} (b_i)_n, \quad n \in \mathbb{N}_0,$$

where the *Pochhammer polynomial* $(x)_n$ is defined by $(x)_0 = 1$ and [43, 18:12]

$$(x)_n = \prod_{j=0}^{n-1} (x + j), \quad n \in \mathbb{N}. \tag{2}$$

From (2), we immediately obtain the recurrence

$$(x)_{n+1} = (x + n)(x)_n, \quad n \in \mathbb{N}_0. \tag{3}$$

On the other hand, if we consider the analytic continuation with $\alpha \in \mathbb{C}$, then

$$(x)_\alpha = \frac{\Gamma(x + \alpha)}{\Gamma(x)}, \quad -(x + \alpha) \notin \mathbb{N}_0, \tag{4}$$

where $\Gamma(x)$ denotes the Gamma function [44, Chapter 5.].

In this work, we deal with linear functionals L defined on $\mathbb{C}[x]$, the linear space of polynomials with complex coefficients. Indeed,

$$L[r] = \sum_{x=0}^{\infty} r(x)\,\varrho(x), \quad r \in \mathbb{C}[x], \tag{5}$$

where the weight function $\varrho : \mathbb{N}_0 \to \mathbb{C}$ is

$$\varrho(x) = \frac{(\mathbf{a})_x}{(\mathbf{b}+1)_x} \frac{z^x}{x!}. \tag{6}$$

Using (3), we see that the weight function $\varrho(x)$ satisfies the *Pearson equation*

$$\frac{\varrho(x+1)}{\varrho(x)} = \frac{\eta(x)}{\sigma(x+1)}, \tag{7}$$

where the polynomials $\eta(x)$, $\sigma(x)$ are

$$\eta(x) = z(x+\mathbf{a}), \quad \sigma(x) = x(x+\mathbf{b}). \tag{8}$$

Note that from (1) we have $\deg(\eta) = p$ and $\deg(\sigma) = q+1$.

The linear functional L (and also ϱ) is said to be *semiclassical*. The class of L (and also ϱ) is the non-negative integer number

$$s = \max\{\deg(\sigma) - 2,\ \deg(\sigma - \eta) - 1\}.$$

The linear functionals of class $s = 0$ are called *discrete classical* [24] linear functionals.

The number s depends on p, q, and z. It follows from (8) that we have four cases to consider:

1. $p > q+1$. In this case, $\deg(\sigma - \eta) - 1 = p - 1 > q$ and $s = \max\{q-1,\ p-1\} = p-1$.
2. $p < q+1$. In this case, $\deg(\sigma - \eta) - 1 = q$ and $s = \max\{q-1,\ q\} = q$.
3. $q+1 = p$, $z \neq 1$. In this case, $\deg(\sigma - \eta) - 1 = q$ and $s = \max\{q-1,\ q\} = q$.
4. $q+1 = p$, $z = 1$. In this case, $\deg(\sigma - \eta) - 1 = q - 1$ and $s = \max\{q-1,\ q-1\} = q-1$.

Thus, we conclude that

$$s = \begin{cases} p-1, & p > q+1, \\ q, & p < q+1, \\ q, & p = q+1, \quad z \neq 1, \\ q-1, & p = q+1, \quad z = 1. \end{cases} \tag{9}$$

Note that we can also write s as

$$s = \begin{cases} \max\{\deg(\sigma) - 1,\ \deg(\sigma - \eta) - 1\}, & \deg(\sigma - \eta) \neq \deg(\sigma) - 1, \\ \deg(\sigma) - 2, & \deg(\sigma - \eta) = \deg(\sigma) - 1. \end{cases} \tag{10}$$

The *Stieltjes transform* of the linear functional L is defined by

$$S(t) = L\left[\frac{1}{t - x}\right]. \tag{11}$$

Here the linear functional is acting on x. The connection between $\eta(x), \sigma(x)$ and $S(t)$ is given by the following result.

Lemma 1 *The Stieltjes transform $S(t)$ of the linear functional L satisfies the difference equation*

$$\sigma(t + 1) S(t + 1) - \eta(t) S(t) = \xi(t), \tag{12}$$

where $\xi(t)$ is a polynomial of degree s that is the class of the linear functional.

Proof From (11), we have

$$\sigma(t + 1) S(t + 1) - \eta(t) S(t) = \sigma(t + 1) \sum_{x=0}^{\infty} \frac{\varrho(x)}{t + 1 - x} - \eta(t) \sum_{x=0}^{\infty} \frac{\varrho(x)}{t - x}$$

$$= \frac{\sigma(t + 1)}{t + 1} + \sigma(t + 1) \sum_{x=1}^{\infty} \frac{\varrho(x)}{t + 1 - x} - \eta(t) \sum_{x=0}^{\infty} \frac{\varrho(x)}{t - x}$$

$$= \frac{\sigma(t + 1)}{t + 1} + \sigma(t + 1) \sum_{x=0}^{\infty} \frac{\varrho(x + 1)}{t - x} - \eta(t) \sum_{x=0}^{\infty} \frac{\varrho(x)}{t - x}.$$

Using the Pearson equation (7), we get

$$\sigma(t + 1) S(t + 1) - \eta(t) S(t) = \frac{\sigma(t + 1)}{t + 1}$$

$$+\sigma(t + 1) \sum_{x=0}^{\infty} \frac{\eta(x)}{\sigma(x + 1)} \frac{\varrho(x)}{t - x} - \eta(t) \sum_{x=0}^{\infty} \frac{\varrho(x)}{t - x}$$

$$= \frac{\sigma(t + 1)}{t + 1} + \sum_{x=0}^{\infty} \frac{\sigma(t + 1)\eta(x) - \eta(t)\sigma(x + 1)}{\sigma(x + 1)} \frac{\varrho(x)}{t - x}.$$

But

$$\lambda_x(t) = \frac{\sigma(t+1)\,\eta(x) - \eta(t)\,\sigma(x+1)}{t-x}$$

is a polynomial in t (with coefficients depending on x) of degree $\deg(\sigma - \eta) - 1$. Since $\eta(x) \neq \sigma(x+1)$ (otherwise ϱ is a constant function), we have $\deg(\sigma - \eta) \neq \deg(\sigma) - 1$.

Therefore,

$$\xi(t) = \frac{\sigma(t+1)}{t+1} + \sum_{x=0}^{\infty} \lambda_x(t)\,\frac{\varrho(x)}{\sigma(x+1)} \tag{13}$$

is a polynomial in t of degree $\max\{\deg(\sigma) - 1,\ \deg(\sigma - \eta) - 1\}$. From (10) we conclude that $\deg(\xi) = s$. □

Let the *falling factorial polynomials* be defined by $\phi_0(x) = 1$ and

$$\phi_n(x) = \prod_{j=0}^{n-1} (x - j), \quad n \in \mathbb{N}. \tag{14}$$

Comparing (14) with (2), we see that

$$\phi_n(x) = (-1)^n (-x)_n = (x - n + 1)_n, \, n \in \mathbb{N}_0, \tag{15}$$

or, using (4),

$$\phi_n(x) = \frac{\Gamma(x+1)}{\Gamma(x-n+1)} = \frac{x!}{(x-n)!} = n!\binom{x}{n}, \quad n \in \mathbb{N}_0. \tag{16}$$

We define the *moments* of L with respect to the basis $\{\phi_n(x)\}_{n\geq 0}$ by

$$\nu_n = L[\phi_n], \quad n \in \mathbb{N}_0.$$

From (6) and (16), we have

$$\nu_n = \sum_{x=n}^{\infty} \frac{(\mathbf{a})_x}{(\mathbf{b}+1)_x} \frac{z^x}{(x-n)!} = \sum_{x=0}^{\infty} \frac{(\mathbf{a})_{x+n}}{(\mathbf{b}+1)_{x+n}} \frac{z^{x+n}}{x!}.$$

If we use the recurrence [43, 18:5:12]

$$(x)_{n+m} = (x)_n (x+n)_m, \quad n, m \in \mathbb{N}_0, \tag{17}$$

we get

$$\nu_n = z^n \frac{(\mathbf{a})_n}{(\mathbf{b}+1)_n} \sum_{x=0}^{\infty} \frac{(\mathbf{a}+n)_x}{(\mathbf{b}+1+n)_x} \frac{z^x}{x!} = z^n \frac{(\mathbf{a})_n}{(\mathbf{b}+1)_n} \, {}_pF_q\left(\begin{matrix} \mathbf{a}+n \\ \mathbf{b}+1+n \end{matrix} ; z \right), \tag{18}$$

where ${}_pF_q$ denotes the *generalized hypergeometric function* defined by [44, 16.2]

$$ {}_pF_q\left(\begin{matrix} a_1, \ldots, a_p \\ b_1, \ldots, b_q \end{matrix} ; z \right) = \sum_{k=0}^{\infty} \frac{(a_1)_k \cdots (a_p)_k}{(b_1)_k \cdots (b_q)_k} \frac{z^k}{k!}. \tag{19}$$

The moments of discrete semiclassical orthogonal polynomials of class $s \le 1$ were studied in [18].

Remark 1 The convergence of the series (19) depends on the values of p and q. Three different situations appear [44, 16.2]:

1. If $p < q+1$, then ${}_pF_q$ is an entire function of z.
2. If $p = q+1$, then ${}_pF_q$ is analytic inside the unit circle, $|z| < 1$, and can be extended by analytic continuation to the cut plane $\mathbb{C} \setminus [1, \infty)$. Let

$$\gamma = b_1 + \cdots + b_q - (a_1 + \cdots + a_{q+1}). \tag{20}$$

   On the unit circle $|z| = 1$, the series (19) is

   (a) absolutely convergent if $\mathrm{Re}(\gamma) > 0$,
   (b) convergent except at $z = 1$ if $\mathrm{Re}(\gamma) \in (-1, 0]$, and
   (c) divergent if $\mathrm{Re}(\gamma) \le -1$.

3. If $p > q+1$, then ${}_pF_q$ diverges for all $z \ne 0$, up to $a_i = -N$, with $N \in \mathbb{N}$ for some $1 \le i \le p$. In this case, ${}_pF_q$ becomes a polynomial of degree N.

3 Modification of Functionals

In this section, we describe three ways to change the class of a linear functional.

3.1 Rational Spectral Transformations

Let L be a linear functional and $S(t)$ denote its Stieltjes transform introduced in (11). A *rational spectral transformation* of $S(t)$ was defined by Zhedanov [50]

$$\widetilde{L}\left[\frac{1}{t-x}\right] = \widetilde{S}(t) = \frac{A(t)\,S(t)+B(t)}{C(t)\,S(t)+D(t)},$$

where $A(t)$, $B(t)$, $C(t)$, $D(t)$ are polynomials such that $A(z)D(z)-B(z)C(z)\neq 0$. See also [13, 23, 32, 45, 48], and [50], among others.

It was shown in [50] what are the families of semiclassical orthogonal polynomials related by *spectral linear transformations* (with $C=0$). They can be written as a composition of two basic transformations:

1. The *Christoffel transformation*, (see [12–14, 46]) defined by

$$\frac{L_C[r]}{\lambda_C} = L[(x-\omega)\,r(x)], \quad r\in\mathbb{C}[x], \tag{21}$$

or, equivalently, by

$$\frac{S_C(z)}{\lambda_C} = (z-\omega)\,S(z) - L[1], \tag{22}$$

where $L[x-\omega]\neq 0$, $n\geq 0$, and

$$\lambda_C = \frac{L_C[1]}{L[x-\omega]}.$$

2. The *Geronimus transformation* (see [15, 16, 26, 27, 32, 38]) defined by

$$\frac{L_G[r]}{\lambda_G} = L\left[\frac{r(x)}{x-\omega}\right] + M r(\omega), \quad r\in\mathbb{C}[x], \tag{23}$$

or, equivalently, by

$$(z-\omega)\,\frac{S_G(z)}{\lambda_G} = S(z) - S(\omega) + M, \tag{24}$$

with

$$\lambda_G = \frac{L_G[1]}{M-S(\omega)},$$

where we assume that $S(z)$ is analytic at $z=\omega$ and $M-S(\omega)\neq 0$.

On the other hand, the *Uvarov transformation* (see [4, 33, 48]) is defined by

$$\frac{L_U[r]}{\lambda_U} = L[r] + Mr(\omega), \quad r \in \mathbb{C}[x], \tag{25}$$

or by

$$\frac{S_U(z)}{\lambda_U} = S(z) + \frac{M}{z-\omega}, \tag{26}$$

where $M + L[1] \neq 0$ and

$$\lambda_U = \frac{L_U[1]}{L[1] + M}.$$

Note that since

$$r(x) = \sum_{k=0}^{\deg(r)} \frac{r^{(k)}(\omega)}{k!}(x-\omega)^k, \quad r \in \mathbb{C}[x],$$

we can rewrite (23) as

$$\frac{L_G[r]}{\lambda_G} = [M - S(\omega)]\, r(\omega) + \sum_{k=1}^{\deg(r)} \frac{r^{(k)}(\omega)}{k!} L\left[(x-\omega)^k\right],$$

which is well defined for all $r \in \mathbb{C}[x]$.

We denote by $L_{T_2T_1}, S_{T_2T_1}, \varrho_{T_2T_1}$, etc., the object obtained by applying the transformation T_1 followed by the transformation T_2 (in other words, by the composition $T_2 \circ T_1$).

Let us assume that $\lambda_U = \lambda_C = \lambda_G = 1$. If we apply a Geronimus transformation followed by a Christoffel transformation, we get

$$L_{CG}[r] = L_G[(x-\omega)\, r(x)] = L\left[\frac{(x-\omega)\, r(x)}{x-\omega}\right] + M\,[(x-\omega)\, r(x)]_{x=\omega} = L[r]. \tag{27}$$

On the other hand, changing the order in the composition, we obtain

$$\begin{aligned} L_{GC}[r] &= L_C\left[\frac{r(x)}{x-\omega}\right] + Mr(\omega) = L\left[\frac{(x-\omega)\, r(x)}{x-\omega}\right] + Mr(\omega) \\ &= L[r] + Mr(\omega) = L_U[r]. \end{aligned} \tag{28}$$

Thus, the Christoffel and Geronimus transformations are almost inverses of each other in the sense that L_{CG} is the identity and L_{GC} is an Uvarov transformation.

3.1.1 Uvarov Transformations

Let L be given by (5) and consider the linear functional L_U defined by (25) with

$$L_U[1] = L[1] + M \neq 0.$$

It follows that the moments of L_U with respect to the basis $\{\phi_n(x)\}_{n\geq 0}$ are

$$\nu_n^U = \nu_n + M\phi_n(\omega), \quad n \in \mathbb{N}_0. \tag{29}$$

Using (26) in (12), we obtain

$$\sigma(t+1)\left[S_U(t+1) - \frac{M}{t+1-\omega}\right] - \eta(t)\left[S_U(t) - \frac{M}{t-\omega}\right] = \xi(t). \tag{30}$$

Therefore, the Stieltjes transform of L_U satisfies the difference equation

$$(t+1-\omega)(t-\omega)\sigma(t+1)S_U(t+1) - (t+1-\omega)(t-\omega)\eta(t)S_U(t) = \xi_U(t),$$

where

$$\xi_U(t) = (t-\omega)(t-\omega+1)\xi(t) + M[(t-\omega)\sigma(t+1) - (t-\omega+1)\eta(t)],$$

and the modified weight function $\varrho_U(x)$ satisfies the Pearson equation

$$\frac{\varrho_U(x+1)}{\varrho_U(x)} = \frac{(x-\omega)(x+1-\omega)\eta(x)}{(x-\omega)(x+1-\omega)\sigma(x+1)}.$$

We conclude that if L was of class s, then L_U will be of class at most $s+2$.

Notice that we have two special cases to consider:

1. If $\sigma(\omega) = 0$ and $\eta(\omega) \neq 0$, with $\sigma(x) = (x-\omega)\sigma_1(x)$, then

$$\frac{\varrho_U(x+1)}{\varrho_U(x)} = \frac{(x-\omega)(x+1-\omega)\eta(x)}{(x-\omega)(x+1-\omega)\sigma(x+1)} = \frac{(x-\omega)\eta(x)}{(x-\omega)\sigma(x+1)},$$

$$\sigma(t+1)S_U(t+1) - \eta(t)S_U(t) = \xi(t) + M\left[\sigma_1(t+1) - \frac{\eta(t)}{t-\omega}\right], \tag{31}$$

and L_U will be of class $s+1$.

2. If $\sigma(\omega) \neq 0$ and $\eta(\omega) = 0$, with $\eta(x) = (x-\omega)\eta_1(x)$, then

$$\frac{\varrho_U(x+1)}{\varrho_U(x)} = \frac{(x-\omega)(x+1-\omega)\eta(x)}{(x-\omega)(x+1-\omega)\sigma(x+1)} = \frac{(x+1-\omega)\eta(x)}{(x+1-\omega)\sigma(x+1)},$$

$$\sigma(t+1)\, S_U(t+1) - \eta(t)\, S_U(t) = \xi(t) + M\left[\frac{\sigma(t+1)}{t+1-\omega} - \eta_1(t)\right],$$

and L_U will be of class $s+1$.

Perturbations of discrete semiclassical functionals by Dirac masses were studied in [4, 6, 24, 28], as well as in [2, 5], and [3].

3.1.2 Christoffel Transformations

Let L be given by (5), and consider the functional L_C defined by (21), with

$$L_C[1] = L[x-\omega] \neq 0.$$

It follows that the moments of L_C with respect to the basis $\{\phi_n(x)\}_{n\geq 0}$ are

$$\nu_n^C = \nu_{n+1} + (n-\omega)\,\nu_n, \quad n \in \mathbb{N}_0 \tag{32}$$

since we see from (14) that

$$\phi_{n+1}(x) = (x-n)\,\phi_n(x) = (x-\omega)\,\phi_n(x) + (\omega-n)\,\phi_n(x). \tag{33}$$

Using (22) in (12), we obtain

$$\sigma(t+1)\,\frac{S_C(t+1)+\nu_0}{t+1-\omega} - \eta(t)\,\frac{S_C(t)+\nu_0}{t-\omega} = \xi(t).$$

Therefore, the Stieltjes transform of L_C satisfies the difference equation

$$(t-\omega)\,\sigma(t+1)\, S_C(t+1) - (t+1-\omega)\,\eta(t)\, S_C(t) = \xi_C(t),$$

where

$$\xi_C(t) = (t-\omega)(t-\omega+1)\,\xi(t) - [(t-\omega)\,\sigma(t+1) - (t-\omega+1)\,\eta(t)]\,\nu_0,$$

and the modified weight function $\varrho_C(x)$ satisfies the Pearson equation

$$\frac{\varrho_C(x+1)}{\varrho_C(x)} = \frac{(x+1-\omega)\,\eta(x)}{(x-\omega)\,\sigma(x+1)}.$$

We conclude that if L was of class s, then L_C will be of class at most $s+1$.

Note that we have two special cases to consider:

1. If $\sigma(\omega) = 0$ and $\eta(\omega) \neq 0$, with $\sigma(x) = (x-\omega)\sigma_1(x)$, then

$$\frac{\varrho_C(x+1)}{\varrho_C(x)} = \frac{\eta(x)}{(x-\omega)\sigma_1(x+1)},$$

$$\begin{aligned}&(t-\omega)\sigma_1(t+1)S_C(t+1) - \eta(t)S_C(t)\\&= (t-\omega)\xi(t) - [(t-\omega)\sigma_1(t+1) - \eta(t)]\nu_0,\end{aligned}$$

and L_C could be of class s.

2. If $\sigma(\omega) \neq 0$ and $\eta(\omega) = 0$, with $\eta(x) = (x-\omega)\eta_1(x)$, then

$$\frac{\varrho_C(x+1)}{\varrho_C(x)} = \frac{(x+1-\omega)\eta_1(x)}{\sigma(x+1)},$$

$$\begin{aligned}&\sigma(t+1)S_C(t+1) - (t+1-\omega)\eta_1(t)S_C(t)\\&= (t-\omega+1)\xi(t) - [\sigma(t+1) - (t-\omega+1)\eta_1(t)]\nu_0,\end{aligned}$$

and L_C could be of class s.

Christoffel transformations in the discrete case have been studied in [46].

3.1.3 Geronimus Transformations

Let L be given by (5), and consider the functional L_G defined by (23), with

$$L_G[1] = M - S(\omega) \neq 0.$$

Since

$$S(\omega) = \sum_{x=0}^{\infty} \frac{1}{\omega - x}\frac{(\mathbf{a})_x}{(\mathbf{b}+1)_x}\frac{z^x}{x!}$$

is a meromorphic function, we need $\omega \notin \mathbb{N}_0$ for $L_G[1]$ to be well defined.

It follows from (23) that the moments of L_G with respect to the basis $\{\phi_n(x)\}_{n\geq 0}$ satisfy

$$\nu_{n+1}^G + (n-\omega)\nu_n^G = \nu_n, \quad n \in \mathbb{N}_0, \tag{34}$$

since from (33) we see that

$$L_G\left[\phi_{n+1} - (\omega - n)\,\phi_n\right] = L_G\left[(x - \omega)\,\phi_n(x)\right] = L\left[\phi_n\right].$$

As it is well known, the general solution of the initial value problem

$$y_{n+1} = c_n y_n + g_n, \quad y_{n_0} = y_0,$$

is [20, 1.2.4]

$$y_n = y_0 \prod_{j=n_0}^{n-1} c_j + \sum_{k=n_0}^{n-1} \left(g_k \prod_{j=k+1}^{n-1} c_j \right).$$

Solving (34) with $\nu_0^G = M - S(\omega)$, we get

$$\nu_n^G = \phi_n(\omega) \left[\nu_0^G + \sum_{k=0}^{n-1} \frac{\nu_k}{\phi_{k+1}(\omega)} \right],$$

since from (14) we have

$$\prod_{j=0}^{n-1} (\omega - j) = \phi_n(\omega), \quad \prod_{j=k+1}^{n-1} (\omega - j) = \frac{\phi_n(\omega)}{\phi_{k+1}(\omega)}.$$

Replacing (24) in (12), we obtain

$$\sigma(t+1)\left[(t+1-\omega)\,S_G(t+1) - \nu_0^G\right] - \eta(t)\left[(t-\omega)\,S_G(t) - \nu_0^G\right] = \xi(t).$$

Therefore, the Stieltjes transform of L_G satisfies the difference equation

$$(t+1-\omega)\,\sigma(t+1)\,S_G(t+1) - (t-\omega)\,\eta(t)\,S_G(t) = \xi_G(t),$$

where

$$\xi_G(t) = \xi(t) + \left[\sigma(t+1) - \eta(t)\right]\nu_0^G,$$

and the modified weight function $\varrho_G(x)$ satisfies the Pearson equation

$$\frac{\varrho_G(x+1)}{\varrho_G(x)} = \frac{(x-\omega)\,\eta(x)}{(x+1-\omega)\,\sigma(x+1)}.$$

We conclude that if L was of class s, then L_G will be of class at most $s+1$.

Remark 2 For all these transformations, it is understood that

$$\eta(\omega) = 0 \Rightarrow \frac{\eta(x)}{x-\omega} \times (x-\omega) = \eta(x),$$

$$\sigma(\omega) = 0 \Rightarrow \frac{x+1-\omega}{\sigma(x+1)} \times \frac{1}{x+1-\omega} = \frac{1}{\sigma(x+1)}.$$

Thus, if we write

$$\frac{\varrho_G(x+1)}{\varrho_G(x)} = \frac{(x-\omega)\eta(x)}{(x+1-\omega)\sigma(x+1)} = \frac{\eta_G(x)}{\sigma_G(x+1)},$$

then clearly $\eta_G(\omega) = 0 = \sigma_G(\omega)$. Hence,

$$\frac{\varrho_{CG}(x+1)}{\varrho_{CG}(x)} = \frac{\eta_G(x)}{\sigma_G(x+1)} \times \frac{x-\omega}{x+1-\omega} = \frac{\eta(x)}{\sigma(x+1)} = \frac{\varrho(x+1)}{\varrho(x)},$$

in agreement with (27).

On the other hand, if we write

$$\frac{\varrho_C(x+1)}{\varrho_C(x)} = \frac{(x+1-\omega)\eta(x)}{(x-\omega)\sigma(x+1)} = \frac{\eta_C(x)}{\sigma_C(x+1)},$$

then $\eta_G(\omega)\sigma_G(\omega) \neq 0$, and therefore

$$\begin{aligned}\frac{\varrho_{GC}(x+1)}{\varrho_{GC}(x)} &= \frac{\eta_C(x)}{\sigma_C(x+1)} \times \frac{x+1-\omega}{x-\omega} \\ &= \frac{(x-\omega)(x+1-\omega)\eta(x)}{(x-\omega)(x+1-\omega)\sigma(x+1)} = \frac{\varrho_U(x+1)}{\varrho_U(x)},\end{aligned}$$

in agreement with (28).

3.2 Truncated Linear Functionals

Let $N \in \mathbb{N}_0$ and define the *truncated functional* L_T of the linear functional L by

$$L_T[r] = \sum_{x=0}^{N} r(x)\varrho(x), \quad r \in \mathbb{C}[x], \tag{35}$$

where $\varrho(x)$ satisfies (7) with $\eta(N) \neq 0$. We define the truncated weight function $\varrho_T(x)$ by

$$\varrho_T(x) = \varrho(x)[1 - H(x-N)], \tag{36}$$

where $H(x)$ is the Heaviside function [44, 1.16(iv)] defined by

$$H(x) = \begin{cases} 1, & x > 0 \\ 0, & x \leq 0 \end{cases}.$$

Multiplying (36) by $\frac{1}{(N-x)!}$, we obtain

$$\frac{1}{(N-x)!}\varrho_T(x) = \frac{1}{(N-x)!}\varrho(x), \quad x \in \mathbb{N}_0. \tag{37}$$

Using (37) in (7), we get

$$(N-x)\frac{\varrho_T(x+1)}{\varrho_T(x)} = \frac{\frac{1}{(N-x-1)!}\varrho_T(x+1)}{\frac{1}{(N-x)!}\varrho_T(x)} = \frac{\frac{1}{(N-x-1)!}\varrho(x+1)}{\frac{1}{(N-x)!}\varrho(x)} = (N-x)\frac{\eta(x)}{\sigma(x+1)}.$$

We conclude that the modified weight function $\varrho_T(x)$ satisfies the Pearson equation

$$\frac{\varrho_T(x+1)}{\varrho_T(x)} = \frac{\eta(x)(x-N)}{\sigma(x+1)(x-N)}, \tag{38}$$

and L_T will be of class $s+1$.

The moments of L_T with respect to the basis $\{\phi_n(x)\}_{n\geq 0}$ are

$$\nu_n^T(z) = \sum_{x=0}^{N}\phi_n(x)\frac{(\mathbf{a})_x}{(\mathbf{b}+1)_x}\frac{z^x}{x!}, \quad n \in \mathbb{N}_0.$$

A similar calculation leading to (18), gives

$$\nu_n^T(z) = z^n\frac{(\mathbf{a})_n}{(\mathbf{b}+1)_n}\sum_{x=0}^{N-n}\frac{(\mathbf{a}+n)_x}{(\mathbf{b}+1+n)_x}\frac{z^x}{x!}, \quad n \in \mathbb{N}_0. \tag{39}$$

Using the identity [44, 16.2.4]

$$\sum_{x=0}^{K}\frac{(\mathbf{a})_x}{(\mathbf{b}+1)_x}\frac{z^x}{x!} = \frac{(\mathbf{a})_K}{(\mathbf{b}+1)_K}\frac{z^K}{K!}\,{}_{q+2}F_p\left[\begin{matrix}-K,1,-K-\mathbf{b} \\ 1-K-\mathbf{a}\end{matrix};\frac{(-1)^{p+q+1}}{z}\right], \quad K \in \mathbb{N},$$

we can rewrite (39) as

$$\nu_n^T(z) = \frac{(\mathbf{a})_N}{(\mathbf{b}+1)_N}\frac{z^N}{(N-n)!}\,{}_{q+2}F_p\left[\begin{matrix}n-N,1,-N-\mathbf{b} \\ 1-N-\mathbf{a}\end{matrix};\frac{(-1)^{p+q+1}}{z}\right]. \tag{40}$$

3.3 Symmetrized Functionals

Let $\varrho(x)$ be given by (6) and $m \in \mathbb{N}$. We consider the symmetric weight function $\varrho_{\varsigma}(x)$ defined by

$$\varrho_{\varsigma}(x) = C\varrho(x+m)\,\varrho(-x+m), \tag{41}$$

where C is a constant factor to be determined. Replacing (7) in (41), we get

$$\frac{\varrho_{\varsigma}(x+1)}{\varrho_{\varsigma}(x)} = \frac{\varrho(x+1+m)}{\varrho(x+m)}\frac{\varrho(-x-1+m)}{\varrho(-x+m)} = \frac{\eta(x+m)}{\sigma(x+m+1)}\frac{\sigma(-x+m)}{\eta(-x-1+m)}.$$

Therefore,

$$\frac{\varrho_{\varsigma}(x+1)}{\varrho_{\varsigma}(x)} = \frac{x+m+\mathbf{a}}{(x+m+\mathbf{b}+1)(x+m+1)}\frac{(m+\mathbf{b}-x)(m-x)}{m+\mathbf{a}-x-1},$$

or

$$\frac{\varrho_{\varsigma}(x+1)}{\varrho_{\varsigma}(x)} = \frac{(-1)^{p+q+1}(x+m+\mathbf{a})(x-m-\mathbf{b})(x-m)}{(x+1-m-\mathbf{a})(x+1+m+\mathbf{b})(x+1+m)}. \tag{42}$$

Note that, from (42), we have $\varrho_{\varsigma}(x) = \varrho_1(x+m)$, with

$$\frac{\varrho_1(x+1)}{\varrho_1(x)} = \frac{(-1)^{p+q+1}(x+\mathbf{a})(x-2m-\mathbf{b})(x-2m)}{(x+1-2m-\mathbf{a})(x+1+\mathbf{b})(x+1)}.$$

If we write $N = 2m$, then

$$\frac{\varrho_1(x+1)}{\varrho_1(x)} = \frac{(-1)^{p+q+1}(x+\mathbf{a})(x-N-\mathbf{b})(x-N)}{(x+1-N-\mathbf{a})(x+1+\mathbf{b})(x+1)},$$

and it follows that $\varrho_1(x)$ (and ϱ_{ς}) is of class s_1 with

$$s_1 = \begin{cases} p+q, & \text{if } q+p \text{ is even} \\ p+q-1, & \text{if } q+p \text{ is odd} \end{cases}.$$

If $\mathbf{a} = (\mathbf{a}_1, -N)$, then

$$\frac{\varrho_1(x+1)}{\varrho_1(x)} = \frac{(-1)^{q+1+p}(x+\mathbf{a}_1)}{(x+1+\mathbf{b})(x+1)}\frac{x-N}{x+1}\frac{(x-N-\mathbf{b})(x-N)}{x+1-N-\mathbf{a}_1},$$

and since the term $\frac{x-N}{x+1}$ is repeated, we consider the reduced Pearson equation

$$\begin{aligned}\frac{\varrho_2(x+1)}{\varrho_2(x)} &= \frac{(-1)^{q+1+p}(x+\mathbf{a}_1)}{(x+1+\mathbf{b})(x+1)} \frac{(x-N-\mathbf{b})(x-N)}{x+1-N-\mathbf{a}_1} \\ &= \frac{(-1)^{q+1+p}(x+\mathbf{a})}{(x+1+\mathbf{b})} \frac{(x-N-\mathbf{b})}{x+1-N-\mathbf{a}},\end{aligned}$$

and $\varrho_\varsigma(x) = \varrho_2(x+m)$. In this case $\varrho_2(x)$ (and ϱ_ς) is of class s_2 with

$$s_2 = \begin{cases} p+q-1, & \text{if } q+p \text{ is even} \\ p+q-2, & \text{if } q+p \text{ is odd} \end{cases}.$$

As a conclusion, if $\varrho(x)$ satisfies (7), then we have $\varrho_\varsigma(x) = \rho(x+m)$ where

$$\rho(x) = \begin{cases} \frac{(\mathbf{a})_x}{(\mathbf{b}+1)_x} \frac{(z_0)^x}{x!} \frac{(-N)_x(-N-\mathbf{b})_x}{(1-N-\mathbf{a})_x}, & \eta(N) \neq 0, \\ \frac{(\mathbf{a})_x}{(\mathbf{b}+1)_x}(z_0)^x \frac{(-N-\mathbf{b})_x}{(1-N-\mathbf{a})_x}, & \eta(N) = 0, \end{cases}$$

and

$$z_0 = (-1)^{p+q+1}, \quad N = 2m.$$

We define the *symmetrized functional* L_ς of L_ρ by

$$L_\varsigma[r] = \sum_{x=-m}^{m} r(x)\varrho_\varsigma(x) = \sum_{x=0}^{N} r(x-m)\rho(x) = L_\rho[r(x-m)].$$

Hence, the moments of L_ς on the basis

$$\overline{\phi}_n(x) = \phi_n(x+m), \quad n \in \mathbb{N}_0,$$

are

$$\nu_n^\varsigma = \sum_{x=-m}^{m} \overline{\phi}_n(x)\varrho_\varsigma(x) = \sum_{x=0}^{N} \phi_n(x)\rho(x) = L_\rho[\phi_n], \quad n \in \mathbb{N}_0, \tag{43}$$

and the Stieltjes transform of L_ς is

$$S_\varsigma(t) = \sum_{x=-m}^{m} \frac{\varrho_\varsigma(x)}{t-x} = \sum_{x=0}^{N} \frac{\rho(x)}{t+m-x} = S_\rho(t+m).$$

It follows that if the Stieltjes transform of L_ρ satisfies the difference equation

$$\sigma_\rho (t+1) S_\rho (t+1) - \eta_\rho (t) S_\rho (t) = \xi_\rho (t),$$

then

$$\sigma_\rho (t+m+1) S_\rho (t+m+1) - \eta_\rho (t+m) S_\rho(t+m) = \xi_\rho (t+m).$$

Therefore

$$\sigma_\varsigma (t+1) S_\varsigma (t+1) - \eta_\varsigma (t) S_\varsigma(t) = \xi_\rho (t+m), \tag{44}$$

since

$$\frac{\eta_\varsigma (x)}{\sigma_\varsigma (x+1)} = \frac{\varrho_\varsigma (x+1)}{\varrho_\varsigma (x)} = \frac{\rho (x+m+1)}{\rho (x+m)} = \frac{\eta_\rho (x+m)}{\sigma_\rho (x+m+1)}.$$

Symmetric orthogonal polynomials of a discrete variable were studied in [7, 8, 40, 41].

3.4 Summary

Let's list all the transformations that we will use in the sequel. For simplicity, we will not consider compositions of them.

1. Uvarov transformation: $s \to s+2$

$$\frac{\varrho_U (x+1)}{\varrho_U (x)} = \frac{\eta (x)}{\sigma (x+1)} \times \frac{(x-\omega)(x+1-\omega)}{(x-\omega)(x+1-\omega)}, \quad \eta (\omega) \sigma (\omega) \neq 0.$$

2. Reduced Uvarov transformation: $s \to s+1$

(a)

$$\frac{\varrho_U (x+1)}{\varrho_U (x)} = \frac{\eta (x)}{\sigma (x+1)} \times \frac{x+1-\omega}{x+1-\omega}, \quad \eta (\omega) = 0.$$

(b)

$$\frac{\varrho_U (x+1)}{\varrho_U (x)} = \frac{\eta (x)}{\sigma (x+1)} \times \frac{x-\omega}{x-\omega}, \quad \sigma (\omega) = 0.$$

3. Christoffel transformation: $s \to s+1$

$$\frac{\varrho_C(x+1)}{\varrho_C(x)} = \frac{\eta(x)}{\sigma(x+1)} \times \frac{x+1-\omega}{x-\omega}, \quad \eta(\omega)\sigma(\omega) \neq 0.$$

4. Geronimus transformation: $s \to s+1$

$$\frac{\varrho_G(x+1)}{\varrho_G(x)} = \frac{\eta(x)}{\sigma(x+1)} \times \frac{x-\omega}{x+1-\omega}, \quad \eta(\omega-1)\sigma(\omega+1) \neq 0.$$

5. Truncation at $x = N : s \to s+1$

$$\frac{\varrho_T(x+1)}{\varrho_T(x)} = \frac{\eta(x)}{\sigma(x+1)} \times \frac{x-N}{x-N}, \quad \eta(N) \neq 0.$$

6. Symmetrization on the interval $[-m, m]$:

(a)

$$\frac{\varrho_\varsigma(x+1)}{\varrho_\varsigma(x)} = \frac{\eta(m+x)}{\sigma(m+x+1)} \times \frac{\sigma(m-x)}{\eta(m-x-1)}, \quad \eta(2m) \neq 0.$$

(b)

$$\frac{\varrho_\varsigma(x+1)}{\varrho_\varsigma(x)} = \frac{\eta(m+x)}{\sigma(m+x+1)} \times \frac{\sigma_1(m-x)}{\eta_1(m-x-1)}, \quad \eta(2m) = 0, \tag{45}$$

where

$$\eta(x) = (x-2m)\,\eta_1(x), \quad \sigma(x) = x\sigma_1(x).$$

4 Semiclassical Polynomials of Class 0 (Classical Polynomials)

In this section, we consider the families of discrete classical polynomials. We have three main cases, corresponding to $(p, q) = (0, 0), (1, 0), (2, 1)$. There are also three symmetrized subcases.

We use the notation $(p, q; N)$ to denote the family such that one of the parameters in the numerator is a non-negative integer, and $(p, q; N, 1)$ if in addition the value of z is equal to 1.

For each polynomial, we list the linear functional, the Pearson equation satisfied by the weight function, the moments computed from (18), and the difference equation satisfied by the Stieltjes transform, using (12) and (13).

4.1 $(0, 0)$*: Charlier Polynomials*

Linear functional

$$L[r] = \sum_{x=0}^{\infty} r(x) \frac{z^x}{x!}.$$

Pearson equation

$$\frac{\varrho(x+1)}{\varrho(x)} = \frac{z}{x+1}.$$

Moments

$$\nu_n(z) = z^n \, {}_0F_0\left[\begin{matrix} - \\ - \end{matrix}; z\right] = z^n e^z, \quad n \in \mathbb{N}_0.$$

Stieltjes transform difference equation

$$(t+1) S(t+1) - zS(t) = \nu_0.$$

4.2 $(1, 0)$*: Meixner Polynomials*

Linear functional

$$L[r] = \sum_{x=0}^{\infty} r(x) \ (a)_x \frac{z^x}{x!}.$$

Pearson equation

$$\frac{\varrho(x+1)}{\varrho(x)} = \frac{z(x+a)}{x+1}.$$

Moments

$$\nu_n(z) = z^n (a)_n \ {}_1F_0\left[\begin{matrix} a+n \\ - \end{matrix}; z\right] = z^n (a)_n (1-z)^{-a-n}, \quad n \in \mathbb{N}_0,$$

where we choose the principal branch $z \in \mathbb{C} \setminus [1, \infty)$.

Stieltjes transform difference equation

$$(t+1) S(t+1) - z(t+a) S(t) = (1-z) \nu_0.$$

4.3 (1, 0; *N*): *Krawtchouk Polynomials*

These polynomials are a particular case of the Meixner polynomials, with $-a = N \in \mathbb{N}$.

Linear functional

$$L[r] = \sum_{x=0}^{N} r(x)\ (-N)_x\ \frac{z^x}{x!}.$$

Pearson equation

$$\frac{\varrho(x+1)}{\varrho(x)} = \frac{z(x-N)}{x+1}.$$

Moments

$$\nu_n(z) = z^n (-N)_n (1-z)^{N-n}, \quad z \neq 1, \quad n \in \mathbb{N}_0. \tag{46}$$

Stieltjes transform difference equation

$$(t+1) S(t+1) - z(t-N) S(t) = (1-z)\nu_0.$$

Remark 3 Let's consider the symmetrized Krawtchouk polynomials. Since $\eta(N) = 0$, we use (45) and obtain

$$\frac{\varrho_\varsigma(x+1)}{\varrho_\varsigma(x)} = \frac{z(x-m)}{x+m+1}\frac{(x+m+1)(-x+m)}{(x-m)z(-x-1-m)} = \frac{x-m}{x+m+1}.$$

Hence, the symmetrized Krawtchouk polynomials are shifted Krawtchouk polynomials with $z = 1$. But from (43) and (46) we see that

$$\nu_n^\varsigma = \nu_n(1) = 0, \quad n \in \mathbb{N}_0,$$

and, therefore, we need to discard this example.

4.3.1 Symmetrized Charlier Polynomials

Special values

$$N \to 2m, \quad z \to -1, \quad x \to x+m.$$

Weight function

$$\varrho_{\varsigma}(x) = \varrho(x+m) = (-2m)_{x+m}\frac{(-1)^{x+m}}{(x+m)!}$$
$$= (-2m)_m \frac{(-1)^m}{m!}\frac{(-m)_x}{(m+1)_x}(-1)^x = \binom{2m}{m}\frac{(-m)_x}{(m+1)_x}(-1)^x .$$

Linear functional

$$L_{\varsigma}[r] = \binom{2m}{m}\sum_{x=-m}^{m} r(x)\frac{(-m)_x}{(m+1)_x}(-1)^x .$$

Pearson equation

$$\frac{\varrho_{\varsigma}(x+1)}{\varrho_{\varsigma}(x)} = \frac{-(x-m)}{x+m+1}. \tag{47}$$

Moments on the basis $\{\overline{\phi}_n(x)\}_{n\geq 0}$

$$\nu_n^{\varsigma} = L_{\varsigma}[\phi(x+m)] = (-1)^n(-2m)_n\, 2^{2m-n} = \nu_n(-1), \quad n\in\mathbb{N}_0, \tag{48}$$

where $\nu_n(z)$ are the moments of the Krawtchouk polynomials defined in (46).

Stieltjes transform difference equation

$$(t+m+1)\,S_{\varsigma}(t+1) + (t-m)\,S_{\varsigma}(t) = 2\nu_0^{\varsigma}.$$

Remark 4 These polynomials were studied in [7]. The Pearson equation (47) is the same as equation (11) in that paper, with $m = c$. The authors used the weight function

$$\rho(x) = \frac{1}{\Gamma(x+m+1)\,\Gamma(-x+m+1)} = \frac{1}{(m!)^2}\frac{(-m)_x}{(m+1)_x}(-1)^x ,$$

and, therefore,

$$\varrho_{\varsigma}(x) = (2m)!\rho(x). \tag{49}$$

The moments (48) are the same as those appearing in Eq. (15), if we use (15) and write

$$\nu_n^{\varsigma} = (-1)^n(-2m)_n\, 2^{2m-n} = 2^{2m-n}\phi_n(2m),$$

after taking the scaling (49) into account.

4.4 (2, 1; *N*, 1): *Hahn Polynomials*

Linear functional

$$L[r] = \sum_{x=0}^{N} r(x) \, \frac{(a)_x (-N)_x}{(b+1)_x} \frac{1}{x!}.$$

Pearson equation

$$\frac{\eta(x)}{\sigma(x+1)} = \frac{\varrho(x+1)}{\varrho(x)} = \frac{(x+a)(x-N)}{(x+b+1)(x+1)}.$$

Moments

$$\nu_n = \frac{(-N)_n (a)_n}{(b+1)_n} \, {}_2F_1\left(\begin{matrix} -N+n, a+n \\ b+1+n \end{matrix} ; 1 \right), n \in \mathbb{N}_0.$$

Using the Chu–Vandermonde Identity [44, 15.4.24],

$${}_2F_1\left(\begin{matrix} -n, b \\ c \end{matrix} ; 1 \right) = \frac{(c-b)_n}{(c)_n},$$

we get

$$\nu_n = \frac{(-N)_n (a)_n}{(b+1)_n} \frac{(b+1-a)_{N-n}}{(b+1+n)_{N-n}}, \quad n \in \mathbb{N}_0.$$

Stieltjes transform difference equation

$$(t+1)(t+b+1) S(t+1) - (t-N)(t+a) S(t) = (b+1-a+N) \nu_0.$$

4.4.1 Symmetrized Meixner Polynomials

Special values

$$b \to -N-a, \quad N \to 2m, \quad x \to x+m.$$

Linear functional

$$L_\varsigma[r] = \binom{2m}{m} \frac{(a)_m}{(a+m)_m} \sum_{x=-m}^{m} r(x) \, \frac{(a+m)_x (-m)_x}{(-m-a+1)_x (m+1)_x}.$$

Pearson equation

$$\frac{\varrho_{\varsigma}(x+1)}{\varrho_{\varsigma}(x)} = \frac{(x+a+m)(x-m)}{(x+1-m-a)(x+1+m)}. \tag{50}$$

Moments

$$\nu_n^{\varsigma} = \frac{(-N)_n (a)_n}{(1-N-a)_n} \frac{(1-N-2a)_{N-n}}{(n+1-N-a)_{N-n}}, \quad n \in \mathbb{N}_0.$$

Stieltjes transform difference equation

$$(t+m+1)(t-m-a+1) S_{\varsigma}(t+1) - (t-m)(t+m+a) S_{\varsigma}(t) = (1-2a)\nu_0^{\varsigma}.$$

Remark 5 If we use (4) and (41) to write the weight function $\varrho_{\varsigma}(x)$ in terms of Gamma functions, we have

$$\varrho_{\varsigma}(x) = \varrho_{\varsigma}(0) \left[\frac{m!}{(a)_m}\right]^2 \frac{(a)_{x+m}}{(x+m)!} \frac{(a)_{-x+m}}{(-x+m)!} = \frac{(2m)!}{(a)_{2m}} \frac{\Gamma(x+m+a)\Gamma(-x+m+a)}{\Gamma(x+m+1)\Gamma(-x+m+1)}.$$

This agrees (up to a normalization factor) with the weight function considered by the authors in [7, equation 27], if we set $\delta_1 = -m$ and $\delta_2 = m+a$. Note that the condition $\delta_2 > -\delta_1$ is satisfied if $a > 0$ (the positive-definite case).

4.4.2 Symmetrized Generalized Charlier Polynomials

See Sect. 5.1 for a definition of the Generalized Charlier polynomials.

Special values

$$a \to -N-b, \quad N \to 2m, \quad x \to x+m.$$

Linear functional

$$L_{\varsigma}[r] = \binom{2m}{m} \frac{(b+1+m)_m}{(b+1)_m} \sum_{x=-m}^{m} r(x) \frac{(-m-b)_x (-m)_x}{(b+1+m)_x (m+1)_x}.$$

Pearson equation

$$\frac{\varrho_{\varsigma}(x+1)}{\varrho_{\varsigma}(x)} = \frac{(x-m-b)(x-m)}{(x+m+b+1)(x+m+1)}.$$

Moments

$$\nu_n^{\varsigma} = \frac{(-2m)_n\,(-2m-b)_n}{(b+1)_n}\frac{(2m+2b+1)_{2m-n}}{(b+1+n)_{2m-n}}, \qquad n \in \mathbb{N}_0.$$

Stieltjes transform difference equation

$$(t+m+1)\,(t+m+b+1)\,S_{\varsigma}\,(t+1) - (t-m)\,(t-m-b)\,S_{\varsigma}(t)$$
$$= (2b+1+4m)\,\nu_0^{\varsigma}.$$

Remark 6 If we use (4) and (41) to write the weight function $\varrho_{\varsigma}\,(x)$ in terms of Gamma functions, we have

$$\varrho_{\varsigma}\,(x) = \varrho_{\varsigma}\,(0)\left[(b+1)_m\,m!\right]^2 \frac{1}{(b+1)_{x+m}\,(x+m)!}\frac{1}{(b+1)_{-x+m}\,(-x+m)!}$$
$$= \frac{(b+1)_{2m}\,(2m)!}{\Gamma\,(x+m+b+1)\,\Gamma\,(x+m+1)\,\Gamma\,(-x+m+b+1)\,\Gamma\,(-x+m+1)}.$$

This agrees (up to a normalization factor) with the weight function considered by the authors in [7, equation 26].

5 Semiclassical Polynomials of Class 1

In this section, we consider all families of polynomials of class 1. We have five main cases, corresponding to

$$(p,q) = (0,1)\,,\,(1,1)\,,\,(2,0)\,,\,(2,1)\,,\,(3,2)\,.$$

There are also 14 subcases.

5.1 **(0, 1)***: Generalized Charlier Polynomials*

Linear functional

$$L\,[r] = \sum_{x=0}^{\infty} r\,(x)\;\frac{1}{(b+1)_x}\frac{z^x}{x!}.$$

Pearson equation

$$\frac{\varrho(x+1)}{\varrho(x)} = \frac{z}{(x+b+1)(x+1)}.$$

Moments

$$\nu_n(z) = \frac{z^n}{(b+1)_n}\, {}_0F_1\left[\begin{matrix} - \\ b+1+n \end{matrix}; z\right], \quad n \in \mathbb{N}_0.$$

Stieltjes transform difference equation

$$(t+1)(t+b+1)S(t+1) - zS(t) = (t+b+1)\nu_0 + \nu_1.$$

5.2 (1, 1): *Generalized Meixner Polynomials*

Linear functional

$$L[r] = \sum_{x=0}^{\infty} r(x) \frac{(a)_x}{(b+1)_x} \frac{z^x}{x!}.$$

Pearson equation

$$\frac{\varrho(x+1)}{\varrho(x)} = \frac{z(x+a)}{(x+1)(x+b+1)}.$$

Moments

$$\nu_n(z) = z^n \frac{(a)_n}{(b+1)_n}\, {}_1F_1\left[\begin{matrix} a+n \\ b+1+n \end{matrix}; z\right], \quad n \in \mathbb{N}_0.$$

Stieltjes transform difference equation

$$(t+1)(t+b+1)S(t+1) - z(t+a)S(t) = (t+b+1-z)\nu_0 + \nu_1. \quad (51)$$

5.2.1 Reduced-Uvarov Charlier Polynomials

Let $\omega = 0$.

Special values

$$a \to 0, \quad b \to -1.$$

Linear functional

$$L_U [r] = \sum_{x=0}^{\infty} r (x) \frac{z^x}{x!} + Mr (0) .$$

Pearson equation

$$\frac{\varrho_U (x + 1)}{\varrho_U (x)} = \frac{zx}{(x + 1) x}.$$

Moments

$$\nu_n^U (z) = z^n e^z + M\phi_n (0) , \quad n \in \mathbb{N}_0.$$

Stieltjes transform difference equation

$$t [(t + 1) S_U (t + 1) - zS_U (t)] = (t - z) \nu_0^U + \nu_1^U . \tag{52}$$

Since

$$(t - z) \nu_0^U + \nu_1^U = (t - z) \left(e^z + M\right) + ze^z = t\nu_0 + M (t - z) ,$$

where $\nu_0 = e^z$ is the first moment of the Charlier polynomials, we can rewrite (52) in the M-dependent form

$$(t + 1) S_U (t + 1) - zS_U (t) = \nu_0 + M \left(1 - \frac{z}{t}\right),$$

which agrees with (31).

5.2.2 Christoffel Charlier Polynomials

Let $\omega \notin \{0, z\}$.

Special values

$$a \to -\omega + 1, \quad b \to -\omega - 1.$$

Linear functional

$$L_C [r] = \sum_{x=0}^{\infty} r (x) (x - \omega) \frac{z^x}{x!}.$$

Pearson equation

$$\frac{\varrho_C(x+1)}{\varrho_C(x)} = \frac{z(x+1-\omega)}{(x+1)(x-\omega)}.$$

Moments

$$\nu_n^C(z) = \nu_{n+1} + (n-\omega)\nu_n = (n+z-\omega) z^n e^z, \quad n \in \mathbb{N}_0. \tag{53}$$

Stieltjes transform difference equation

$$(t-\omega)(t+1) S_C(t+1) - (t+1-\omega) z S_C(t) = \nu_0 \left[(z-\omega)t + \left(\omega^2 - z\omega + z\right)\right].$$

Note that if we use (53), we have

$$\nu_0^C = (z-\omega)\nu_0, \quad \nu_1^C = (1+z-\omega) z\nu_0,$$

and therefore

$$(t-\omega)(t+1) S_C(t+1) - (t+1-\omega) z S_C(t) = (t-\omega-z)\nu_0^C + \nu_1^C,$$

which agrees with (51) as $a \to 1-\omega, \quad b \to -\omega - 1$.

5.2.3 Geronimus Charlier Polynomials

Let $\omega \notin \{-1\} \cup \mathbb{N}_0$.

Special values

$$a \to -\omega, \quad b \to -\omega.$$

Linear functional

$$L_G[r] = \sum_{x=0}^{\infty} \frac{r(x)}{x-\omega} \frac{z^x}{x!} + Mr(\omega).$$

Pearson equation

$$\frac{\varrho_G(x+1)}{\varrho_G(x)} = \frac{z(x-\omega)}{(x+1)(x+1-\omega)}.$$

Moments

$$\nu_n^G(z) = \phi_n(\omega) \left[M - S(\omega) + e^z \sum_{k=0}^{n-1} \frac{z^k}{\phi_{k+1}(\omega)} \right], \quad n \in \mathbb{N}_0.$$

Stieltjes transform difference equation

$$(t+1-\omega)\,\sigma\,(t+1)\,S_G\,(t+1)-(t-\omega)\,\eta\,(t)\,S_G\,(t)=\nu_0+\nu_0^G\,(t+1-z)\,.$$

Note that if we use (34), we have

$$\nu_1^G-\omega\nu_0^G=\nu_0,$$

and therefore we can write

$$(t+1)\,(t+1-\omega)\,S_G\,(t+1)-z\,(t-\omega)\,S_G\,(t)=(t+1-\omega-z)\,\nu_0^G+\nu_1^G,$$

which agrees with (51) as $a\to-\omega,\quad b\to-\omega.$

5.2.4 Truncated Charlier Polynomials

Special values

$$a\to-N,\quad b\to-N-1.$$

Linear functional

$$L_T\,[r]=\sum_{x=0}^{N} r\,(x)\,\frac{z^x}{x!}.$$

Pearson equation

$$\frac{\varrho_T\,(x+1)}{\varrho_T\,(x)}=\frac{z\,(x-N)}{(x+1)\,(x-N)}.$$

Moments

$$\nu_n^T\,(z)=\frac{z^N}{(N-n)!}\,{}_2F_0\left[\begin{matrix} n-N,1\\ -\end{matrix};-\frac{1}{z}\right],\qquad n\in\mathbb{N}_0.$$

Stieltjes transform difference equation

$$(t-N)\,[(t+1)\,S_T\,(t+1)-zS_T\,(t)]=(t-N-z)\,\nu_0^T+\nu_1^T.$$

Remark 7 The performance of the modified Chebyshev algorithm used to compute the moments of these polynomials was studied in [25] (example 4.3).

5.3 (2, 0; N): *Generalized Krawtchouk Polynomials*

Linear functional

$$L[r] = \sum_{x=0}^{N} r(x)\ (a)_x\, (-N)_x\, \frac{z^x}{x!}.$$

Pearson equation

$$\frac{\varrho(x+1)}{\varrho(x)} = \frac{z(x+a)(x-N)}{x+1}.$$

Moments

$$\nu_n(z) = z^n\,(-N)_n\,(a)_n\ {}_2F_0\left[\begin{matrix} -N+n,\, a+n \\ - \end{matrix};z\right], \qquad n \in \mathbb{N}_0.$$

Stieltjes transform difference equation

$$(t+1)\,S(t+1) - z\,(t+a)\,(t-N)\,S(t) = [1-(t+a-N)\,z]\,\nu_0 - z\nu_1.$$

5.4 (2, 1): *Generalized Hahn Polynomials of Type I*

Linear functional

$$L[r] = \sum_{x=0}^{\infty} r(x)\ \frac{(a_1)_x\,(a_2)_x}{(b+1)_x}\,\frac{z^x}{x!}.$$

Pearson equation

$$\frac{\varrho(x+1)}{\varrho(x)} = \frac{z(x+a_1)(x+a_2)}{(x+b+1)(x+1)}.$$

Moments

$$\nu_n(z) = z^n\,\frac{(a_1)_n\,(a_2)_n}{(b+1)_n}\ {}_2F_1\left[\begin{matrix} a_1+n,\, a_2+n \\ b+1+n \end{matrix};z\right], \qquad n \in \mathbb{N}_0,$$

where we choose the principal branch $z \in \mathbb{C} \setminus [1,\infty)$. Note that since

$$b+1+n-(a_1+n+a_2+n) \leq 0, \quad n \geq b-a_1-a_2+1,$$

$\nu_n(1)$ is undefined for $n \geq b-a_1-a_2+1$.

Stieltjes transform difference equation

$$(t+1)(t+b+1)S(t+1) - z(t+a_1)(t+a_2)S(t)$$
$$= [t+b+1-z(t+a_1+a_2)]\nu_0 + (1-z)\nu_1.$$

5.4.1 Reduced-Uvarov Meixner Polynomials

Let $(\omega, \Omega) \in \{(0,0), (-a, a+1)\}$.

Special values

$$a_1 \to \Omega, \quad a_2 \to a, \quad b \to \Omega - 1.$$

Linear functional

$$L_U[r] = \sum_{x=0}^{\infty} r(x)\ (a)_x \frac{z^x}{x!} + Mr(\omega).$$

Pearson equation

$$\frac{\varrho_U(x+1)}{\varrho_U(x)} = \frac{z(x+\Omega)(x+a)}{(x+\Omega)(x+1)}.$$

Moments

$$\nu_n^U(z) = z^n (a)_n (1-z)^{-a-n} + M\phi_n(\omega), \quad n \in \mathbb{N}_0.$$

Stieltjes transform difference equation

$$(t+\Omega)[(t+1)S_U(t+1) - z(t+a)S_U(t)]$$
$$= [(1-z)(t+\Omega) - az]\nu_0^U + (1-z)\nu_1^U.$$

Stieltjes transform M-dependent difference equation

$$(t+1)S_U(t+1) - z(t+a)S_U(t) = (1-z)\nu_0 + M\left[\frac{t+1}{t-\omega+1} - \frac{z(t+a)}{t-\omega}\right].$$

5.4.2 Christoffel Meixner Polynomials

Let $\omega \notin \{-a, 0, az(1-z)^{-1}\}$.

Special values

$$a_1 \to a, \quad a_2 \to -\omega + 1, \quad b \to -\omega - 1.$$

Linear functional

$$L_C[r] = \sum_{x=0}^{\infty} r(x)(x-\omega)(a)_x \frac{z^x}{x!}.$$

Pearson equation

$$\frac{\varrho_C(x+1)}{\varrho_C(x)} = \frac{z(x+a)(x+1-\omega)}{(x+1)(x-\omega)}.$$

Moments

$$\nu_n^C(z) = \nu_{n+1} + (n-\omega)\nu_n = z^n (a)_n (1-z)^{-a-n-1}(az+\omega z+n-\omega), \quad n \in \mathbb{N}_0. \tag{54}$$

Stieltjes transform difference equation

$$(t-\omega)(t+1)S_C(t+1) - (t+1-\omega)z(t+a)S_C(t)$$
$$= \nu_0\left[(az-\omega+\omega z)t + \left(\omega^2 + az + \omega z - az\omega - z\omega^2\right)\right].$$

Note that if we use (54), we have

$$\nu_0^C = \frac{az+\omega z-\omega}{1-z}\nu_0, \quad \nu_1^C = \frac{za(az+\omega z+1-\omega)}{(1-z)^2}\nu_0,$$

and therefore we can also write

$$(t-\omega)(t+1)S_C(t+1) - (t+1-\omega)z(t+a)S_C(t)$$
$$= [t-\omega-z(t+a-\omega+1)]\nu_0^C + (1-z)\nu_1^C.$$

5.4.3 Geronimus Meixner Polynomials

Let $\omega \notin \{-1, 1-a\} \cup \mathbb{N}_0$.

Special values

$$a_1 \to a, \quad a_2 \to -\omega, \quad b \to -\omega.$$

Linear functional

$$L_G[r] = \sum_{x=0}^{\infty} \frac{r(x)}{x-\omega}(a)_x \frac{z^x}{x!} + Mr(\omega).$$

Pearson equation

$$\frac{\varrho_G(x+1)}{\varrho_G(x)} = \frac{z(x+a)(x-\omega)}{(x+1)(x+1-\omega)}.$$

Moments

$$\nu_n^G(z) = \phi_n(\omega)\left[M - S(\omega) + (1-z)^{-a}\sum_{k=0}^{n-1}\frac{z^k (a)_k (1-z)^{-k}}{\phi_{k+1}(\omega)}\right], \quad n \in \mathbb{N}_0.$$

Stieltjes transform difference equation

$$(t+1)(t-\omega+1) S_G(t+1) - z(t+a)(t-\omega) S_G(t)$$
$$= (1-z)\nu_0 + \nu_0^G[(1-z)t + (1-az)].$$

Note that if we use (34), we have

$$\nu_1^G - \omega\nu_0^G = \nu_0,$$

and therefore we can also write

$$(t+1)(t-\omega+1) S_G(t+1) - z(t+a)(t-\omega) S_G(t)$$
$$= [(1-z)t + z\omega - \omega - az + 1]\nu_0^G + (1-z)\nu_1^G.$$

5.4.4 Truncated Meixner Polynomials

Special values

$$a_1 \to a, \quad a_2 \to -N, \quad b \to -N-1.$$

Linear functional

$$L_T[r] = \sum_{x=0}^{N} r(x)\,(a)_x\,\frac{z^x}{x!}.$$

Moments

$$\nu_n^T = (a)_N \frac{z^N}{(N-n)!}\,{}_2F_1\left[\begin{matrix} n-N, 1 \\ 1-N-a \end{matrix}; \frac{1}{z}\right], \quad n \in \mathbb{N}_0.$$

Pearson equation

$$\frac{\varrho_T(x+1)}{\varrho_T(x)} = \frac{z(x+a)(x-N)}{(x-N)(x+1)}.$$

Stieltjes transform difference equation

$$\begin{aligned}&(t-N)\left[(t+1)S_T(t+1) - z(t+a)S_T(t)\right]\\&= \left[(t-N)(1-z) - az\right]\nu_0^T + (1-z)\nu_1^T.\end{aligned}$$

5.4.5 Symmetrized Generalized Krawtchouk Polynomials

Special values

$$a_1 \to a, \quad a_2 \to -2m, \quad b \to -2m-a, \quad z \to -1, \quad x \to x+m.$$

Linear functional

$$L_\varsigma[r] = \varrho_\varsigma(0) \sum_{x=-m}^{m} r(x) \frac{(a+m)_x\,(-m)_x}{(-m-a+1)_x\,(m+1)_x}(-1)^x,$$

where

$$\varrho_\varsigma(0) = (-1)^m \binom{2m}{m} \frac{(a)_m}{(a+m)_m}.$$

Pearson equation

$$\frac{\varrho_\varsigma(x+1)}{\varrho_\varsigma(x)} = \frac{-(x+a+m)(x-m)}{(x+1-a-m)(x+1+m)}.$$

Moments on the basis $\{\overline{\phi}_n(x)\}_{n\geq 0}$

$$\nu_n^\varsigma = (-1)^n \frac{(a)_n(-2m)_n}{(-2m-a)_n}\,{}_2F_1\left[\begin{matrix}-2m+n,\ a+n\\ -2m-a+n\end{matrix}; -1\right], \quad n \in \mathbb{N}_0.$$

Stieltjes transform difference equation

$$\begin{aligned}&(t+m+1)(t-m-a+1)S_\varsigma(t+1) + (t-m)(t+m+a)S_\varsigma(t)\\&= (2t-2m+1)\nu_0^\varsigma + 2\nu_1^\varsigma.\end{aligned}$$

5.5 (3, 2; N, 1): *Generalized Hahn Polynomials of Type II*

Linear functional

$$L[r] = \sum_{x=0}^{N} r(x) \frac{(a_1)_x (a_2)_x (-N)_x}{(b_1+1)_x (b_2+1)_x} \frac{1}{x!}.$$

Pearson equation

$$\frac{\varrho(x+1)}{\varrho(x)} = \frac{(x+a_1)(x+a_2)(x-N)}{(x+b_1+1)(x+b_2+1)(x+1)}.$$

Moments

$$\nu_n = \frac{(-N)_n (a_1)_n (a_2)_n}{(b_1+1)_n (b_2+1)_n} \, {}_3F_2\left[\begin{matrix} -N+n, a_1+n, a_2+n \\ b_1+1+n, b_2+1+n \end{matrix}; 1\right], \qquad n \in \mathbb{N}_0.$$

Stieltjes transform difference equation

$$\begin{aligned}
&(t+b_1+1)(t+b_2+1)(t+1)\,S(t+1) - (t+a_1)(t+a_2)(t-N)\,S(t) \\
&= [(N+2+b_1+b_2-a_1-a_2)\,t + b_1 + b_2 + b_1b_2 - a_1 - a_2 - a_1a_2 \\
&\quad + Na_1 + Na_2 + 1]\,\nu_0 \\
&+ (N+b_1+b_2-a_1-a_2+1)\,\nu_1.
\end{aligned}$$

5.5.1 Reduced-Uvarov Hahn Polynomials

Let $(\omega, \Omega) \in \{(0,0), (-a, a+1), (-b, b), (N, -N+1)\}$.

Special values

$$a_1 \to a, \quad a_2 \to \Omega, \quad b_1 \to \Omega - 1, \quad b_2 \to a.$$

Linear functional

$$L_U[r] = \sum_{x=0}^{N} r(x) \frac{(a)_x (-N)_x}{(b+1)_x} \frac{1}{x!} + Mr(\omega).$$

Pearson equation

$$\frac{\varrho_U(x+1)}{\varrho_U(x)} = \frac{(x+a)(x+\Omega)(x-N)}{(x+b+1)(x+\Omega)(x+1)}.$$

Moments

$$\nu_n^U = \frac{(-N)_n\,(a)_n}{(b+1)_n}\frac{(b+1-a)_{N-n}}{(b+1+n)_{N-n}} + M\phi_n(\omega), \quad n \in \mathbb{N}_0.$$

Stieltjes transform difference equation

$$(t+\Omega)\left[(t+b+1)(t+1)\,S_U(t+1) - (t+a)(t-N)\right] S_U(t)$$
$$= \left[(N-a+b+1)t + (N-a+b+1)\,\Omega + aN\right]\nu_0^U + (N+b-a)\,\nu_1^U.$$

Stieltjes transform M-dependent difference equation

$$(t+b+1)(t+1)\,S_U(t+1) - (t-N)(t+a)\,S_U(t)$$
$$= (N+b-a+1)\,\nu_0 + M\left[\frac{(t+b+1)(t+1)}{t-\omega+1} - \frac{(t-N)(t+a)}{t-\omega}\right].$$

5.5.2 Christoffel Hahn Polynomials

Let $\omega \notin \left\{-a, -b, 0, N, -\frac{aN}{b-a+N}\right\}$.

Special values

$$a_1 \to a, \quad a_2 \to -\omega+1, \quad b_1 \to b, \quad b_2 \to -\omega-1.$$

Linear functional

$$L_C[r] = \sum_{x=0}^{\infty} r(x)\,(x-\omega)\,\frac{(a)_x\,(-N)_x}{(b+1)_x}\frac{1}{x!}.$$

Pearson equation

$$\frac{\varrho_C(x+1)}{\varrho_C(x)} = \frac{(x+a)(x-N)(x+1-\omega)}{(x+b+1)(x+1)(x-\omega)}.$$

Moments

$$\nu_n^C(z) = \nu_{n+1} + (n-\omega)\,\nu_n = \left[n-\omega-\frac{(a+n)(N-n)}{b-a+N-n}\right]\nu_n, \quad n \in \mathbb{N}_0. \tag{55}$$

Stieltjes transform difference equation

$$(t-\omega)(t+b+1)(t+1)\,S_C(t+1) - (t-\omega+1)(t-N)(t+a)\,S_C(t)$$
$$= \left[(a\omega - N\omega - b\omega - Na)\,t + a\omega - N\omega + \omega^2 + N\omega^2 - a\omega^2 + b\omega^2 - Na + Na\omega\right]\nu_0.$$

Note that if we use (55), we have

$$\nu_0^C = -\frac{(a+\omega)N + (b-a)\omega}{b-a+N}\nu_0, \qquad \nu_1^C = \frac{(Na + N\omega - a\omega + b\omega - b - \omega)Na}{(b-a+N)(b-a+N-1)}\nu_0,$$

and therefore we can also write

$$\begin{aligned}&(t-\omega)(t+b+1)(t+1)S_C(t+1) - (t-\omega+1)(t-N)(t+a)S_C(t)\\ &= [(N-a+b)t + Na - N\omega + a\omega - b\omega + N - a - \omega]\nu_0^C + (N-a+b-1)\nu_1^C.\end{aligned}$$

5.5.3 Geronimus Hahn Polynomials

Let $\omega \notin \{-b-1, -1, 1-a, N+1\} \cup \mathbb{N}_0$.

Special values

$$a_1 \to a, \quad a_2 \to -\omega, \quad b \to -\omega.$$

Linear functional

$$L_G[r] = \sum_{x=0}^{\infty} \frac{r(x)}{x-\omega} \frac{(a)_x (-N)_x}{(b+1)_x} \frac{1}{x!} + Mr(\omega).$$

Pearson equation

$$\frac{\varrho_G(x+1)}{\varrho_G(x)} = \frac{(x+a)(x-N)(x-\omega)}{(x+b+1)(x+1)(x+1-\omega)}.$$

Moments

$$\nu_n^G(z) = \phi_n(\omega)\left[M - S(\omega) + \sum_{k=0}^{n-1} \frac{\nu_k}{\phi_{k+1}(\omega)}\right], \qquad n \in \mathbb{N}_0.$$

Stieltjes transform difference equation

$$\xi_G(t) = \xi(t) + [\sigma(t+1) - \eta(t)]\nu_0^G,$$

$$\begin{aligned}&(t-\omega+1)(t+b+1)(t+1)S_G(t+1) - (t-\omega)(t-N)(t+a)S_G(t)\\ &= (b+1-a+N)\nu_0 + [(N-a+b+2)t + b + Na + 1]\nu_0^G.\end{aligned}$$

Note that if we use (34), we have

$$\nu_1^G - \omega \nu_0^G = \nu_0,$$

and therefore we can also write

$$\begin{aligned}&(t-\omega+1)(t+b+1)(t+1)\,S_G(t+1)-(t-\omega)(t-N)(t+a)\,S_G(t)\\&=[(N-a+b+2)t+Na-N\omega+a\omega-b\omega+b-\omega+1]\,\nu_0^G+(b-1-a+N)\,\nu_1^G.\end{aligned}$$

5.5.4 Symmetrized Hahn Polynomials

Special values

$$a_1 \to a, \quad a_2 \to -N-b, \quad b_1 \to b, \quad b_2 \to -N-a, \quad N \to 2m, \quad x \to x+m.$$

Linear functional

$$L_\varsigma[r] = \varrho_\varsigma(0) \sum_{x=-m}^{m} r(x) \frac{(m+a)_x (-m-b)_x (-m)_x}{(1-m-a)_x (m+b+1)_x (m+1)_x},$$

where

$$\varrho_\varsigma(0) = (-1)^m \binom{2m}{m} \frac{(a)_m}{(a+m)_m} \frac{(b+1+m)_m}{(b+1)_m}.$$

Pearson equation

$$\frac{\varrho_\varsigma(x+1)}{\varrho_\varsigma(x)} = \frac{(x+m+a)(x-m-b)(x-m)}{(x+1-m-a)(x+1+m+b)(x+1+m)}.$$

Moments on the basis $\{\overline{\phi}_n(x)\}_{n\geq 0}$

$$\nu_n^\varsigma = \frac{(-2m)_n (a)_n (-2m-b)_n}{(b+1)_n (-2m-a+1)_n} \, {}_3F_2\left[\begin{matrix} -2m+n, a+n, -2m-b+n \\ b+1+n, -2m-a+1+n \end{matrix}; 1\right], \quad n \in \mathbb{N}_0.$$

Stieltjes transform difference equation

$$\begin{aligned}&(t+1-m-a)(t+1+m+b)(t+1+m)\,S_\varsigma(t+1)\\&-(t+m+a)(t-m-b)(t-m)\,S_\varsigma(t)\end{aligned}$$

$$= \left[2(1+b-a+m)t + 2am - 2bm - 2m^2 - a + b + 1 \right] \nu_0^{\varsigma}$$
$$+ (1+2m-2a+2b)\,\nu_1^{\varsigma}.$$

5.5.5 Symmetrized Polynomials of Type (3, 0; N)

For a definition of the polynomials of type $(3, 0; N)$, see Sect. 6.4.

Special values

$$b_1 \to -N - a_1, \quad b_2 \to -N - a_2, \quad N \to 2m, \quad x \to x + m.$$

Linear functional

$$L_{\varsigma}[r] = \varrho_{\varsigma}(0) \sum_{x=-m}^{m} r(x) \frac{(a_1+m)_x (a_2+m)_x (-m)_x}{(1-a_1-m)_x (1-a_2-m)_x (m+1)_x},$$

where

$$\varrho_{\varsigma}(0) = (-1)^m \binom{2m}{m} \frac{(a_1)_m}{(a_1+m)_m} \frac{(a_2)_m}{(a_2+m)_m}.$$

Pearson equation

$$\frac{\varrho_{\varsigma}(x+1)}{\varrho_{\varsigma}(x)} = \frac{(x+a_1+m)(x+a_2+m)(x-m)}{(x+1-a_1-m)(x+1-a_2-m)(x+1+m)}.$$

Moments on the basis $\{\overline{\phi}_n(x)\}_{n\geq 0}$

$$\nu_n^{\varsigma} = \frac{(-2m)_n (a_1)_n (a_2)_n}{(1-a_1-2m)_n (1-a_2-2m)_n} \, {}_3F_2\left[\begin{matrix} -2m+n, a_1+n, a_2+n \\ 1-a_1-2m+n, 1-a_2-2m+n \end{matrix}; 1 \right],$$

$n \in \mathbb{N}_0$.

Stieltjes transform difference equation

$$(t+1-a_1-m)(t+1-a_2-m)(t+1+m)\,S_{\varsigma}(t+1)$$
$$-(t+a_1+m)(t+a_2+m)(t-m)\,S_{\varsigma}(t)$$
$$= \left[-2(m+a_1+a_2-1)t + 2m^2 - 2m + 1 + (2m-1)(a_1+a_2) \right] \nu_0^{\varsigma}$$
$$+ (-2m+1-2a_1-2a_2)\,\nu_1^{\varsigma}.$$

6 Semiclassical Polynomials of Class 2

In this section, we consider all families of polynomials of class 2. We have seven main cases, corresponding to

$$(p, q) = (0, 2), (1, 2), (2, 2), (3, 0), (3, 1), (3, 2), (4, 3).$$

There are also 25 subcases.

6.1 Polynomials of Type (0,2)

Linear functional

$$L[r] = \sum_{x=0}^{\infty} r(x) \frac{1}{(b_1+1)_x (b_2+1)_x} \frac{z^x}{x!}.$$

Pearson equation

$$\frac{\varrho(x+1)}{\varrho(x)} = \frac{z}{(x+b_1+1)(x+b_2+1)(x+1)}.$$

Moments

$$\nu_n(z) = \frac{z^n}{(b_1+1)_n (b_2+1)_n} \, {}_0F_2 \left[\begin{matrix} - \\ b_1+1+n, b_2+1+n \end{matrix} ; z \right], \quad n \in \mathbb{N}_0.$$

Stieltjes transform difference equation

$$\begin{aligned} &(t+b_1+1)(t+b_2+1)(t+1) S(t+1) - zS(t) \\ &= (t+b_1+1)(t+b_2+1)\nu_0 + (t+b_1+b_2+2)\nu_1 + \nu_2. \end{aligned}$$

6.2 Polynomials of Type (1,2)

Linear functional

$$L[r] = \sum_{x=0}^{\infty} r(x) \frac{(a)_x}{(b_1+1)_x (b_2+1)_x} \frac{z^x}{x!}.$$

Pearson equation

$$\frac{\varrho(x+1)}{\varrho(x)} = \frac{z(x+a)}{(x+b_1+1)(x+b_2+1)(x+1)}.$$

Moments

$$\nu_n(z) = z^n \frac{(a)_n}{(b_1+1)_n (b_2+1)_n} \, {}_1F_2\left[\begin{matrix} a+n \\ b_1+1+n, b_2+1+n \end{matrix}; z\right], \quad n \in \mathbb{N}_0.$$

Stieltjes transform difference equation

$$\begin{aligned}&(t+b_1+1)(t+b_2+1)(t+1)S(t+1) - z(t+a)S(t)\\ &= [(t+b_1+1)(t+b_2+1) - z]\nu_0 + (t+b_1+b_2+2)\nu_1 + \nu_2.\end{aligned}$$

6.2.1 Reduced-Uvarov Generalized Charlier Polynomials

Let $\omega \in \{0, -b\}$.

Special values

$$a \to -\omega, \quad b_1 \to b, \quad b_2 \to -\omega - 1.$$

Linear functional

$$L_U[r] = \sum_{x=0}^{\infty} r(x) \frac{1}{(b+1)_x} \frac{z^x}{x!} + Mr(\omega).$$

Pearson equation

$$\frac{\varrho_U(x+1)}{\varrho_U(x)} = \frac{z(x-\omega)}{(x+b+1)(x-\omega)(x+1)}.$$

Moments

$$\nu_n^U(z) = \frac{z^n}{(b+1)_n} \, {}_0F_1\left[\begin{matrix} - \\ b+1+n \end{matrix}; z\right] + M\phi_n(\omega), \quad n \in \mathbb{N}_0.$$

Stieltjes transform difference equation

$$\begin{aligned}&(t-\omega)[(t+b+1)(t+1)S_U(t+1) - zS_U(t)]\\ &= [(t+b+1)(t+b) - z]\nu_0^U + (t+2b+1)\nu_1^U + \nu_2^U.\end{aligned}$$

Stieltjes transform M-dependent difference equation

$$\begin{aligned}&(t+b+1)(t+1)\,S_U(t+1) - zS_U(t)\\ &= (t+b+1)\,\nu_0 + \nu_1 + M\left[\frac{(t+b+1)(t+1)}{t-\omega+1} - \frac{z}{t-\omega}\right].\end{aligned}$$

6.2.2 Christoffel Generalized Charlier Polynomials

Let $\omega \notin \left\{-b, 0, \frac{\nu_1}{\nu_0}\right\}$.

Special values

$$a \to -\omega+1, \quad b_1 \to b, \quad b_2 \to -\omega-1.$$

Linear functional

$$L_C[r] = \sum_{x=0}^{\infty} r(x)\,(x-\omega)\,\frac{1}{(b+1)_x}\,\frac{z^x}{x!}.$$

Pearson equation

$$\frac{\varrho_C(x+1)}{\varrho_C(x)} = \frac{z(x+1-\omega)}{(x+b+1)(x+1)(x-\omega)}.$$

Moments

$$\nu_n^C(z) = \nu_{n+1} + (n-\omega)\,\nu_n, \quad n \in \mathbb{N}_0.$$

Stieltjes transform difference equation

$$\begin{aligned}&(t-\omega)(t+b+1)(t+1)\,S_C(t+1) - z(t-\omega+1)\,S_C(t)\\ &= \left[-\omega t^2 + \left(z-\omega-b\omega+\omega^2\right)t + \left(z-z\omega+\omega^2+b\omega^2\right)\right]\nu_0\\ &+ (t-\omega)(t-\omega+1)\,\nu_1.\end{aligned}$$

6.2.3 Geronimus Generalized Charlier Polynomials

Let $\omega \notin \{-b-1, -1\} \cup \mathbb{N}_0$.

Special values

$$a \to -\omega, \quad b_1 \to b, \quad b_2 \to -\omega.$$

Linear functional

$$L_G[r] = \sum_{x=0}^{\infty} \frac{r(x)}{x-\omega} \frac{1}{(b+1)_x} \frac{z^x}{x!} + Mr(\omega).$$

Pearson equation

$$\frac{\varrho_G(x+1)}{\varrho_G(x)} = \frac{z(x-\omega)}{(x+b+1)(x+1)(x+1-\omega)}.$$

Moments

$$\nu_n^G(z) = \phi_n(\omega)\left[M - S(\omega) + \sum_{k=0}^{n-1} \frac{\nu_k}{\phi_{k+1}(\omega)}\right], \quad n \in \mathbb{N}_0.$$

Stieltjes transform difference equation

$$\begin{aligned}&(t-\omega+1)(t+b+1)(t+1)S_G(t+1) - (t-\omega)zS_G(t)\\&= (t+b+1)\nu_0 + \nu_1 + [(t+b+1)(t+1) - z]\nu_0^G.\end{aligned}$$

6.2.4 Truncated Generalized Charlier Polynomials

Special values

$$a \to -N, \quad b_1 \to b, \quad b_2 \to -N-1.$$

Linear functional

$$L_T[r] = \sum_{x=0}^{N} r(x) \frac{1}{(b+1)_x} \frac{z^x}{x!}.$$

Pearson equation

$$\frac{\varrho_T(x+1)}{\varrho_T(x)} = \frac{z(x-N)}{(x+b+1)(x-N)(x+1)}.$$

Moments

$$\nu_n^T(z) = \frac{1}{(b+1)_N} \frac{z^N}{(N-n)!} \, {}_3F_0\left[\begin{matrix} n-N, 1, -N-b \\ - \end{matrix}; \frac{1}{z}\right], \quad n \in \mathbb{N}_0.$$

Stieltjes transform difference equation

$$(t-N)\left[(t+b+1)(t+1)S_T(t+1)-zS_T(t)\right]$$
$$=\left[(t+b+1)(t-N)-z\right]\nu_0^T+(t+b+1-N)\nu_1^T+\nu_2^T.$$

6.3 Polynomials of Type (2,2)

Linear functional

$$L[r]=\sum_{x=0}^{\infty}r(x)\,\frac{(a_1)_x\,(a_2)_x}{(b_1+1)_x\,(b_2+1)_x}\frac{z^x}{x!}.$$

Pearson equation

$$\frac{\varrho(x+1)}{\varrho(x)}=\frac{z(x+a_1)(x+a_2)}{(x+b_1+1)(x+b_2+1)(x+1)}.$$

Moments

$$\nu_n(z)=z^n\frac{(a_1)_n\,(a_2)_n}{(b_1+1)_n\,(b_2+1)_n}\,{}_2F_2\left[\begin{matrix}a_1+n,a_2+n\\ b_1+1+n,b_2+1+n\end{matrix};z\right],\quad n\in\mathbb{N}_0.$$

Stieltjes transform difference equation

$$(t+b_1+1)(t+b_2+1)(t+1)S(t+1)-z(t+a_1)(t+a_2)S(t)$$
$$=\left[(t+b_1+1)(t+b_2+1)-(t+a_1+a_2)z\right]\nu_0+(t+b_1+b_2+2-z)\nu_1+\nu_2.$$

6.3.1 Uvarov Charlier Polynomials

Let $\omega\neq 0$.

Special values

$$a_1\to-\omega,\quad a_2\to-\omega+1,\quad b_1\to-\omega-1,\quad b_2\to-\omega.$$

Linear functional

$$L_U[r]=\sum_{x=0}^{\infty}r(x)\,\frac{z^x}{x!}+Mr(\omega).$$

Pearson equation

$$\frac{\varrho_U(x+1)}{\varrho_U(x)} = \frac{z(x-\omega)(x-\omega+1)}{(x-\omega)(x-\omega+1)(x+1)}.$$

Moments

$$\nu_n^U(z) = z^n e^z + M\phi_n(\omega), \quad n \in \mathbb{N}_0.$$

Stieltjes transform difference equation

$$\begin{aligned}&(t-\omega)(t-\omega+1)\left[(t+1)S_U(t+1) - zS_U(t)\right]\\&= \left[(t-\omega)(t-\omega+1) - (t-2\omega+1)z\right]\nu_0^U + (t-z-2\omega+1)\nu_1^U + \nu_2^U.\end{aligned}$$

Stieltjes transform M-dependent difference equation

$$(t+1)S_U(t+1) - zS_U(t) = \nu_0 + M\left(\frac{t+1}{t+1-\omega} - \frac{z}{t-\omega}\right).$$

6.3.2 Reduced-Uvarov Generalized Meixner Polynomials

Let $(\omega, \Omega) \in \{(-a, a+1), (-b, b), (0, 0)\}$.

Special values

$$a_1 \to a, \quad a_2 \to \Omega, \quad b_1 \to b, \quad b_2 \to \Omega - 1.$$

Linear functional

$$L_U[r] = \sum_{x=0}^{\infty} r(x)\frac{(a)_x}{(b+1)_x}\frac{z^x}{x!} + Mr(\omega).$$

Pearson equation

$$\frac{\varrho_U(x+1)}{\varrho_U(x)} = \frac{z(x+a)(x+\Omega)}{(x+b+1)(x+\Omega)(x+1)}.$$

Moments

$$\nu_n^U(z) = z^n\frac{(a)_n}{(b+1)_n}\,{}_1F_1\left[\begin{matrix}a+n\\b+1+n\end{matrix}; z\right] + M\phi_n(\omega), \quad n \in \mathbb{N}_0.$$

Stieltjes transform difference equation

$$(t+\Omega)\left[(t+b+1)(t+1)S_U(t+1)-z(t+a)S_U(t)\right]$$
$$=\left[(t+b+1)(t+\Omega)-(t+a+\Omega)z\right]\nu_0^U+(t+b+1+\Omega-z)\nu_1^U+\nu_2^U.$$

Stieltjes transform M-dependent difference equation

$$(t+b+1)(t+1)S_U(t+1)-z(t+a)S_U(t)$$
$$=(t+b+1-z)\nu_0+\nu_1+M\left[\frac{(t+b+1)(t+1)}{t+1-\omega}-\frac{z(t+a)}{t-\omega}\right].$$

6.3.3 Christoffel Generalized Meixner Polynomials

Let $\omega\notin\left\{-a,-b,0,\frac{\nu_1}{\nu_0}\right\}$.

Special values

$$a_1\to a,\quad a_2\to-\omega+1,\quad b_1\to b,\quad b_2\to-\omega-1.$$

Linear functional

$$L_C[r]=\sum_{x=0}^{\infty}r(x)(x-\omega)\frac{(a)_x}{(b+1)_x}\frac{z^x}{x!}.$$

Pearson equation

$$\frac{\varrho_C(x+1)}{\varrho_C(x)}=\frac{z(x+a)(x+1-\omega)}{(x+1)(x+b+1)(x-\omega)}.$$

Moments

$$\nu_n^C(z)=\nu_{n+1}+(n-\omega)\nu_n,\quad n\in\mathbb{N}_0.$$

Stieltjes transform difference equation

$$(t-\omega)(t+b+1)(t+1)S_C(t+1)-z(t-\omega+1)(t+a)S_C(t)$$
$$=\left[-\omega t^2+\left(z\omega-b\omega-\omega+\omega^2+az\right)t+z\omega+\omega^2+b\omega^2-z\omega^2+az-az\omega\right]\nu_0$$
$$+(t-\omega)(t-\omega+1)\nu_1.$$

6.3.4 Geronimus Generalized Meixner Polynomials

Let $\omega \notin \{-b-1, -1, 1-a\} \cup \mathbb{N}_0$.

Special values

$$a_1 \to a, \quad a_2 \to -\omega, \quad b_1 \to b, \quad b_2 \to -\omega.$$

Linear functional

$$L_G[r] = \sum_{x=0}^{\infty} \frac{r(x)}{x-\omega} \frac{(a)_x}{(b+1)_x} \frac{z^x}{x!} + Mr(\omega).$$

Pearson equation

$$\frac{\varrho_G(x+1)}{\varrho_G(x)} = \frac{z(x+a)(x-\omega)}{(x+1)(x+b+1)(x+1-\omega)}.$$

Moments

$$\nu_n^G(z) = \phi_n(\omega)\left[M - S(\omega) + \sum_{k=0}^{n-1} \frac{\nu_k}{\phi_{k+1}(\omega)}\right], \quad n \in \mathbb{N}_0.$$

Stieltjes transform difference equation

$$\begin{aligned}&(t-\omega+1)(t+b+1)(t+1)S_G(t+1) - z(t-\omega)(t+a)S_G(t)\\&= (t+b+1-z)\nu_0 + \nu_1 + [(t+b+1)(t+1) - z(t+a)]\nu_0^G.\end{aligned}$$

6.3.5 Truncated Generalized Meixner Polynomials

Special values

$$a_1 \to a, \quad a_2 \to -N, \quad b_1 \to b, \quad b_2 \to -N-1.$$

Linear functional

$$L_T[r] = \sum_{x=0}^{N} r(x) \frac{(a)_x}{(b+1)_x} \frac{z^x}{x!}.$$

Pearson equation

$$\frac{\varrho_T(x+1)}{\varrho_T(x)} = \frac{z(x+a)(x-N)}{(x+b+1)(x-N)(x+1)}.$$

Moments

$$\nu_n^T(z) = \frac{(a)_N}{(b+1)_N} \frac{z^N}{(N-n)!} \, {}_3F_1\left[\begin{matrix} n-N, 1, -N-b \\ 1-N-a \end{matrix} ; \frac{1}{z} \right], \quad n \in \mathbb{N}_0.$$

Stieltjes transform difference equation

$$(t-N)\left[(t+b+1)(t+1)S_T(t+1) - z(t+a)S_T(t)\right]$$
$$= \left[(t+b+1)(t-N) - (t+a-N)z\right]\nu_0^T + (t+b-N+1-z)\nu_1^T + \nu_2^T.$$

6.4 Polynomials of Type (3,0;N)

Linear functional

$$L[r] = \sum_{x=0}^{N} r(x) \, (a_1)_x (a_2)_x (-N)_x \frac{z^x}{x!}.$$

Pearson equation

$$\frac{\varrho(x+1)}{\varrho(x)} = \frac{z(x+a_1)(x+a_2)(x-N)}{x+1}.$$

Moments

$$\nu_n(z) = z^n (-N)_n (a_1)_n (a_2)_n \, {}_3F_0\left[\begin{matrix} -N+n, a_1+n, a_2+n \\ - \end{matrix} ; z \right], \quad n \in \mathbb{N}_0.$$

Stieltjes transform difference equation

$$(t+1)S(t+1) - z(t+a_1)(t+a_2)(t-N)S(t)$$
$$= \left[-zt^2 - (a_1+a_2-N)zt + z(Na_1+Na_2-a_1a_2) + 1\right]\nu_0$$
$$-(t+a_1+a_2+1-N)z\nu_1 - z\nu_2.$$

6.5 Polynomials of Type (3,1;N)

Linear functional

$$L[r] = \sum_{x=0}^{N} r(x) \, \frac{(a_1)_x (a_2)_x (-N)_x}{(b+1)_x} \frac{z^x}{x!}.$$

Pearson equation

$$\frac{\varrho(x+1)}{\varrho(x)} = \frac{z(x+a_1)(x+a_2)(x-N)}{(x+b+1)(x+1)}.$$

Moments

$$\nu_n(z) = z^n \frac{(-N)_n (a_1)_n (a_2)_n}{(b+1)_n} \, {}_3F_1\left[\begin{matrix} -N+n, a_1+n, a_2+n \\ b+1+n \end{matrix}; z\right], \quad n \in \mathbb{N}_0.$$

Stieltjes transform difference equation

$$\begin{aligned}&(x+b+1)(t+1)S(t+1) - z(t+a_1)(t+a_2)(t-N)S(t)\\ &= \left[-zt^2 + t - (a_1+a_2-N)zt + (Na_1+Na_2-a_1a_2)z + b + 1\right]\nu_0\\ &+ [1-(t+a_1+a_2+1-N)z]\nu_1 - z\nu_2.\end{aligned}$$

6.5.1 Reduced-Uvarov Generalized Krawtchouk Polynomials

Let $(\omega, \Omega) \in \{(-a, a+1), (0,0), (N, -N+1)\}$.

Special values

$$a_1 \to a, \quad a_2 \to \Omega, \quad b \to \Omega - 1.$$

Linear functional

$$L_U[r] = \sum_{x=0}^{N} r(x)\,(a)_x (-N)_x \frac{z^x}{x!} + Mr(\omega).$$

Pearson equation

$$\frac{\varrho_U(x+1)}{\varrho_U(x)} = \frac{z(x+a)(x+\Omega)(x-N)}{(x+\Omega)(x+1)}.$$

Moments

$$\nu_n^U(z) = z^n (-N)_n (a)_n \, {}_2F_0\left[\begin{matrix} -N+n, a+n \\ - \end{matrix}; z\right] + M\phi_n(\omega), \quad n \in \mathbb{N}_0.$$

Stieltjes transform difference equation

$$(t+\Omega)\left[(t+1)\,S_U(t+1)-z\,(t+a)\,(t-N)\,S_U(t)\right]$$
$$=\left[-zt^2+t-(a+\Omega-N)\,zt+(Na+N\Omega-a\Omega)\,z+\Omega\right]\nu_0^U$$
$$+\left[1-(t+a+\Omega+1-N)\,z\right]\nu_1^U-z\nu_2^U.$$

Stieltjes transform M-dependent difference equation

$$(t+1)\,S_U(t+1)-z\,(t+a)\,(t-N)\,S_U(t)$$
$$=\left[1-(t+a-N)\,z\right]\nu_0-z\nu_1+M\left[\frac{t+1}{t-\omega+1}-\frac{z\,(t+a)\,(t-N)}{t-\omega}\right].$$

6.5.2 Christoffel Generalized Krawtchouk Polynomials

Let $\omega\notin\left\{-a,0,N,\dfrac{\nu_1}{\nu_0}\right\}$.

Special values

$$a_1\to a,\quad a_2\to-\omega+1,\quad b\to-\omega-1.$$

Linear functional

$$L_C[r]=\sum_{x=0}^{\infty}r(x)\,(x-\omega)\,(a)_x\,(-N)_x\,\frac{z^x}{x!}.$$

Pearson equation

$$\frac{\varrho_C(x+1)}{\varrho_C(x)}=\frac{z\,(x+a)\,(x-N)\,(x+1-\omega)}{(x+1)\,(x-\omega)}.$$

Moments

$$\nu_n^C(z)=\nu_{n+1}+(n-\omega)\,\nu_n,\quad n\in\mathbb{N}_0.$$

Stieltjes transform difference equation

$$(t-\omega)\,(t+1)\,S_C(t+1)-z\,(t-\omega+1)\,(t+a)\,(t-N)\,S_C(t)$$
$$=\left[z\omega t^2-\left(\omega-z\omega+z\omega^2+Nz\omega-az\omega+Naz\right)t\right]\nu_0$$
$$+\left(\omega^2-Nz\omega+az\omega+Nz\omega^2-az\omega^2-Naz+Naz\omega\right)\nu_0$$
$$-z\,(t-\omega)\,(t-\omega+1)\,\nu_1.$$

6.5.3 Geronimus Generalized Krawtchouk Polynomials

Let $\omega \notin \{-1, 1-a, 1+N\} \cup \mathbb{N}_0$.

Special values

$$a_1 \to a, \quad a_2 \to -\omega, \quad b \to -\omega.$$

Linear functional

$$L_G[r] = \sum_{x=0}^{\infty} \frac{r(x)}{x-\omega} (a)_x (-N)_x \frac{z^x}{x!} + M r(\omega).$$

Pearson equation

$$\frac{\varrho_G(x+1)}{\varrho_G(x)} = \frac{z(x+a)(x-N)(x-\omega)}{(x+1)(x+1-\omega)}.$$

Moments

$$\nu_n^G(z) = \phi_n(\omega) \left[M - S(\omega) + \sum_{k=0}^{n-1} \frac{\nu_k}{\phi_{k+1}(\omega)} \right], \quad n \in \mathbb{N}_0.$$

Stieltjes transform difference equation

$$(t-\omega+1)(t+1) S_G(t+1) - z(t-\omega)(t+a)(t-N) S_G(t)$$
$$= (-zt + Nz - az + 1)\nu_0 - z\nu_1 + [t+1 - z(t+a)(t-N)]\nu_0^G.$$

6.6 *Polynomials of Type (3,2)*

Linear functional

$$L[r] = \sum_{x=0}^{\infty} r(x) \frac{(a_1)_x (a_2)_x (a_3)_x}{(b_1+1)_x (b_2+1)_x} \frac{z^x}{x!}.$$

Pearson equation

$$\frac{\varrho(x+1)}{\varrho(x)} = \frac{z(x+a_1)(x+a_2)(x+a_3)}{(x+b_1+1)(x+b_2+1)(x+1)}.$$

Moments

$$\nu_n(z) = z^n \frac{(a_1)_n (a_2)_n (a_3)_n}{(b_1+1)_n (b_2+1)_n} {}_3F_2\left[\begin{matrix} a_1+n, a_2+n, a_3+n \\ b_1+1+n, b_2+1+n \end{matrix}; z\right], \quad n \in \mathbb{N}_0,$$

where we choose the principal branch $z \in \mathbb{C} \setminus [1, \infty)$. Note that since

$$b_1+1+n+b_2+1+n-(a_1+n+a_2+n+a_3+n) \leq 0, \quad n \geq b_1+b_2-a_2-a_3-a_1+2,$$

$\nu_n(1)$ is undefined for $n \geq b_1+b_2-a_2-a_3-a_1+2$.

Stieltjes transform difference equation

$$(t+b_1+1)(t+b_2+1)(t+1)S(t+1) - z(t+a_1)(t+a_2)(t+a_3)S(t) = \sum_{k=0}^{2} \xi_k t^k,$$

where

$$\xi_2 = (1-z)\nu_0, \quad \xi_1 = [e_1(\mathbf{b}+1) - e_1(\mathbf{a})z]\nu_0 + (1-z)\nu_1,$$
$$\xi_0 = [e_2(\mathbf{b}+1) - e_2(\mathbf{a})z]\nu_0 + [e_1(\mathbf{b}+1) - e_1(\mathbf{a})z - z]\nu_1 + (1-z)\nu_2,$$

and e_1, e_2, e_3 denote the *elementary symmetric polynomials* defined by

$$e_1(\mathbf{x}) = \sum_i x_i, \quad e_2(\mathbf{x}) = \sum_{i<j} x_i x_j, \quad e_3(\mathbf{x}) = \sum_{i<j<k} x_i x_j x_k.$$

6.6.1 Uvarov Meixner Polynomials

Let $\omega \notin \{-a, 0\}$.

Special values

$$a_1 \to a, \quad a_2 \to -\omega, \quad a_3 \to -\omega+1, \quad b_1 \to -\omega-1, \quad b_2 \to -\omega.$$

Linear functional

$$L_U[r] = \sum_{x=0}^{\infty} r(x)\,(a)_x\,\frac{z^x}{x!} + Mr(\omega).$$

Pearson equation

$$\frac{\varrho_U(x+1)}{\varrho_U(x)} = \frac{z(x+a)(x-\omega)(x-\omega+1)}{(x-\omega)(x-\omega+1)(x+1)}.$$

Moments

$$\nu_n^U(z) = z^n (a)_n (1-z)^{-a-n} + M\phi_n(\omega), \quad n \in \mathbb{N}_0.$$

Stieltjes transform difference equation

$$(t-\omega)(t-\omega+1)[(t+1)S_U(t+1) - z(t+a)S_U(t)]$$

$$= \left[(1-z)t^2 + (-az+2\omega z - 2\omega - z + 1)t + \omega(z+\omega - z\omega + 2az - 1) - az\right]\nu_0^{U\varsigma}$$

$$+ [(1-z)t - 2z + 2z\omega - 2\omega - az + 1]\nu_1^U + (1-z)\nu_2^U.$$

Stieltjes transform M-dependent difference equation

$$(t+1)S_U(t+1) - z(t+a)S_U(t) = (1-z)\nu_0 + M\left[\frac{t+1}{t+1-\omega} - \frac{z(t+a)}{t-\omega}\right].$$

6.6.2 Symmetrized Generalized Meixner Polynomials

Special values

$$a_1 \to a, \quad a_2 \to -b-2m, \quad a_3 \to -2m,$$
$$b_1 \to -a-2m, \quad b_2 \to b-2m, \quad z \to -1, \quad x \to x+m.$$

Linear functional

$$L_\varsigma[r] = \varrho_\varsigma(0) \sum_{x=-m}^{m} r(x) \frac{(a+m)_x (-b-m)_x (-m)_x}{(-a-m+1)_x (b+1-m)_x} \frac{(-1)^x}{(m+1)_x},$$

where

$$\varrho_\varsigma(0) = \binom{2m}{m} \frac{(a)_m \ (b+1+m)_m}{(a+m)_m \ (b+1-2m)_m}.$$

Pearson equation

$$\frac{\varrho_\varsigma(x+1)}{\varrho_\varsigma(x)} = \frac{-(x+a+m)(x-b-m)(x-m)}{(x-a-m+1)(x+b+1+m)(x+1+m)}.$$

Moments on the basis $\{\overline{\phi}_n(x)\}_{n\geq 0}$

$$\nu_n^\varsigma = \frac{(-1)^n (a)_n (-b-2m)_n (-2m)_n}{(-a-2m+1)_n (b-2m+1)_n} \, {}_3F_2\left[\begin{matrix} a+n, -b-2m+n, -2m+n \\ -a-2m+1+n, b-2m+1+n \end{matrix}; -1\right],$$

$n \in \mathbb{N}_0$.

Stieltjes transform difference equation

$$(t-a-m+1)(t+b+1+m)(t+1+m)S_{\varsigma}(t+1)$$
$$+(t+a+m)(t-b-m)(t-m)S_{\varsigma}(t)$$
$$=\left[2t^{2}+2(1-m)t+b-a-2ab-4am+1\right]\nu_{0}^{\varsigma}+(2t-4m+3)\nu_{1}^{\varsigma}+2\nu_{2}^{\varsigma}.$$

6.6.3 Reduced-Uvarov Generalized Hahn Polynomials of Type I

Let $\{\omega,\Omega\}\in\{(-a_1,a_1+1),(-a_2,a_2+1),(-b,b),(0,0)\}$.

Special values

$$a_3\to\Omega,\quad b_1\to b,\quad b_2\to\Omega-1.$$

Linear functional

$$L_U[r]=\sum_{x=0}^{\infty}r(x)\frac{(a_1)_x(a_2)_x}{(b+1)_x}\frac{z^x}{x!}+Mr(\omega).$$

Pearson equation

$$\frac{\varrho_U(x+1)}{\varrho_U(x)}=\frac{z(x+a_1)(x+a_2)(x+\Omega)}{(x+b+1)(x+\Omega)(x+1)}.$$

Moments

$$\nu_n^U(z)=z^n\frac{(a_1)_n(a_2)_n}{(b+1)_n}\,{}_2F_1\left[\begin{matrix}a_1+n,a_2+n\\ b+1+n\end{matrix};z\right]+M\phi_n(\omega),\qquad n\in\mathbb{N}_0.$$

Stieltjes transform difference equation

$$(t+\Omega)\left[(t+b+1)(t+1)S_U(t+1)-z(t+a_1)(t+a_2)S_U(t)\right]$$
$$=\Big[(1-z)t^2+(b+\Omega-z\Omega-za_1-za_2+1)t$$
$$-(\Omega a_1+\Omega a_2+a_1a_2)z+(b+1)\Omega\Big]\nu_0^U$$
$$+\left[(1-z)t+1-z(\Omega+a_1+a_2+1)+\Omega+b\right]\nu_1^U+(1-z)\nu_2^U.$$

Stieltjes transform M-dependent difference equation

$$\begin{aligned}&(t+b+1)(t+1)\,S_U(t+1)-z(t+a_1)(t+a_2)\,S_U(t)\\&=[t+b+1-z(t+a_1+a_2)]\,\nu_0\\&+(1-z)\,\nu_1+M\left[\frac{(t+b+1)(t+1)}{t-\omega+1}-\frac{z(t+a_1)(t+a_2)}{t-\omega}\right].\end{aligned}$$

6.6.4 Christoffel Generalized Hahn Polynomials of Type I

Let $\omega \notin \left\{-a_1, -a_2, -b, 0, \frac{\nu_1}{\nu_0}\right\}$.

Special values

$$a_3 \to -\omega+1, \quad b_1 \to b, \quad b_2 \to -\omega-1.$$

Linear functional

$$L_C[r]=\sum_{x=0}^{\infty} r(x)(x-\omega)\,\frac{(a_1)_x\,(a_2)_x}{(b+1)_x}\,\frac{z^x}{x!}.$$

Pearson equation

$$\frac{\varrho_C(x+1)}{\varrho_C(x)}=\frac{z(x+a_1)(x+a_2)(x+1-\omega)}{(x+b+1)(x+1)(x-\omega)}, \qquad n\in\mathbb{N}_0.$$

Moments

$$\nu_n^C(z)=\nu_{n+1}+(n-\omega)\,\nu_n.$$

Stieltjes transform difference equation

$$\begin{aligned}&(t-\omega)(t+b+1)(t+1)\,S_C(t+1)-z(t-\omega+1)(t+a_1)(t+a_2)\,S_C(t)\\&=\left[(z-1)\,\omega t^2-\left(\omega+b\omega-z\omega-\omega^2+z\omega^2-z\omega a_1-z\omega a_2-za_1a_2\right)t\right]\nu_0\\&+\left[\omega^2(b+1)-z(\omega-1)(\omega a_1+\omega a_2+a_1a_2)\right]\nu_0-(z-1)(t-\omega)(t-\omega+1)\,\nu_1.\end{aligned}$$

6.6.5 Geronimus Generalized Hahn Polynomials of Type I

Let $\omega \notin \{-b-1, -1, 1-a_1, 1-a_2\}\cup\mathbb{N}_0$.

Special values

$$a_3 \to -\omega, \quad b_1 \to b, \quad b_2 \to -\omega.$$

Linear functional

$$L_G[r] = \sum_{x=0}^{\infty} \frac{r(x)}{x-\omega} \frac{(a_1)_x (a_2)_x}{(b+1)_x} \frac{z^x}{x!} + Mr(\omega).$$

Pearson equation

$$\frac{\varrho_G(x+1)}{\varrho_G(x)} = \frac{z(x+a_1)(x+a_2)(x-\omega)}{(x+b+1)(x+1)(x+1-\omega)}.$$

Moments

$$\nu_n^G(z) = \phi_n(\omega)\left[M - S(\omega) + \sum_{k=0}^{n-1} \frac{\nu_k}{\phi_{k+1}(\omega)}\right], \quad n \in \mathbb{N}_0.$$

Stieltjes transform difference equation

$$\begin{aligned}&(t-\omega+1)(t+b+1)(t+1)\,S_G(t+1) - z(t-\omega)(t+a_1)(t+a_2)\,S_G(t)\\&= [(1-z)t + (b - za_1 - za_2 + 1)]\,\nu_0 + (1-z)\,\nu_1\\&+ [(t+b+1)(t+1) - z(t+a_1)(t+a_2)]\,\nu_0^G.\end{aligned}$$

6.6.6 Truncated Generalized Hahn Polynomials of Type I

Special values

$$a_3 \to -N, \quad b_1 \to b, \quad b_2 \to -N-1.$$

Linear functional

$$L_T[r] = \sum_{x=0}^{N} r(x) \frac{(a_1)_x (a_2)_x}{(b+1)_x} \frac{z^x}{x!}.$$

Pearson equation

$$\frac{\varrho_T(x+1)}{\varrho_T(x)} = \frac{z(x+a_1)(x+a_2)(x-N)}{(x+b+1)(x-N)(x+1)}.$$

Moments

$$\nu_n^T(z) = \frac{(a_1)_N\,(a_2)_N}{(b+1)_N} \frac{z^N}{(N-n)!}\,{}_3F_2\left[\begin{matrix} n-N,1,\,-N-b \\ 1-N-\,a_1, 1-N-\,a_2\end{matrix};\frac{1}{z}\right], \quad n \in \mathbb{N}_0.$$

Stieltjes transform difference equation

$$(t-N)\left[(t+b+1)(t+1)\,S_U(t+1)-z(t+a_1)(t+a_2)\,S_U(t)\right]$$
$$=\Big[(1-z)t^2+(b-N+zN-za_1-za_2+1)\,t$$
$$-(-Na_1-Na_2+a_1a_2)\,z-(b+1)N\Big]\,\nu_0^U$$
$$+\left[(1-z)t+1-z\,(-N+a_1+a_2+1)-N+b\right]\nu_1^U+(1-z)\,\nu_2^U.$$

6.6.7 Symmetrized Polynomials of Type (0, 2)

Special values

$$a_1\to -b_1-2m,\quad a_2\to -b_2-2m,\quad a_3\to -2m,\quad z\to -1,\quad x\to x+m.$$

Linear functional

$$L_\varsigma[r]=\varrho_\varsigma(0)\sum_{x=-m}^{m} r(x)\,\frac{(-b_1-m)_x\,(-b_2-m)_x\,(-m)_x}{(b_1+1+m)_x\,(b_2+1+m)_x}\,\frac{(-1)^x}{(m+1)_x},$$

where

$$\varrho_\varsigma(0)=\binom{2m}{m}\frac{(b_1+1+m)_m\,(b_1+1+m)_m}{(b_1+1)_m\,(b_2+1)_m}.$$

Pearson equation

$$\frac{\varrho_\varsigma(x+1)}{\varrho_\varsigma(x)}=\frac{-(x-b_1-m)(x-b_2-m)(x-m)}{(x+b_1+1+m)(x+b_2+1+m)(x+1+m)}.$$

Moments on the basis $\{\overline{\phi}_n(x)\}_{n\geq 0}$.
For $n\in\mathbb{N}_0$

$$\nu_n^\varsigma=\frac{(-1)^n\,(-2m)_n\,(-b_1-2m)_n\,(-b_2-2m)_n}{(b_1+1)_n\,(b_2+1)_n}\;{}_3F_2\left[\begin{matrix}-2m+n,\,-b_1-2m,\,-b_2-2m+n,\\ b_1+1+n,\,b_2+1+n\end{matrix};-1\right].$$

Stieltjes transform difference equation

$$(t+b_1+1+m)(t+b_2+1+m)(t+1+m)\,S_\varsigma(t+1)$$
$$+(t-b_1-m)(t-b_2-m)(t-m)\,S_\varsigma(t)$$

$$= \left[2t^2 + 2(1-m)t + 8m^2 + 2m(2b_1 + 2b_2 + 1) + b_1 + b_2 + 2b_1 b_2 + 1\right]\nu_0^{\varsigma}$$
$$+ (2t - 4m + 3)\nu_1^{\varsigma} + 2\nu_2^{\varsigma}.$$

6.7 Polynomials of Type (4,3;N,1)

Linear functional

$$L[r] = \sum_{x=0}^{N} r(x) \frac{(a_1)_x (a_2)_x (a_3)_x (-N)_x}{(b_1+1)_x (b_2+1)_x (b_3+1)_x} \frac{1}{x!}.$$

Pearson equation

$$\frac{\varrho(x+1)}{\varrho(x)} = \frac{(x+a_1)(x+a_2)(x+a_3)(x-N)}{(x+b_1+1)(x+b_2+1)(x+b_3+1)(x+1)}.$$

Moments. For $n \in \mathbb{N}_0$

$$\nu_n = \frac{(-N)_n (a_1)_n (a_2)_n (a_3)_n}{(b_1+1)_n (b_2+1)_n (b_3+1)_n} \, {}_4F_3 \left[\begin{matrix} -N+n, a_1+n, a_2+n, a_3+n \\ b_1+1+n, b_2+1+n, b_3+1+n \end{matrix} ; 1 \right].$$

Stieltjes transform difference equation

$$(t+b_1+1)(t+b_2+1)(t+b_3+1)(t+1)S(t+1)$$
$$-(t+a_1)(t+a_2)(t+a_3)(t-N)S(t) = \sum_{k=0}^{2} \xi_k t^k,$$

where

$$\xi_2 = [e_1(\mathbf{b}+1) - e_1(\mathbf{a})]\nu_0, \quad \xi_1 = [e_2(\mathbf{b}+1) - e_2(\mathbf{a})]\nu_0$$
$$+ [e_1(\mathbf{b}+1) - e_1(\mathbf{a}) - 1]\nu_1,$$
$$\xi_0 = [e_3(\mathbf{b}+1) - e_3(\mathbf{a})]\nu_0 + [e_2(\mathbf{b}+1) - e_2(\mathbf{a}) - e_1(\mathbf{a}) - 1]\nu_1$$
$$+ [e_1(\mathbf{b}+1) - e_1(\mathbf{a}) - 2]\nu_2,$$

with $a_4 = -N$.

If we use the relations

$$e_1(\mathbf{a}, -N) = e_1(\mathbf{a}) - N, \quad e_2(\mathbf{a}, -N) = e_2(\mathbf{a}) - Ne_1(\mathbf{a}),$$
$$e_3(\mathbf{a}, -N) = e_3(\mathbf{a}) - Ne_2(\mathbf{a}),$$

we obtain expressions for the coefficients ξ_k explicitly depending on N

$$\xi_2 = [e_1(\mathbf{b}+1) - e_1(\mathbf{a}) + N]\,\nu_0,$$

$$\xi_1 = [e_2(\mathbf{b}+1) - e_2(\mathbf{a}) + Ne_1(\mathbf{a})]\,\nu_0 + [e_1(\mathbf{b}+1) - e_1(\mathbf{a}) + N - 1]\,\nu_1,$$

$$\xi_0 = [e_3(\mathbf{b}+1) - e_3(\mathbf{a}) + Ne_2(\mathbf{a})]\,\nu_0 + [e_2(\mathbf{b}+1) - e_2(\mathbf{a}) + (N-1)(e_1(\mathbf{a})+1)]\,\nu_1 + [e_1(\mathbf{b}+1) - e_1(\mathbf{a}) + N - 2]\,\nu_2.$$

6.7.1 Uvarov Hahn Polynomials

Let $\omega \notin \{-a, -b, 0, N\}$.

Special values

$$a_1 \to a, \quad a_2 \to -\omega, \quad a_3 \to -\omega + 1,$$
$$b_1 \to b, \quad b_2 \to -\omega - 1, \quad b_3 \to -\omega.$$

Linear functional

$$L_U[r] = \sum_{x=0}^{N} r(x)\,\frac{(a)_x(-N)_x}{(b+1)_x}\,\frac{1}{x!} + Mr(\omega).$$

Pearson equation

$$\frac{\varrho_U(x+1)}{\varrho_U(x)} = \frac{(x+a)(x-\omega)(x-\omega+1)(x-N)}{(x+b+1)(x-\omega)(x-\omega+1)(x+1)}.$$

Moments

$$\nu_n^U = \frac{(-N)_n(a)_n}{(b+1)_n}\,\frac{(b+1-a)_{N-n}}{(b+1+n)_{N-n}} + M\phi_n(\omega), \quad n \in \mathbb{N}_0.$$

Stieltjes transform difference equation

$$(t-\omega)(t-\omega+1)[(t+b+1)(t+1)S_U(t+1) - (t+a)(t-N)S_U(t)]$$
$$= \left[(N-a+b+1)t^2 + (N-a+b-2\omega-2N\omega+2a\omega-2b\omega+Na+1)t\right]\nu_0^{U\varsigma}$$
$$+ \left(a\omega - N\omega - \omega - b\omega + \omega^2 + N\omega^2 - a\omega^2 + b\omega^2 + Na - 2Na\omega\right)\nu_0^{U\varsigma}$$
$$+ [(N-a+b)t + 2N - 2a + b - 2N\omega + 2a\omega - 2b\omega + Na - 1]\,\nu_1^U$$
$$+ (-a-1+b+N)\,\nu_2^U.$$

Stieltjes transform M-dependent difference equation

$$(t+b+1)(t+1)\,S_U(t+1)-(t+a)(t-N)\,S_U(t)$$
$$=(b+1-a+N)\,\nu_0+M\left[\frac{(t+b+1)(t+1)}{t+1-\omega}-\frac{(t+a)(t-N)}{t-\omega}\right].$$

6.7.2 Symmetrized Generalized Hahn Polynomials of Type I

Special values

$$a_3\to -b-N,\quad b_1\to -a_1-N,\quad b_2\to -a_2-N,$$
$$b_3\to b,\quad N\to 2m,\quad x\to x+m.$$

Linear functional

$$L_\varsigma[r]=\varrho_\varsigma(0)\sum_{x=-m}^{m} r(x)\,\frac{(a_1+m)_x\,(a_2+m)_x\,(-b-m)_x\,(-m)_x}{(-a_1-m+1)_x\,(-a_2-m+1)_x\,(b+1+m)_x}\,\frac{1}{(m+1)_x},$$

where

$$\varrho_\varsigma(0)=\binom{2m}{m}\frac{(a_1)_m\,(a_2)_m\,(b+1+m)_m}{(a_1+m)_m\,(a_2+m)_m\,(b+1)_m}.$$

Pearson equation

$$\frac{\varrho_\varsigma(x+1)}{\varrho_\varsigma(x)}=\frac{(x+a_1+m)(x+a_2+m)(x-b-m)(x-m)}{(x+1-a_1-m)(x+1-a_2-m)(x+1+b+m)(x+1+m)}.$$

Moments on the basis $\{\overline{\phi}_n(x)\}_{n\geq 0}$.
For $n\in\mathbb{N}_0$

$$\nu_n^\varsigma=\frac{(a_1)_n\,(a_2)_n\,(-b-2m)_n\,(-2m)_n}{(-a_1-2m+1)_n\,(-a_2-2m+1)_n\,(b+1)_n}$$
$$\times\ {}_4F_3\left[\begin{matrix}-2m+n,a_1+n,a_2+n,-b-2m+n\\ -a_1-2m+1+n,-a_2-2m+1+n,b+1+n\end{matrix};1\right].$$

Stieltjes transform difference equation

$$(t+1-a_1-m)(t+1-a_2-m)(t+1+b+m)(t+1+m)\,S_\varsigma(t+1)$$
$$-(t+a_1+m)(t+a_2+m)(t-b-m)(t-m)\,S_\varsigma(t)$$
$$=[2(b-a_1-a_2+1)(t+1-m)t+t(t+1)]\,\nu_0^\varsigma$$

$$+\left[b-m-2bm-m^2+1-(b+1)(a_1+a_2)+(2b+4m+1)\,a_1a_2\right]\nu_0^{\varsigma}$$
$$+\left[(2t+3-4m)(b-a_1-a_2)+2(t+1-m)\right]\nu_1^{\varsigma}+(2b+1-2a_1-2a_2)\,\nu_2^{\varsigma}.$$

6.7.3 Reduced-Uvarov Generalized Hahn Polynomials of Type II

Let $\{\omega, \Omega\} \in \{(-a_1, a_1+1), (-a_2, a_2+1), (-b_1, b_1), (-b_2, b_2), (0, 0), (N, -N+1)\}$.

Special values

$$a_3 \to \Omega, \quad b_3 \to \Omega - 1.$$

Linear functional

$$L_U[r] = \sum_{x=0}^{N} r(x) \frac{(-N)_x (a_1)_x (a_2)_x}{(b_1+1)_x (b_2+1)_x} \frac{1}{x!} + Mr(\omega).$$

Pearson equation

$$\frac{\varrho_U(x+1)}{\varrho_U(x)} = \frac{(x+a_1)(x+a_2)(x+\Omega)(x-N)}{(x+b_1+1)(x+b_2+1)(x+\Omega)(x+1)}.$$

Moments

$$\nu_n^U = \frac{(-N)_n (a_1)_n (a_2)_n}{(b_1+1)_n (b_2+1)_n}\, {}_3F_2\left[\begin{matrix} -N+n, a_1+n, a_2+n \\ b_1+1+n, b_2+1+n \end{matrix}; 1\right] + M\phi_n(\omega), \qquad n \in \mathbb{N}_0.$$

Stieltjes transform difference equation

$$(t+\Omega)\left[(t+b_1+1)(t+b_2+1)(t+1)\,S_U(t+1)\right.$$
$$\left.-(t+a_1)(t+a_2)(t-N)\,S_U(t)\right]$$
$$=\left[(N-a_1-a_2+b_1+b_2+2)(t+\Omega)\,t+(b_1+1)(b_2+1)t\right.$$
$$\left.+N(a_1+a_2)\,t-a_1a_2t\right]\nu_0^U$$
$$+\left[(b_1+1)(b_2+1)\Omega+a_1a_2(N-\Omega)+(a_1+a_2)\,\Omega N\right]\nu_0^U$$
$$+\left[(N-a_1-a_2+b_1+b_2+1)(t+\Omega)+N+b_1+b_2+b_1b_2\right.$$
$$\left.+(N-1)(a_1+a_2)-a_1a_2\right]\nu_1^U$$
$$+(N+b_1+b_2-a_1-a_2)\,\nu_2^U.$$

Stieltjes transform M-dependent difference equation

$$(t+b_1+1)(t+b_2+1)(t+1)\,S_U(t+1)-(t+a_1)(t+a_2)(t-N)\,S_U(t)$$
$$=\left[(N+2+b_1+b_2-a_1-a_2)\,t+b_1+b_2+b_1b_2+(N-1)(a_1+a_2)-a_1a_2+1\right]\nu_0$$
$$+(N+b_1+b_2-a_1-a_2+1)\,\nu_1$$
$$+M\left[\frac{(t+b_1+1)(t+b_2+1)(t+1)}{t-\omega+1}-\frac{(t+a_1)(t+a_2)(t-N)}{t-\omega}\right].$$

6.7.4 Christoffel Generalized Hahn Polynomials of Type II

Let $\omega\notin\left\{-a_1,-a_2,-b_1,-b_2,0,N,\frac{\nu_1}{\nu_0}\right\}$.

Special values

$$a_3\to-\omega+1,\quad b_3\to-\omega-1.$$

Linear functional

$$L_C[r]=\sum_{x=0}^{\infty}r(x)(x-\omega)\frac{(a_1)_x(a_2)_x(-N)_x}{(b_1+1)_x(b_2+1)_x}\frac{1}{x!}.$$

Pearson equation

$$\frac{\varrho_C(x+1)}{\varrho_C(x)}=\frac{(x+a_1)(x+a_2)(x-N)(x+1-\omega)}{(x+b_1+1)(x+b_2+1)(x+1)(x-\omega)}.$$

Moments

$$\nu_n^C(z)=\nu_{n+1}+(n-\omega)\,\nu_n,\quad n\in\mathbb{N}_0.$$

Stieltjes transform difference equation

$$(t-\omega)(t+b_1+1)(t+b_2+1)(t+1)\,S_C(t+1)$$
$$-(t-\omega+1)(t+a_1)(t+a_2)(t-N)\,S_C(t)$$
$$=-\left[\omega(N+b_1+b_2+1)-(\omega-1)(a_1+a_2)\right]t^2\nu_0+\omega(\omega-1)$$
$$(N+b_1+b_2+1)\,t\nu_0$$
$$+\left[(-N+\omega)\,a_1a_2+\left(3\omega-N\omega-\omega^2-1\right)(a_1+a_2)+\omega(\omega-b_1b_2)\right]t\nu_0$$
$$+\left[\omega(\omega-1)(N-1)(a_1+a_2)+(\omega-1)(N-\omega)\,a_1a_2+\omega^2(b_2+1)(b_1+1)\right]\nu_0$$
$$+(N-a_1-a_2+b_1+b_2+1)(t-\omega)(t-\omega+1)\,\nu_1.$$

6.7.5 Geronimus Generalized Hahn Polynomials of Type II

Let $\omega \notin \{-b_1 - 1, -b_2 - 1, -1, 1 - a_1, 1 - a_2, 1 + N\} \cup \mathbb{N}_0$.

Special values

$$a_3 \to -\omega, \quad b_3 \to -\omega.$$

Linear functional

$$L_G[r] = \sum_{x=0}^{\infty} \frac{r(x)}{x - \omega} \frac{(a_1)_x (a_2)_x (-N)_x}{(b_1 + 1)_x (b_2 + 1)_x} \frac{1}{x!} + Mr(\omega).$$

Pearson equation

$$\frac{\varrho_G(x+1)}{\varrho_G(x)} = \frac{(x + a_1)(x + a_2)(x - N)(x - \omega)}{(x + b_1 + 1)(x + b_2 + 1)(x + 1)(x + 1 - \omega)}.$$

Moments

$$\nu_n^G(z) = \phi_n(\omega) \left[M - S(\omega) + \sum_{k=0}^{n-1} \frac{\nu_k}{\phi_{k+1}(\omega)} \right], \quad n \in \mathbb{N}_0.$$

Stieltjes transform difference equation

$$\begin{aligned}
&(t - \omega + 1)(t + b_1 + 1)(t + b_2 + 1)(t + 1) S_G(t + 1) \\
&- (t - \omega)(t + a_1)(t + a_2)(t - N) S_G(t) \\
&= [(N + 2 + b_1 + b_2 - a_1 - a_2) t + (b_1 + 1)(b_2 + 1) + (N - 1)(a_1 + a_2) - a_1 a_2] \nu_0 \\
&+ (N + b_1 + b_2 - a_1 - a_2 + 1) \nu_1 \\
&+ \left[(N - a_1 - a_2 + b_1 + b_2 + 3) t^2 + (2b_1 + 2b_2 + Na_1 + Na_2 - a_1 a_2 + b_1 b_2 + 3) t \right] \nu_0^G \\
&+ (b_1 + b_2 + b_1 b_2 + Na_1 a_2 + 1) \nu_0^G.
\end{aligned}$$

6.7.6 Symmetrized Polynomials of Type (1, 2)

Special values

$$\begin{aligned}
&a_1 \to a, \quad a_2 \to -b_1 - N, \quad a_3 \to -b_2 - N, \quad b_3 \to -a - N, \\
&N \to 2m, \quad x \to x + m.
\end{aligned}$$

Linear functional

$$L_{\varsigma}[r]=\varrho_{\varsigma}(0)\sum_{x=-m}^{m}r(x)\,\frac{(a+m)_x(-b_1-m)_x(-b_2-m)_x(-m)_x}{(b_1+1+m)_x(b_2+1+m)_x(-a-m+1)_x}\frac{1}{(m+1)_x},$$

where

$$\varrho_{\varsigma}(0)=\binom{2m}{m}\frac{(a)_m\;(b_1+1+m)_m\,(b_1+1+m)_m}{(a+m)_m\;(b_1+1)_m\,(b_2+1)_m}.$$

Pearson equation

$$\frac{\varrho_{\varsigma}(x+1)}{\varrho_{\varsigma}(x)}=\frac{(x+a+m)(x-b_1-m)(x-b_2-m)(x-m)}{(x+b_1+1+m)(x+b_2+1+m)(x+1-a-m)(x+1+m)}.$$

Moments on the basis $\{\overline{\phi}_n(x)\}_{n\geq 0}$

$$\nu_n^{\varsigma}=\frac{(-2m)_n(a)_n(-b_1-2m)_n(-b_2-2m)_n}{(1-a-2m)_n(b_1+1)_n(b_2+1)_n}$$
$$\times\;{}_4F_3\left[\begin{matrix}-2m+n,a+n,-b_1-2m,-b_2-2m+n,\\1-a-2m+n,b_1+1+n,b_2+1+n\end{matrix};1\right],\quad n\in\mathbb{N}_0.$$

Stieltjes transform difference equation

$$\begin{aligned}&(t+1-a-m)(t+b_1+1+m)(t+b_2+1+m)(t+1+m)S_{\varsigma}(t+1)\\&\quad-(t+a+m)(t-b_1-m)(t-b_2-m)(t-m)S_{\varsigma}(t)\\&=\left[(4m+3+2b_1+2b_2-2a)t^2+2(m-1)(a-b_1-b_2)t\right]\nu_0^{\varsigma}\\&\quad+\left[\left(-4m^2+2m+3\right)t-(18a+4ab_1+4ab_2+1)m\right.\\&\quad\left.-(b_1+b_2+1)(a-1)-b_1b_2(2a-1)\right]\nu_0^{\varsigma}\\&\quad+\left[(2t-4m+3)(b_1+b_2-a)+2(t+2mt+1)+4m(1-2m)\right]\nu_1^{\varsigma}\\&\quad+(4m+1+2b_1+2b_2-2a)\nu_2^{\varsigma}.\end{aligned}$$

7 Conclusions

We have considered all solutions of the Pearson equation

$$\frac{\varrho(x+1)}{\varrho(x)}=\frac{\eta(x)}{\sigma(x+1)},$$

with deg $(\sigma) \leq 4$ and deg $(\sigma - \eta) \leq 3$. We deduce 15 canonical cases, from which 42 subcases can be obtained and we relate those to rational spectral transformations (Uvarov, Christoffel, and Geronimus), truncations, and symmetrizations.

For each family, we have listed a representation of the moments and the first order linear difference equations satisfied by their Stieltjes transform.

Extensions to the class $s = 3$ are possible, but the complexity of the formulas will demand new forms of notation in order to be able to type the results in a comprehensive way.

Acknowledgments This work of the first author was done while visiting the Johannes Kepler Universität Linz and supported by the strategic program "Innovatives OÖ– 2010 plus" from the Upper Austrian Government. The second author acknowledges financial support of Dirección general de Investigación, Ministerio de Economía, Industria y Competitividad of Spain, grant MTM2015-65888. C4- 2-P.

Finally, we also wish to express our gratitude to the anonymous referee, who provided us with invaluable suggestions and comments that greatly improved our first draft of the paper.

References

1. Abdelkarim, F., Maroni, P.: The D_ω-classical orthogonal polynomials. Results Math. **32**(1–2), 1–28 (1997)
2. Álvarez Nodarse, R., Marcellán, F.: Difference equation for modifications of Meixner polynomials. J. Math. Anal. Appl. **194**(1), 250–258 (1995)
3. Álvarez Nodarse, R., Marcellán, F.: The modification of classical Hahn polynomials of a discrete variable. Integral Transform. Spec. Funct. **3**(4), 243–262 (1995)
4. Álvarez Nodarse, R., Petronilho, J.: On the Krall-type discrete polynomials. J. Math. Anal. Appl. **295**(1), 55–69 (2004)
5. Álvarez Nodarse, R., García, A.G., Marcellán, F.: On the properties for modifications of classical orthogonal polynomials of discrete variables. In: Proceedings of the International Conference on Orthogonality, Moment Problems and Continued Fractions (Delft, 1994), vol. 65, pp. 3–18 (1995)
6. Álvarez Nodarse, R., Arvesú, J., Marcellán, F.: Modifications of quasi-definite linear functionals via addition of delta and derivatives of delta Dirac functions. Indag. Math. (N.S.) **15**(1), 1–20 (2004)
7. Area, I., Godoy, E., Ronveaux, A., Zarzo, A.: Classical symmetric orthogonal polynomials of a discrete variable. Integral Transforms Spec. Funct. **15**(1), 1–12 (2004)
8. Area, I., Dimitrov, D.K., Godoy, E., Paschoa, V.: Bounds for the zeros of symmetric Kravchuk polynomials. Numer. Algorithms **69**(3), 611–624 (2015)
9. Belmehdi, S.: On semi-classical linear functionals of class $s = 1$. Classification and integral representations. Indag. Math. (N.S.) **3**(3), 253–275 (1992)
10. Boelen, L., Filipuk, G., Van Assche, W.: Recurrence coefficients of generalized Meixner polynomials and Painlevé equations. J. Phys. A **44**(3), 035202, 19 (2011)
11. Boelen, L., Filipuk, G., Smet, C., Van Assche, W., Zhang, L.: The generalized Krawtchouk polynomials and the fifth Painlevé equation. J. Differ. Equ. Appl. **19**(9), 1437–1451 (2013)
12. Bueno, M.I., Dopico, F.M.: A more accurate algorithm for computing the Christoffel transformation. J. Comput. Appl. Math. **205**(1), 567–582 (2007)
13. Bueno, M.I., Marcellán, F.: Darboux transformation and perturbation of linear functionals. Linear Algebra Appl. **384**, 215–242 (2004)

14. Christoffel, E.B.: Über die Gaußische Quadratur und eine Verallgemeinerung derselben. J. Reine Angew. Math. **55**, 61–82 (1858)
15. Derevyagin, M., Marcellán, F.: A note on the Geronimus transformation and Sobolev orthogonal polynomials. Numer. Algorithms **67**(2), 271–287 (2014)
16. Derevyagin, M., García-Ardila, J.C., Marcellán, F.: Multiple Geronimus transformations. Linear Algebra Appl. **454**, 158–183 (2014)
17. Dominici, D., Marcellán, F.: Discrete semiclassical orthogonal polynomials of class one. Pacific J. Math. **268**(2), 389–411 (2014)
18. Dominici, D.: Polynomial sequences associated with the moments of hypergeometric weights. SIGMA Symmetry Integrability Geom. Methods Appl. **12**, 18, Paper No. 044 (2016)
19. Dominici, D.: Laguerre-Freud equations for generalized Hahn polynomials of type I. J. Differ. Equ. Appl. **24**(6), 916–940 (2018)
20. Elaydi, S.: An Introduction to Difference Equations. Undergraduate Texts in Mathematics, 3rd edn. Springer, New York (2005)
21. Filipuk, G., Van Assche, W.: Recurrence coefficients of a new generalization of the Meixner polynomials. SIGMA Symmetry Integrability Geom. Methods Appl. **7**, 11, Paper 068 (2011)
22. Filipuk, G., Van Assche, W.: Discrete orthogonal polynomials with hypergeometric weights and Painlevé VI. SIGMA Symmetry Integrability Geom. Methods Appl. **14**, 19, Paper No. 088 (2018)
23. Galant, D.: Algebraic methods for modified orthogonal polynomials. Math. Comput. **59**(200), 541–546 (1992)
24. García, A.G., Marcellán, F., Salto, L.: A distributional study of discrete classical orthogonal polynomials. In: Proceedings of the Fourth International Symposium on Orthogonal Polynomials and their Applications (Evian-Les-Bains, 1992), vol. 57, pp. 147–162 (1995)
25. Gautschi, W.: On generating orthogonal polynomials. SIAM J. Sci. Statist. Comput. **3**(3), 289–317 (1982)
26. Geronimus, J.: On polynomials orthogonal with regard to a given sequence of numbers. Commun. Inst. Sci. Math. Méc. Univ. Kharkoff [Zapiski Inst. Mat. Mech.] (4) **17**, 3–18 (1940)
27. Geronimus, J.: Sur quelques propriétés des polynomes orthogonaux généralisés. Rec. Math. [Mat. Sbornik] N.S. **9**(51), 121–135 (1941)
28. Godoy, E., Marcellán, F., Salto, L., Zarzo, A.: Perturbations of discrete semiclassical functionals by Dirac masses. Integral Transform. Spec. Funct. **5**(1–2), 19–46 (1997)
29. Hounkonnou, M.N., Hounga, C., Ronveaux, A.: Discrete semi-classical orthogonal polynomials: generalized Charlier. J. Comput. Appl. Math. **114**(2), 361–366 (2000)
30. Ismail, M.E.H.: Classical and Quantum Orthogonal Polynomials in One Variable. Encyclopedia of Mathematics and its Applications, vol. 98. Cambridge University Press, Cambridge (2005)
31. Koekoek, R., Lesky, P.A., Swarttouw, R.F.: Hypergeometric Orthogonal Polynomials and their q-analogues. Springer Monographs in Mathematics. Springer, Berlin (2010)
32. Lee, J.H., Kwon, K.H.: Division problem of moment functionals. Rocky Mountain J. Math. **32**(2), 739–758 (2002). Conference on Special Functions (Tempe, AZ, 2000)
33. Marcellán, F., Maroni, P.: Sur l'adjonction d'une masse de Dirac à une forme régulière et semi-classique. Ann. Mat. Pura Appl. (4) **162**, 1–22 (1992)
34. Marcellán, F., Salto, L.: Discrete semi-classical orthogonal polynomials. J. Differ. Equations Appl. **4**(5), 463–496 (1998)
35. Marcellán, F., Sghaier, M., Zaatra, M.: On semiclassical linear functionals of class $s = 2$: classification and integral representations. J. Differ. Equ. Appl. **18**(6), 973–1000 (2012)
36. Maroni, P.: Une caractérisation des polynômes orthogonaux semi-classiques. C. R. Acad. Sci. Paris Sér. I Math. **301**(6), 269–272 (1985)
37. Maroni, P.: Prolégomènes à l'étude des polynômes orthogonaux semi-classiques. Ann. Mat. Pura Appl. (4) **149**, 165–184 (1987)
38. Maroni, P.: Sur la suite de polynômes orthogonaux associée à la forme $u = \delta_c + \lambda(x-c)^{-1}L$. Period. Math. Hungar. **21**(3), 223–248 (1990)

39. Maroni, P.: Une théorie algébrique des polynômes orthogonaux. Application aux polynômes orthogonaux semi-classiques. In: Orthogonal Polynomials and their Applications (Erice, 1990). IMACS Ann. Comput. Appl. Math., vol. 9, pp. 95–130. Baltzer, Basel (1991)
40. Masjed-Jamei, M., Area, I.: A basic class of symmetric orthogonal polynomials of a discrete variable. J. Math. Anal. Appl. **399**(1), 291–305 (2013)
41. Masjed-Jamei, M., Area, I.: A symmetric generalization of Sturm-Liouville problems in discrete spaces. J. Difference Equ. Appl. **19**(9), 1544–1562 (2013)
42. Nikiforov, A.F., Suslov, S.K., Uvarov, V.B.: Classical Orthogonal Polynomials of a Discrete Variable. Springer Series in Computational Physics. Springer, Berlin (1991)
43. Oldham, K., Myland, J., Spanier, J.: An Atlas of Functions, 2nd edn. Springer, New York (2009)
44. Olver, F.W.J., Lozier, D.W., Boisvert, R.F., Clark, C.W. (eds.): NIST Handbook of Mathematical Functions. U.S. Department of Commerce National Institute of Standards and Technology, Washington (2010)
45. Peherstorfer, F.: Finite perturbations of orthogonal polynomials. J. Comput. Appl. Math. **44**(3), 275–302 (1992)
46. Ronveaux, A., Salto, L.: Discrete orthogonal polynomials—polynomial modification of a classical functional. J. Differ. Equations Appl. **7**(3), 323–344 (2001)
47. Sghaier, M., Zaatra, M.: A family of discrete semi-classical orthogonal polynomials of class one. Period. Math. Hungar. **76**(1), 68–87 (2018)
48. Uvarov, V.B.: The connection between systems of polynomials that are orthogonal with respect to different distribution functions. Ž. Vyčisl. Mat. i Mat. Fiz. **9**, 1253–1262 (1969)
49. Van Assche, W.: Orthogonal Polynomials and Painlevé Equations. Australian Mathematical Society Lecture Series, vol. 27. Cambridge University Press, Cambridge (2018)
50. Zhedanov, A.: Rational spectral transformations and orthogonal polynomials. J. Comput. Appl. Math. **85**(1), 67–86 (1997)

Stable Equilibria for the Roots of the Symmetric Continuous Hahn and Wilson Polynomials

Jan Felipe van Diejen

Abstract We show that the gradient flows associated with a recently found family of Morse functions converge exponentially to the roots of the symmetric continuous Hahn polynomials. By symmetry reduction the rate of the exponential convergence can be improved, which is clarified by comparing with corresponding gradient flows for the roots of the Wilson polynomials.

Keywords Hypergeometric polynomials · Roots of orthogonal polynomials · Electrostatic interpretation · Gradient flow · Stability

1 Introduction

For $\alpha, \beta > -1$, the Jacobi polynomials (cf. e.g. [19, Chapter 9.8])

$$P_n^{(\alpha,\beta)}(x) := \frac{(\alpha+1)_n}{n!} \, {}_2F_1\left[\begin{matrix} -n, n+\alpha+\beta \\ \alpha+1 \end{matrix}; \frac{1-x}{2}\right] \tag{1a}$$

satisfy the orthogonality relation

$$\int_{-1}^{1} (1-x)^\alpha (1+x)^\beta P_n^{(\alpha,\beta)}(x) P_m^{(\alpha,\beta)}(x) \tag{1b}$$

$$= \begin{cases} \frac{2^{\alpha+\beta+1}\Gamma(n+\alpha+1)\Gamma(n+\beta+1)}{(2n+\alpha+\beta+1)\Gamma(n+\alpha+\beta+1)n!} & \text{if } m = n, \\ 0 & \text{if } m \neq n, \end{cases}$$

J. F. van Diejen (✉)
Instituto de Matemática y Física, Universidad de Talca, Talca, Chile
e-mail: diejen@inst-mat.utalca.cl

F. Marcellán, E. J. Huertas (eds.), *Orthogonal Polynomials: Current Trends and Applications*, SEMA SIMAI Springer Series 22,
https://doi.org/10.1007/978-3-030-56190-1_6

for $n, m = 0, 1, 2, \ldots$. Here $\Gamma(\cdot)$ refers to the gamma function, and we have adopted standard conventions for the Pochhammer symbol

$$(a)_n := \prod_{0\leq k<n} (a+k), \quad (a_1, a_2, \ldots, a_r)_n := (a_1)_n (a_2)_n \cdots (a_r)_n,$$

and for the hypergeometric series

$${}_{r+1}F_r\left[\begin{matrix} a_1, \ldots, a_{r+1} \\ b_1, \ldots, b_r \end{matrix} ; z\right] := \sum_{k\geq 0} \frac{(a_1, \ldots, a_{r+1})_k}{(1, b_1, \ldots, b_r)_k} z^k .$$

As an nth degree orthogonal polynomial, $P_n^{(\alpha,\beta)}(x)$ has n simple roots inside the interval of orthogonality:

$$-1 < x_1^{(n)} < x_2^{(n)} < \cdots < x_n^{(n)} < 1.$$

In a landmark paper [30], Stieltjes demonstrated that these roots $x_1^{(n)}, \ldots, x_n^{(n)}$ provide the coordinates of the minimum of the potential

$$V^{(\alpha,\beta)}(x_1, \ldots, x_n) = - \sum_{1\leq j<k\leq n} \log |x_j - x_k| - \sum_{1\leq j\leq n} \left(\frac{\alpha+1}{2} \log |x_j - 1| + \frac{\beta+1}{2} \log |x_j + 1| \right) \tag{2}$$

in the domain $\{(x_1, \ldots, x_n) \in \mathbb{R}^n \mid -1 < x_1 < x_2 < \cdots < x_n < 1\}$. Since (by Gauss' law) the strength of an electric field $E = -\nabla V$ caused by an infinite, straight, uniformly charged wire is proportional to the charge density and inversely proportional to the distance, the potential in Eq. (2) models the total energy of n movable parallel wires with unit charge densities, aligned perpendicular to the real axis between two additional wires that are fixed at $x = 1$ and $x = -1$ with positive charge densities $\frac{1}{2}(\alpha + 1)$ and $\frac{1}{2}(\beta + 1)$, respectively. Following Stieltjes' footsteps, electrostatic models of this type turn out to provide a very fruitful tool for the study of the roots of classical orthogonal polynomials and their relatives, cf. e.g. [2, 7, 9–16, 21, 28, 31] and references therein.

As a member of the classical (hypergeometric) orthogonal families, the Jacobi polynomials satisfy a second-order differential equation [19, Chapter 9.8]:

$$A(x) \frac{\mathrm{d}^2}{\mathrm{d}x^2} P_n^{(\alpha,\beta)}(x) + B(x) \frac{\mathrm{d}}{\mathrm{d}x} P_n^{(\alpha,\beta)}(x) = \lambda_n P_n^{(\alpha,\beta)}(x), \tag{3}$$

where $A(x) = x^2 - 1$, $B(x) = (\alpha + 1)(x + 1) + (\beta + 1)(x - 1)$, and $\lambda_n = n(n + \alpha + \beta + 1)$. In fact, such Sturm-Liouville type differential equations play a fundamental role in the proof of Stieltjes' electrostatic interpretation for the roots.

Moreover, in recent work of Steinerberger [29] it was pointed out that the differential equations at issue also ensure that the roots can be recovered via a dynamical system of the form:

$$\begin{aligned}\frac{\mathrm{d}x_j}{\mathrm{d}t} &= -\, B(x_j) - A(x_j) \sum_{\substack{1\leq k\leq n\\ k\neq j}} \frac{2}{x_j - x_k} \\ &= 2A(x_j)\partial_{x_j} V^{(\alpha,\beta)}(x_1, \ldots, x_n)\end{aligned} \tag{4}$$

($j = 1, \ldots, n$). More specifically, as a special instance of [29, Theorem 2], it follows that for any initial condition $-1 < x_1(0) < \cdots < x_n(0) < 1$ the solution $(x_1(t), \ldots, x_n(t))$, $t \geq 0$ of the system in Eq. (4) converges exponentially to the roots of the Jacobi polynomial $P_n^{(\alpha,\beta)}(x)$:

$$|x_j(t) - x_j^{(n)}| \leq K \exp(-\kappa_n t) \qquad (j = 1, \ldots, n), \tag{5}$$

where $\kappa_n = \lambda_n - \lambda_{n-1} = 2n + \alpha + \beta$ and K denotes a positive constant whose value depends on the parameters and on the initial condition.

Hypergeometric orthogonal polynomials beyond the Jacobi family generally do not fit within Stieltjes' electrostatic framework, for instance because these often satisfy second-order difference equations rather than differential equations. Still, for key examples of such families like the (symmetric) continuous Hahn and Wilson polynomials, it was recently observed that the roots can nevertheless be retrieved from the global minimum of a smooth Morse function, which was found by drawing from ideas developed in the context of the Bethe Ansatz method for the diagonalization of quantum integrable particle models [37]. In the present work it will be shown that the gradient flows associated with the pertinent Morse functions converge again exponentially to the roots in question. The underlying gradient system turns out to conserve parity symmetry (i.e., it preserves configurations that are invariant with respect to coordinate reflection in the origin). Upon performing the corresponding symmetry reduction in the case of the gradient system for the symmetric continuous Hahn polynomial, the rate of the convergence is further improved, as becomes evident by comparing with the gradient system for the roots of the Wilson polynomial.

To avoid confusion, it is imperative to recall at this point that the roots of the (basic) hypergeometric orthogonal polynomials belonging to the Askey scheme can also be characterized alternatively in terms of the equilibria of Calogero-Moser type integrable particle models [3–5, 22–25, 35]. Specifically, the particle systems relevant for the roots of the continuous Hahn and Wilson families were introduced in Refs. [32, 33], and further analyzed at the level of quantum mechanics in [34] and at the level of classical mechanics in [8, 26, 27]. Here, however, we will not pursue this connection and rather try to stick closer to the original electrostatic interpretation of Stieltjes, which—for the polynomials of the symmetric continuous Hahn, Wilson and Askey-Wilson families—turns out to be governed by the Morse functions found in Ref. [37]. A corresponding analysis of the gradient system for the roots of the Askey-Wilson polynomial has recently been carried out in Ref. [36]. A different approach to study the (extremal) roots of the continuous Hahn polynomials and other ${}_3F_2$ reductions of the Wilson polynomials—based on contiguous relations derived via Zeilberger's algorithm—was presented in Ref. [17].

In brief, the plan of the paper is as follows. The exponential convergence of the gradient flows for the roots of the symmetric continuous Hahn polynomials and the Wilson polynomials are established in Sects. 2 and 3, respectively. In Sect. 4 we conclude by pointing out how the convergence rate in the case of the symmetric continuous Hahn polynomials can be improved by symmetry reduction.

2 Symmetric Continuous Hahn Polynomials

2.1 Preliminaries

The monic continuous Hahn polynomial $\mathrm{p}_n(x; a, b)$ [1] of degree n in x is, for generic parameters, given by the terminating hypergeometric series [19, Chapter 9.4]

$$\mathrm{p}_n(x; a, b) = \frac{i^n (a+\overline{a}, a+\overline{b})_n}{(n+a+\overline{a}+b+\overline{b}-1)_n} \times {}_3F_2\left[\begin{matrix} -n, n+a+\overline{a}+b+\overline{b}-1, a+ix \\ a+\overline{a}, a+\overline{b} \end{matrix}; 1\right]. \tag{6}$$

For $\mathrm{Re}(a), \mathrm{Re}(b) > 0$, these polynomials satisfy the orthogonality relations [19, Chapter 9.4]

$$\frac{1}{2\pi}\int_{-\infty}^{\infty} \Delta(x; a, b)\mathrm{p}_n(x; a, b)\mathrm{p}_m(x; a, b)\mathrm{d}x = \tag{7a}$$
$$\begin{cases} \frac{n!\Gamma(a+\overline{a}+b+\overline{b}+n-1)\Gamma(a+\overline{a}+n)\Gamma(a+\overline{b}+n)\Gamma(\overline{a}+b+n)\Gamma(b+\overline{b}+n)}{\Gamma(a+\overline{a}+b+\overline{b}+2n-1)\Gamma(a+\overline{a}+b+\overline{b}+2n)} & \text{if } m = n, \\ 0 & \text{if } m \neq n, \end{cases}$$

where

$$\Delta(x; a, b) = |\Gamma(a+ix)\Gamma(b+ix)|^2 . \tag{7b}$$

Upon assuming that possible non-real parameters a and b arise as a complex conjugate pair, the weight function $\Delta(x; a, b)$ is even. The corresponding orthogonal polynomials are referred to as *symmetric* continuous Hahn polynomials.

2.2 Morse Function

For parameter values belonging to the above orthogonality regime with possible non-real parameters a and b arising as a complex conjugate pair, the roots

$$x_1^{(n)} < x_2^{(n)} < \cdots < x_n^{(n)}$$

of the symmetric continuous Hahn polynomial $\mathrm{p}_n(x; a, b)$ turn out to minimize the Morse function [37, Remarks 5.4, 5.5]

$$V(x_1, \ldots, x_n; a, b) = \sum_{1 \leq j < k \leq n} \int_0^{x_j - x_k} \arctan(\theta) \mathrm{d}\theta \tag{8}$$
$$+ \sum_{1 \leq j \leq n} \left(\int_0^{x_j} \arctan\left(\frac{\theta}{a}\right) + \arctan\left(\frac{\theta}{b}\right) \mathrm{d}\theta + \frac{1}{2}\pi(n+1-2j)x_j \right).$$

To keep our presentation self-contained, let us briefly extract the main ingredients from the proof in [37] sustaining the claim that the coordinates of a critical point of the above Morse function are necessarily given by the roots of the symmetric continuous Hahn polynomial. To this end we first observe that at such a critical point:

$$\arctan\left(\frac{x_j}{a}\right) + \arctan\left(\frac{x_j}{b}\right) + \sum_{\substack{1 \leq k \leq n \\ k \neq j}} \arctan(x_j - x_k) = \frac{1}{2}\pi(2j - n - 1)$$

($j = 1, \ldots, n$), which implies in particular that $x_1 < x_2 < \cdots < x_n$ in this situation. Moreover, since

$$\exp\big(2i \arctan(x)\big) = \frac{i - x}{i + x},$$

it is also clear that then

$$\frac{(ia + x_j)(ib + x_j)}{(ia - x_j)(ib - x_j)} \prod_{\substack{1 \leq k \leq n \\ k \neq j}} \frac{i + x_j - x_k}{i - x_j + x_k} = (-1)^{n+1} \tag{9}$$

for $j = 1, \ldots, n$. The upshot is that the associated polynomial

$$\mathrm{p}_n(x) = (x - x_1)(x - x_2) \cdots (x - x_n)$$

satisfies the following second-order difference equation for the symmetric continuous Hahn polynomials [19, Chapter 9.4]:

$$A(x;a,b)\Big(\mathrm{p}_n(x+i)-\mathrm{p}_n(x)\Big)+ \qquad (10)$$
$$A(-x;a,b)\Big(\mathrm{p}_n(x-i)-\mathrm{p}_n(x)\Big)=\lambda_n\mathrm{p}_n(x),$$

where $A(x;a,b) = (x+ia)(x+ib)$ and $\lambda_n = -n(n+2a+2b-1)$. Indeed, the fact that the basis of symmetric continuous Hahn polynomials $\mathrm{p}_n(x;a,b)$, $n = 0,1,2,\ldots$ satisfies the difference equations in question confirms that the expression at the LHS of Eq. (10) is actually a polynomial of degree n in x with a leading coefficient given by λ_n. It thus suffices to establish the equality of both sides of the difference equation at the nodes $x_1,\ldots,x_n$:

$$A(x_j;a,b)\mathrm{p}_n(x_j+i)+A(-x_j;a,b)\mathrm{p}_n(x_j-i)=0 \quad (j=1,\ldots,n),$$

which amounts in turn to the relations in Eq. (9). Hence $\mathrm{p}_n(x)=\mathrm{p}_n(x;a,b)$ (by the nondegeneracy of the eigenvalues λ_n at the RHS of Eq. (10)) and thus $x_j = x_j^{(n)}$ $(j=1,\ldots,n)$.

It is manifest from the above argument that the roots of the symmetric continuous Hahn polynomial $p_n(x;a,b)$ satisfy the identities in (9), which is in fact an observation going back to [22].

2.3 Gradient Flow

The corresponding gradient flow

$$\frac{\mathrm{d}x_j}{\mathrm{d}t}+\partial_{x_j}V(x_1,\ldots,x_n;a,b)=0 \qquad (11a)$$

becomes of the form

$$\frac{\mathrm{d}x_j}{\mathrm{d}t}+\frac{1}{2}\pi(n+1-2j)+\arctan\left(\frac{x_j}{a}\right)+\arctan\left(\frac{x_j}{b}\right) \qquad (11b)$$
$$+\sum_{\substack{1\le k\le n\\ k\neq j}}\arctan(x_j-x_k)=0,$$

where $j=1,\ldots,n$. The following theorem affirms that the solutions of the gradient system (11a), (11b) converge exponentially to the equilibrium $\big(x_1^{(n)},x_2^{(n)},\ldots,x_n^{(n)}\big)$ given by the symmetric continuous Hahn roots.

Theorem 1 *Let $Re(a), Re(b) > 0$ with possible non-real parameters a and b arising as a complex conjugate pair.*

(a) The unique global minimum in $\mathbb{R}^n$ of the strictly convex, radially unbounded, Morse function $V(x_1, \ldots, x_n; a, b)$ (8) is attained at the symmetric continuous Hahn roots $x_j = x_j^{(n)}$ $(j = 1, \ldots, n)$.

(b) Let $\big(x_1(t), \ldots, x_n(t)\big)$, $t \geq 0$ denote the unique solution of the gradient system (11a), (11b) determined by a choice for the initial condition

$$(x_1(0), \ldots, x_n(0)) \in \mathbb{R}^n,$$

and let us fix any κ in the interval

$$0 < \kappa < \frac{Re(a)}{Re^2(a) + (R_n + |Im(a)|)^2} + \frac{Re(b)}{Re^2(b) + (R_n + |Im(b)|)^2}, \tag{12a}$$

where $R_n := x_n^{(n)}$. Then there exists a constant $K > 0$ (depending on the parameter values, on the initial condition, and on κ) such that

$$|x_j(t) - x_j^{(n)}| \leq K e^{-\kappa t} \qquad (j = 1, \ldots, n) \tag{12b}$$

whenever $t \geq 0$.

2.4 Proof of Theorem 1

Because $V(x_1, \ldots, x_n; a, b)$ (8) is manifestly smooth on $\mathbb{R}^n$, the existence of a global minimum is immediate from the observation that the function under consideration is radially unbounded: $V(x_1, \ldots, x_n; a, b) \to +\infty$ when $x_1^2 + \cdots + x_n^2 \to +\infty$ (cf. [37, Section 5.1] for a detailed verification of this crucial property). The convexity reconfirms the uniqueness of the global minimum. Indeed, the relevant Hessian is of the form

$$H_{j,k}(x_1, \ldots, x_n; a, b) = \partial_{x_j} \partial_{x_k} V(x_1, \ldots, x_n; a, b)$$
$$= \begin{cases} \frac{a}{a^2+x_j^2} + \frac{b}{b^2+x_j^2} + \sum_{l \neq j} \frac{1}{1+(x_j-x_l)^2} & \text{if } j = k, \\ -\frac{1}{1+(x_j-x_k)^2} & \text{if } j \neq k, \end{cases}$$

so

$$\sum_{1\leq j,k\leq n} y_j y_k H_{j,k}(x_1,\ldots,x_n;a,b) = \sum_{1\leq j\leq n}\left(\frac{a}{a^2+x_j^2}+\frac{b}{b^2+x_j^2}\right)y_j^2 + \sum_{1\leq j<k\leq n}\frac{(y_j-y_k)^2}{1+(x_j-x_k)^2} > 0$$

when $y_1^2+\cdots+y_n^2\neq 0$. The fact that this unique global minimum is attained precisely at the roots of the symmetric continuous Hahn polynomial $\mathrm{p}_n(x;a,b)$ follows via [37, Remarks 5.4, 5.5] as outlined above in Sect. 2.2, which completes the proof of part (a) of the theorem.

We conclude from part (a) that $(x_1^{(n)},\ldots,x_n^{(n)})$ constitutes the unique equilibrium of our gradient system (11a), (11b). Given any initial condition in $\mathbb{R}^n$, the smoothness of the Morse function guarantees the existence and uniqueness of the solution $(x_1(t),\ldots,x_n(t))$ locally for small $t\geq 0$. Moreover, since $V(x_1,\ldots,x_n;a,b)$ is radially unbounded and strictly decreasing along the gradient flow:

$$\frac{\mathrm{d}}{\mathrm{d}t}V\big(x_1(t),\ldots,x_n(t);a,b\big) = -\sum_{j=1}^{n}\left|(\partial_{x_j}V)\big(x_1(t),\ldots,x_n(t);a,b\big)\right|^2 < 0$$

(outside the equilibrium), the solution cannot escape to infinity in finite time and thus extends (uniquely) for all $t\geq 0$. In fact, the radial unboundedness of the potential $V(x_1,\ldots,x_n;a,b)$ (8) guarantees that the equilibrium $(x_1^{(n)},\ldots,x_n^{(n)})$ is globally asymptotically stable (cf. e.g. [18, Theorem 4.2]):

$$\lim_{t\to+\infty} x_j(t) = x_j^{(n)} \qquad (j=1,\ldots,n). \tag{13}$$

The rate of the convergence to the equilibrium is determined by the lowest eigenvalue λ of the Jacobian of the linearized system at the equilibrium point. This Jacobian is given by the above Hessian. Since the weight function $\Delta(x;a,b)$ (7b) is even in x, the roots of $\mathrm{p}_n(x;a,b)$ are symmetrically distributed around the origin. Hence $-R_n\leq x_j\leq R_n$ $(j=1,\ldots,n)$ at the equilibrium, and thus

$$\mathrm{Re}\Big(\frac{a}{a^2+x_j^2}\Big) = \frac{1}{2}\left(\frac{\mathrm{Re}(a)}{\mathrm{Re}^2(a)+(x_j+\mathrm{Im}(a))^2}+\frac{\mathrm{Re}(a)}{\mathrm{Re}^2(a)+(x_j-\mathrm{Im}(a))^2}\right) \geq \frac{\mathrm{Re}(a)}{\mathrm{Re}^2(a)+(R_n+|\mathrm{Im}(a)|)^2}$$

in this situation. The upshot is that

$$\sum_{1\leq j,k\leq n} y_j y_k H_{j,k}(x_1^{(n)},\ldots,x_n^{(n)};a,b)$$

$$\geq \left(\frac{\mathrm{Re}(a)}{\mathrm{Re}^2(a)+(R_n+|\mathrm{Im}(a)|)^2}+\frac{\mathrm{Re}(b)}{\mathrm{Re}^2(b)+(R_n+|\mathrm{Im}(b)|)^2}\right)(y_1^2+\cdots+y_n^2)+\sum_{1\leq j<k\leq n}\frac{(y_j-y_k)^2}{1+4R_n^2}$$

$$\geq \left(\frac{\mathrm{Re}(a)}{\mathrm{Re}^2(a)+(R_n+|\mathrm{Im}(a)|)^2}+\frac{\mathrm{Re}(b)}{\mathrm{Re}^2(b)+(R_n+|\mathrm{Im}(b)|)^2}\right)(y_1^2+\cdots+y_n^2),$$

which entails that

$$\lambda = \min_{\substack{(y_1,\ldots,y_n)\in\mathbb{R}^n\\ y_1^2+\cdots+y_n^2\neq 0}} \frac{\sum_{1\leq j,k\leq n} y_j y_k H_{j,k}(x_1^{(n)},\ldots,x_n^{(n)};a,b)}{y_1^2+\cdots+y_n^2}$$

$$\geq \frac{\mathrm{Re}(a)}{\mathrm{Re}^2(a)+(R+|\mathrm{Im}(a)|)^2}+\frac{\mathrm{Re}(b)}{\mathrm{Re}^2(b)+(R+|\mathrm{Im}(b)|)^2}>\kappa.$$

This lower bound on the eigenvalue guarantees (cf. e.g. [6, Corollary 2.78]) that there exists a neighborhood $U\subset\mathbb{R}^n$ around the equilibrium point $\left(x_1^{(n)},\ldots,x_n^{(n)}\right)$ and a constant $C>0$ such that

$$|x_j(t)-x_j^{(n)}|\leq C|x_j(t_0)-x_j^{(n)}|e^{-\kappa(t-t_0)}\qquad (j=1,\ldots,n),$$

whenever $(x_1(t_0),\ldots,x_n(t_0))\in U$ and $t\geq t_0$. To complete the proof of part b), it now suffices to recall that the gradient flow pulls us from any initial condition $(x_1(0),\ldots,x_n(0))\in\mathbb{R}^n$ into U within finite time t (in view of Eq. (13)).

2.5 *Numerical Samples*

For $n=30$ with $a=10$ and $b=\frac{3}{10}$, the continuous Hahn roots become with a precision of 4 decimals:

$$\begin{aligned}
x_1^{(30)}&=-15.6230, & x_6^{(30)}&=-7.5285, & x_{11}^{(30)}&=-2.7503,\\
x_2^{(30)}&=-13.3738, & x_7^{(30)}&=-6.4188, & x_{12}^{(30)}&=-1.9957,\\
x_3^{(30)}&=-11.6001, & x_8^{(30)}&=-5.3956, & x_{13}^{(30)}&=-1.3059,\\
x_4^{(30)}&=-10.0841, & x_9^{(30)}&=-4.4474, & x_{14}^{(30)}&=-0.6907,\\
x_5^{(30)}&=-8.7415, & x_{10}^{(30)}&=-3.5671, & x_{15}^{(30)}&=-0.1919
\end{aligned}$$

(and $x_j^{(30)}=-x_{31-j}^{(30)}$ for $j=16,\ldots,30$). The corresponding trajectories of $x_j^{(n)}(t)$ for $0\leq t\leq 30$ starting from the origin, i.e. with $x_j^{(n)}(0)=0$ ($j=1,\ldots,n$), are

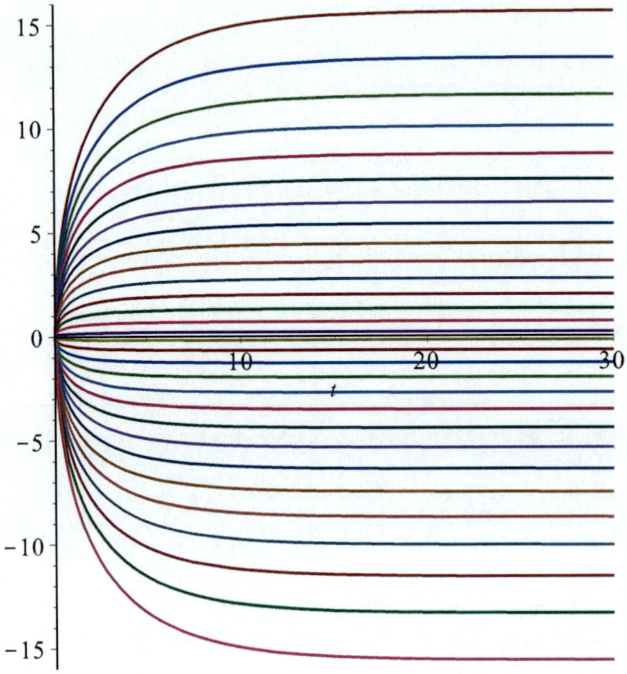

Fig. 1 Continuous Hahn trajectories $x_1^{(30)}(t), \ldots, x_{30}^{(30)}(t)$ corresponding to the initial condition $x_j^{(30)}(0) = 0$ ($j = 1, \ldots, 30$) and the parameter values $a = 10$ and $b = \frac{3}{10}$

exhibited in Fig. 1. The slope of the evolution of the logarithmic error $\log \left|x_j(t) - x_j^{(n)}\right|$ in Fig. 2 confirms that the convergence is exponential with a decay rate that exceeds the not very sharp estimate of $\frac{a}{a^2+R_n^2} + \frac{b}{b^2+R_n^2} \approx 0.030$ guaranteed by Theorem 1.

3 Wilson Polynomials

3.1 Preliminaries

The monic Wilson polynomial $\mathrm{p}_n(x^2; a, b, c, d)$ [38] is an orthogonal polynomial of degree n in x^2 that sits at the top of Askey's hypergeometric scheme [19]. For generic parameter values it is given by the terminating hypergeometric series [19,

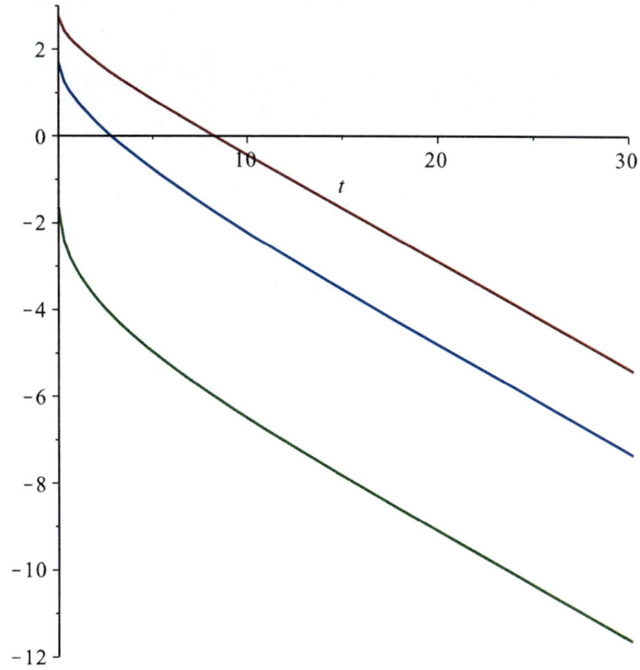

Fig. 2 Evolution of the continuous Hahn logarithmic error $\log\left|x_j(t) - x_j^{(30)}\right|$ for $j = 1$ (top), $j = 8$ (middle) and $j = 15$ (bottom), corresponding to the initial condition $x_j^{(30)}(0) = 0$ ($j = 1, \ldots, 30$) and the parameter values $a = 10$ and $b = \frac{3}{10}$

Chapter 9.1]

$$\mathrm{p}_n(x^2; a, b, c, d) = \frac{(-1)^n (a+b, a+c, a+d)_n}{(n+a+b+c+d-1)_n} \times {}_4F_3\left[\begin{matrix} -n, n+a+b+c+d-1, a+ix, a-ix \\ a+b, a+c, a+d \end{matrix}; 1\right], \tag{14}$$

When $\mathrm{Re}(a), \mathrm{Re}(b), \mathrm{Re}(c), \mathrm{Re}(d) > 0$ with possible non-real parameters arising in complex conjugate pairs, the polynomials in question satisfy the following orthogonality relations [19, Chapter 9.1]

$$\frac{1}{2\pi}\int_0^\infty \Delta(x; a, b, c, d)\mathrm{p}_n(x^2; a, b, c, d)\mathrm{p}_m(x^2; a, b, c, d)\mathrm{d}x = \tag{15a}$$

$$\begin{cases} \frac{n!\Gamma(a+b+c+d+n-1)\Gamma(a+b+n)\Gamma(a+c+n)\Gamma(a+d+n)\Gamma(b+c+n)\Gamma(b+d+n)\Gamma(c+d+n)}{\Gamma(a+b+c+d+2n-1)\Gamma(a+b+c+d+2n)} & \text{if } m = n, \\ 0 & \text{if } m \neq n, \end{cases}$$

where

$$\Delta(x;a,b,c,d)=\left|\frac{\Gamma(a+ix)\Gamma(b+ix)\Gamma(c+ix)\Gamma(d+ix)}{\Gamma(2ix)}\right|^2. \tag{15b}$$

3.2 Morse Function

For parameters values within the above orthogonality regime, the roots

$$0<x_1^{(n)}<x_2^{(n)}<\cdots<x_n^{(n)}$$

of the Wilson polynomial $\mathrm{p}_n(x^2;a,b,c,d)$ minimize the Morse function [37, Remarks 5.4, 5.5]

$$\begin{aligned} V(x_1,\ldots,x_n;a,b,c,d) = &\sum_{1\le j<k\le n}\left(\int_0^{x_j+x_k}\arctan(\theta)\mathrm{d}\theta+\int_0^{x_j-x_k}\arctan(\theta)\mathrm{d}\theta\right)\\ &+\sum_{1\le j\le n}\left(\int_0^{x_j}\arctan\left(\frac{\theta}{a}\right)+\arctan\left(\frac{\theta}{b}\right)\right.\\ &\left.+\arctan\left(\frac{\theta}{c}\right)+\arctan\left(\frac{\theta}{d}\right)\mathrm{d}\theta-\pi jx_j\right). \end{aligned} \tag{16}$$

Indeed, we now have that at the critical points of the Morse function

$$\begin{aligned} &\arctan\left(\frac{x_j}{a}\right)+\arctan\left(\frac{x_j}{b}\right)+\arctan\left(\frac{x_j}{c}\right)+\arctan\left(\frac{x_j}{d}\right)\\ &\quad+\sum_{\substack{1\le k\le n\\k\neq j}}\big(\arctan(x_j+x_k)+\arctan(x_j-x_k)\big)=\pi j \end{aligned}$$

$(j=1,\ldots,n)$, so $0<x_1<x_2<\cdots<x_n$ and (after exponentiation)

$$\frac{(ia+x_j)(ib+x_j)(ic+x_j)(id+x_j)}{(ia-x_j)(ib-x_k)(ic-x_j)(id-x_j)}\prod_{\substack{1\le k\le n\\k\neq j}}\frac{(i+x_j+x_k)(i+x_j-x_k)}{(i-x_j-x_k)(i-x_j+x_k)}=1$$

for $j=1,\ldots,n$. Upon rewriting the latter identities in the form

$$A(x_j;a,b,c,d)\mathrm{p}_n(x_j+i)+A(-x_j;a,b,c,d)\mathrm{p}_n(x_j-i)=0\quad(j=1,\ldots,n),$$

with $A(x; a, b, c, d) = \frac{(x+ia)(x+ib)(x+ic)(x+id)}{2x(2x+i)}$ and

$$\mathrm{p}_n(x) = (x^2 - x_1^2)(x^2 - x_2^2) \cdots (x^2 - x_n^2),$$

it is deduced in the same way as before that the factorized polynomial in question satisfies the second-order difference equation for the Wilson polynomials [19, Chapter 9.1]:

$$A(x; a, b, c, d)\Big(\mathrm{p}_n(x+i) - \mathrm{p}_n(x)\Big) +$$

$$A(-x; a, b, c, d)\Big(\mathrm{p}_n(x-i) - \mathrm{p}_n(x)\Big) = \lambda_n \mathrm{p}_n(x)$$

with $\lambda_n = -n(n+a+b+c+d-1)$. The upshot is that $\mathrm{p}_n(x) = p_n(x^2; a, b, c, d)$ (using again the nondegeneracy of the eigenvalues λ_n at the RHS) and thus $x_j = x_j^{(n)}$ $(j = 1, \ldots, n)$.

Let us stress in this connection that the observation that the roots of Wilson polynomial $p_n(x^2; a, b, c, d)$ satisfy the algebraic identities in the second equation below the Morse function (16) goes again back to [22] (cf. also [35] and [3]).

3.3 Gradient Flow

The corresponding gradient flow now becomes:

$$\frac{\mathrm{d}x_j}{\mathrm{d}t} - \pi j + \arctan\left(\frac{x_j}{a}\right) + \arctan\left(\frac{x_j}{b}\right) + \arctan\left(\frac{x_j}{c}\right) + \arctan\left(\frac{x_j}{d}\right) \tag{17}$$
$$+ \sum_{\substack{1 \le k \le n \\ k \neq j}} \Big(\arctan(x_j + x_k) + \arctan(x_j - x_k)\Big) = 0,$$

$j = 1, \ldots, n$. The following theorem affirms that the solutions of the gradient system (17) converge exponentially to the equilibrium $\big(x_1^{(n)}, \ldots, x_n^{(n)}\big)$ given by the Wilson roots.

Theorem 2 *Let $Re(a), Re(b), Re(c), Re(d) > 0$ with possible non-real parameters arising in complex conjugate pairs.*

(a) The unique global minimum of the strictly convex, radially unbounded, Morse function $V(x_1, \ldots, x_n; a, b, c, d)$ (16) is attained at the Wilson roots $x_j = x_j^{(n)}$ $(j = 1, \ldots, n)$.

(b) Let $\big(x_1(t), \ldots, x_n(t)\big)$, $t \geq 0$ denote the unique solution of the gradient system (17) determined by a choice for the initial condition

$$(x_1(0), \ldots, x_n(0)) \in \mathbb{R}^n,$$

and let us fix any κ *in the interval*

$$0 < \kappa < \tfrac{2(n-1)}{1+4R_n^2} + \tfrac{Re(a)}{Re^2(a)+(R_n+|Im(a)|)^2} + \tfrac{Re(b)}{Re^2(b)+(R_n+|Im(b)|)^2} \\ + \tfrac{Re(c)}{Re^2(c)+(R_n+|Im(c)|)^2} + \tfrac{Re(d)}{Re^2(d)+(R_n+|Im(d)|)^2} \tag{18a}$$

where $R_n := x_n^{(n)}$. *Then there exists a constant* $K > 0$ *(depending on the parameter values, on the initial condition, and on* κ*) such that*

$$|x_j(t) - x_j^{(n)}| \leq K e^{-\kappa t} \qquad (j = 1, \ldots, n) \tag{18b}$$

whenever $t \geq 0$.

3.4 Proof of Theorem 2

The proof runs along the same lines as that of Theorem 1, so we only highlight some of the principal modifications in the corresponding computations.

As before, the first part of the theorem hinges on the considerations in [37, Remarks 5.4, 5.5] with the main points summarized above in Sect. 3.2 (cf. also the proof of [37, Proposition 4.1] for a detailed check that the present Morse function is indeed radially unbounded).

To infer the asserted estimate for the convergence rate, we must again provide the lower bound for the eigenvalues of the pertinent Jacobian evaluated at the equilibrium point. This Jacobian is given by the Hessian

$$H_{j,k}(x_1, \ldots, x_n; a, b, c, d) = \partial_{x_j} \partial_{x_k} V(x_1, \ldots, x_n; a, b, c, d) \tag{19}$$

$$= \begin{cases} \frac{a}{a^2+x_j^2} + \frac{b}{b^2+x_j^2} + \frac{c}{c^2+x_j^2} + \frac{d}{d^2+x_j^2} + \sum_{l \neq j} \left(\frac{1}{1+(x_j+x_l)^2} + \frac{1}{1+(x_j-x_l)^2} \right) & \text{if } j = k, \\ \frac{1}{1+(x_j+x_l)^2} - \frac{1}{1+(x_j-x_l)^2} & \text{if } j \neq k, \end{cases}$$

which confirms the convexity:

$$\sum_{1 \leq j,k \leq n} y_j y_k H_{j,k}(x_1, \ldots, x_n; a, b, c, d) = \sum_{1 \leq j \leq n} \Big(\sum_{\epsilon \in \{a,b,c,d\}} \frac{\epsilon}{\epsilon^2 + x_j^2} \Big) y_j^2 \\ + \sum_{1 \leq j < k \leq n} \left(\frac{(y_j + y_k)^2}{1 + (x_j + x_k)^2} + \frac{(y_j - y_k)^2}{1 + (x_j - x_k)^2} \right).$$

Since $0 < x_j \leq R_n$ $(j = 1, \ldots, n)$ at the equilibrium, we now see that

$$\sum_{1\leq j,k\leq n} y_j y_k H_{j,k}(x_1^{(n)}, \ldots, x_n^{(n)}; a, b, c, d)$$

$$\geq \left(\sum_{\epsilon\in\{a,b,c,d\}} \frac{\mathrm{Re}(\epsilon)}{\mathrm{Re}^2(\epsilon)+(R_n+|\mathrm{Im}(\epsilon)|)^2}\right)(y_1^2 + \cdots + y_n^2)$$

$$+ \sum_{1\leq j<k\leq n} \frac{(y_j + y_k)^2 + (y_j - y_k)^2}{1 + 4R_n^2}$$

$$= \left(\sum_{\epsilon\in\{a,b,c,d\}} \frac{\mathrm{Re}(\epsilon)}{\mathrm{Re}^2(\epsilon)+(R_n+|\mathrm{Im}(\epsilon)|)^2} + \frac{2(n-1)}{1+4R_n^2}\right)(y_1^2 + \cdots + y_n^2).$$

This entails the desired lower bound for the smallest eigenvalue:

$$\lambda = \min_{\substack{(y_1,\ldots,y_n)\in\mathbb{R}^n \\ y_1^2+\cdots+y_n^2\neq 0}} \frac{\sum_{1\leq j,k\leq n} y_j y_k H_{j,k}(x_1^{(n)}, \ldots, x_n^{(n)}; a, b, c, d)}{y_1^2 + \cdots + y_n^2}$$

$$\geq \sum_{\epsilon\in\{a,b,c,d\}} \frac{\mathrm{Re}(\epsilon)}{\mathrm{Re}^2(\epsilon) + (R_n + |\mathrm{Im}(\epsilon)|)^2} + \frac{2(n-1)}{1 + 4R_n^2} > \kappa.$$

3.5 *Numerical Samples*

For $n = 15$ with $a = \frac{17}{3}$, $b = \frac{1}{5}$, $c = 1 + i$, and $d = 1 - i$, the Wilson roots become with a precision of 4 decimals:

$$\begin{aligned}
x_1^{(15)} &= 0.5274, & x_6^{(15)} &= 3.7728, & x_{11}^{(15)} &= 8.5546,\\
x_2^{(15)} &= 1.1194, & x_7^{(15)} &= 4.5787, & x_{12}^{(15)} &= 9.8143,\\
x_3^{(15)} &= 1.7050, & x_8^{(15)} &= 5.4496, & x_{13}^{(15)} &= 11.2449,\\
x_4^{(15)} &= 2.3375, & x_9^{(15)} &= 6.3938, & x_{14}^{(15)} &= 12.9284,\\
x_5^{(15)} &= 3.0266, & x_{10}^{(15)} &= 7.4231, & x_{15}^{(15)} &= 15.0759.
\end{aligned}$$

The corresponding trajectories of $x_j^{(n)}(t)$ for $0 \leq t \leq 30$, with an initial condition of the form $x_j^{(n)}(0) = 0$ $(j = 1, \ldots, n)$, are exhibited in Fig. 3. The slope of the evolution of the logarithmic error $\log\left|x_j(t) - x_j^{(n)}\right|$ in Fig. 4 confirms that the convergence is exponential with a decay rate that exceeds the not very sharp estimate of

$$\sum_{\epsilon\in\{a,b,c,d\}} \frac{\mathrm{Re}(\epsilon)}{\mathrm{Re}^2(\epsilon) + (R_n + |\mathrm{Im}(\epsilon)|)^2} + \frac{2(n-1)}{1 + 4R_n^2} \approx 0.061$$

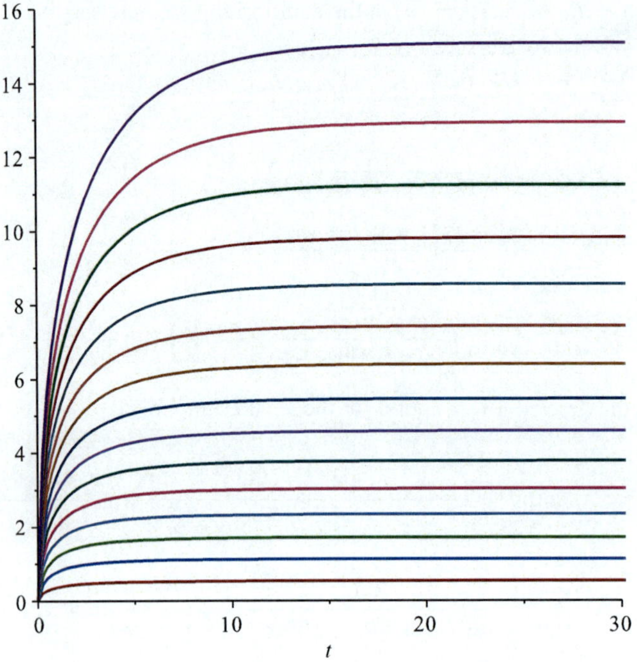

Fig. 3 Wilson trajectories $x_1^{(15)}(t), \ldots, x_{15}^{(15)}(t)$ corresponding to the initial condition $x_j^{(15)}(0) = 0$ $(j = 1, \ldots, 15)$ and the parameter values $a = \frac{17}{3}$, $b = \frac{1}{5}$, $c = 1 + i$, and $d = 1 - i$

guaranteed by Theorem 2.

4 Symmetry Reduction

As noticed above, the weight function $\Delta(x; a, b)$ (7b) is even in x for our parameter regime, so the roots of the corresponding symmetric continuous Hahn polynomial $\mathrm{p}_n(x; a, b)$ are symmetrically distributed around the origin:

$$x_j^{(n)} + x_{n+1-j}^{(n)} = 0 \quad \text{for } j = 1, \ldots, n. \tag{20}$$

It is manifest from the explicit differential equations that the gradient system in Eqs. (11a), (11b) preserves this parity symmetry. More specifically, if the initial condition satisfies

$$x_j(0) + x_{n+1-j}(0) = 0 \quad \text{for } j = 1, \ldots, n, \tag{21a}$$

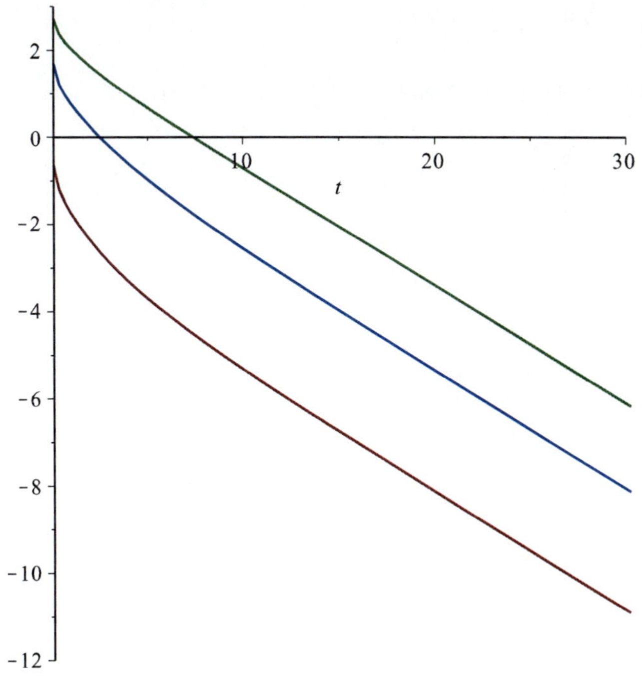

Fig. 4 Evolution of the Wilson logarithmic error $\log |x_j(t) - x_j^{(15)}|$ for $j = 1$ (bottom), $j = 8$ (middle) and $j = 15$ (top), corresponding to the initial condition $x_j^{(15)}(0) = 0$ ($j = 1, \ldots, 15$) and the parameter values $a = \frac{17}{3}$, $b = \frac{1}{5}$, $c = 1 + i$, and $d = 1 - i$

then so does the corresponding gradient flow $\big(x_1(t), \ldots, x_n(t)\big)$, $t \geq 0$:

$$x_j(t) + x_{n+1-j}(t) = 0 \quad \text{for } j = 1, \ldots, n. \tag{21b}$$

If we now perform the reduction of our gradient system to the pertinent $m := \lfloor \frac{n}{2} \rfloor$-dimensional invariant manifold through the substitution

$$(x_1, \ldots, x_n) = \begin{cases} (-y_m, \ldots, -y_1, y_1, \ldots, y_m) & \text{if } n = 2m, \\ (-y_m, \ldots, -y_1, 0, y_1, \ldots, y_m) & \text{if } n = 2m+1, \end{cases} \tag{22}$$

then we arrive at the differential equations

$$\frac{\mathrm{d}y_j}{\mathrm{d}t} - \pi(j - \tfrac{1}{2}) + \arctan\left(\frac{y_j}{a}\right) + \arctan\left(\frac{y_j}{b}\right) + \arctan\big(2y_j\big) + \sum_{\substack{1 \leq k \leq m \\ k \neq j}} \big(\arctan(y_j + y_k) + \arctan(y_j - y_k)\big) = 0 \tag{23a}$$

$(j = 1, \ldots, m)$ when $n = 2m$, and

$$\frac{\mathrm{d}y_j}{\mathrm{d}t} - \pi j + \arctan\left(\frac{y_j}{a}\right) + \arctan\left(\frac{y_j}{b}\right) + \arctan\left(2y_j\right) + \arctan\left(y_j\right) \qquad (23b)$$

$$+ \sum_{\substack{1 \le k \le m \\ k \ne j}} \big(\arctan(y_j + y_k) + \arctan(y_j - y_k)\big) = 0$$

$(j = 1, \ldots, m)$ when $n = 2m + 1$, respectively. By comparing the latter differential equations with the gradient system for the Wilson polynomials in Eq. (17), it is seen that for $n = 2m$ we recover the case $(a, b, c, d) \to (a, b, \frac{1}{2}, 0)$ and for $n = 2m + 1$ we recover the case $(a, b, c, d) \to (a, b, \frac{1}{2}, 1)$. Indeed, the symmetric continuous Hahn polynomials and the Wilson polynomials are known to be related by the following quadratic relations (cf. e.g. [20, Section 2.4]):

$$\mathrm{p}_n(x; a, b) = \begin{cases} \mathrm{p}_m(x^2; a, b, \frac{1}{2}, 0) & \text{if } n = 2m, \\ x\mathrm{p}_m(x^2; a, b, \frac{1}{2}, 1) & \text{if } n = 2m + 1. \end{cases} \qquad (24)$$

The upshot is that for initial conditions respecting the parity invariance, the estimate for the rate of the exponential convergence in the second part of Theorem 1 can be improved as follows.

Theorem 3 *Let $Re(a), Re(b) > 0$ with possible non-real parameters a and b arising as a complex conjugate pair, and let $\big(x_1(t), \ldots, x_n(t)\big)$, $t \ge 0$ denote the unique solution of the gradient system (11a), (11b) determined by a choice for the initial condition $(x_1(0), \ldots, x_n(0)) \in \mathbb{R}^n$ such that*

$$x_j(0) + x_{n+1-j}(0) = 0 \quad \textit{for } j = 1, \ldots, n.$$

Then for any κ in the interval

$$0 < \kappa < \frac{2\lfloor \frac{n}{2} \rfloor}{1+4R_n^2} + \frac{Re(a)}{Re^2(a)+(R_n+|Im(a)|)^2} + \frac{Re(b)}{Re^2(b)+(R_n+|Im(b)|)^2} + \frac{1-(-1)^n}{2+2R_n^2}, \qquad (25a)$$

where $R_n := x_n^{(n)}$, there exists a constant $K > 0$ (depending on the parameter values, on the initial condition, and on κ) such that

$$|x_j(t) - x_j^{(n)}| \le K e^{-\kappa t} \qquad (j = 1, \ldots, n) \qquad (25b)$$

whenever $t \ge 0$.

For $n = 2m + 1$, Theorem 3 is immediate from the reduced differential equation (23b) and the second part of Theorem 2, upon performing the parameter specialization $(a, b, c, d) \to (a, b, \frac{1}{2}, 1)$. The case $n = 2m$ would follow similarly via the reduced differential equation (23a) upon performing the parameter specialization $(a, b, c, d) \to (a, b, \frac{1}{2}, 0)$ in Theorem 2. However, since the parameter specialization $d \to 0$ takes us outside the Wilson parameter domain considered here, for the argument to stick rigorously one formally has to repeat the proof of Theorem 2 for the case of the gradient flow in Eq. (23a).

Since our numerical example in Sect. 2.5 corresponded to an initial condition of the form $x_j(0) = 0$ $(j = 1, \ldots, n)$, the improved exponential convergence of Theorem 3 actually applies in this situation. However, if we break the parity symmetry by moving the initial condition up to $x_j(0) = 3$ $(j = 1, \ldots, n)$—while maintaining the parameter values $a = 10$ and $b = \frac{3}{10}$ (cf. Fig. 5)—then we see from the slopes of the logarithmic error (cf. Fig. 6) that the rate of the exponential convergence indeed slows down considerably. Notice at this point that the downward peaks in the evolution of the logarithmic error reveal that the corresponding limiting values are no longer approached monotonously: the peak detects when the trajectory of $x_j(t)$ overshoots its limiting value $x_j^{(n)}$.

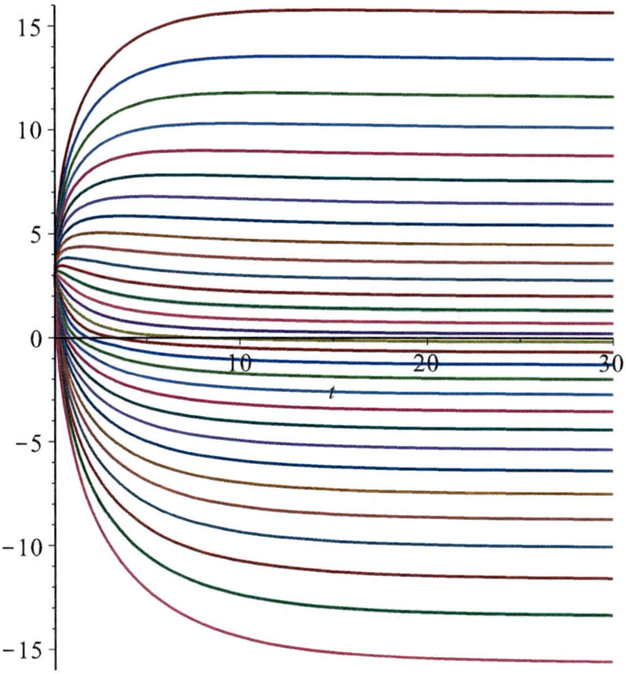

Fig. 5 Continuous Hahn trajectories $x_1^{(30)}(t), \ldots, x_{30}^{(30)}(t)$ corresponding to the initial condition $x_j^{(30)}(0) = 3$ $(j = 1, \ldots, 30)$ and the parameter values $a = 10$ and $b = \frac{3}{10}$

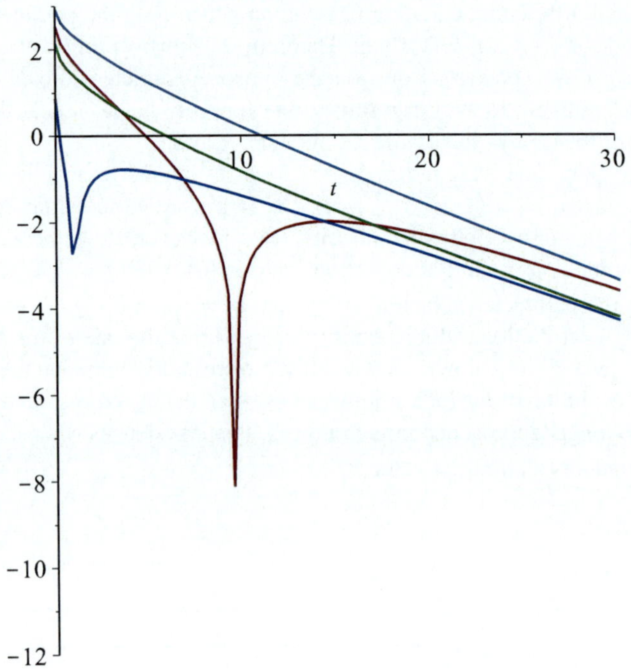

Fig. 6 Evolution of the continuous Hahn logarithmic error $\log|x_j(t) - x_j^{(30)}|$ for $j = 1$ (top), $j = 8$ (second from below), $j = 23$ (bottom), and $j = 30$ (second from above), where the ordering refers to that of the asymptotic tails, with the initial condition $x_j^{(30)}(0) = 3$ ($j = 1, \ldots, 30$) and the parameter values $a = 10$ and $b = \frac{3}{10}$

Acknowledgments It is a pleasure to thank Alexei Zhedanov for emphasizing that the Morse functions from [37], which minimize at the roots of the continuous Hahn, Wilson and Askey-Wilson polynomials, should be viewed as natural analogs of Stieltjes' electrostatic potentials for the roots of the classical orthogonal polynomials. Thanks are also due to an anonymous referee for suggesting some important improvements in the presentation.

References

1. Askey, R., Wilson, J.: A set of hypergeometric orthogonal polynomials. SIAM J. Math. Anal. **13**, 651–655 (1982)
2. Beltrán, C., Marcellán, F., Martínez-Finkelshtein, A.: Some extremal properties of the roots of orthogonal polynomials. Gac. R. Soc. Mat. Esp. **21**, 345–366 (2018)
3. Bihun, O., Calogero, F.: Properties of the zeros of the polynomials belonging to the Askey scheme. Lett. Math. Phys. **104**, 1571–1588 (2014)
4. Bihun, O., Calogero, F.: Properties of the zeros of the polynomials belonging to the q-Askey scheme. J. Math. Anal. Appl. **433**, 525–542 (2016)
5. Calogero, F.: Equilibrium configuration of the one-dimensional n-body problem with quadratic and inversely quadratic pair potentials. Lett. Nuovo Cimento (2) **20**, 251–253 (1977)

6. Chicone, C.: Ordinary Differential Equations with Applications. 2nd edn. Springer, New York (2006)
7. Dimitrov, D.K., Van Assche, W.: Lamé differential equations and electrostatics. Proc. Amer. Math. Soc. **128**, 3621–3628 (2000)
8. Fehér, L., Görbe, T.F.: Duality between the trigonometric BC_n Sutherland system and a completed rational Ruijsenaars-Schneider-van Diejen system. J. Math. Phys. **55**(10), 102704 (2014)
9. Forrester, P.J., Rogers, J.B.: Electrostatics and the zeros of the classical polynomials. SIAM J. Math. Anal. **17**, 461–468 (1986)
10. Grünbaum, F.A.: Variations on a theme of Heine and Stieltjes: an electrostatic interpretation of the zeros of certain polynomials. J. Comput. Appl. Math. **99**, 189–194 (1998)
11. Grünbaum, F.A.: Electrostatic interpretation for the zeros of certain polynomials and the Darboux process. J. Comput. Appl. Math. **133**, 397–412 (2001)
12. Hendriksen, E., van Rossum, H., Electrostatic interpretation of zeros. In: Orthogonal Polynomials and their Applications. Lecture Notes in Mathematics, vol. 1329, pp. 241–250. Springer, Berlin (1988)
13. Horváth, Á.P.: The electrostatic properties of zeros of exceptional Laguerre and Jacobi polynomials and stable interpolation. J. Approx. Theory **194**, 87–107 (2015)
14. Ismail, M.E.H.: An electrostatics model for zeros of general orthogonal polynomials. Pacific J. Math. **193**, 355–369 (2000)
15. Ismail, M.E.H.: More on electrostatic models for zeros of orthogonal polynomials. Numer. Funct. Anal. Optim. **21**, 191–204 (2000)
16. Ismail, M.E.H.: Classical and Quantum Orthogonal Polynomials in One Variable. Cambridge University Press, Cambridge (2005)
17. Jooste, A., Njionou Sadjang, P., Koepf, W.: Inner bounds for the extreme zeros of ${}_3F_2$ hypergeometric polynomials. Integral Transforms Spec. Funct. **28**, 361–373 (2017)
18. Khalil, H.K.: Nonlinear Systems, 3rd edn. Prentice Hall, Upper Saddle River (2002)
19. Koekoek, R., Lesky, P.A., Swarttouw, R.: Hypergeometric Orthogonal Polynomials and their q-Analogues. Springer, Berlin (2010)
20. Koornwinder, T.H.: Quadratic transformations for orthogonal polynomials in one and two variables. In: Representation Theory, Special Functions and Painlevé Equations–RIMS 2015. Advanced Studies in Pure Mathematics, vol. 76, pp. 419–447. Mathematical Society of Japan, Tokyo (2018)
21. Marcellán, F., Martínez-Finkelshtein, A., Martínez-González, P.: Electrostatic models for zeros of polynomials: old, new, and some open problems. J. Comput. Appl. Math. **207**, 258–272 (2007)
22. Odake, S., Sasaki, R.: Equilibria of 'discrete' integrable systems and deformation of classical orthogonal polynomials. J. Phys. A **37**, 11841–11876 (2004)
23. Odake, S., Sasaki, R.: Equilibrium positions, shape invariance and Askey-Wilson polynomials. J. Math. Phys. **46**(6), 063513 (2005)
24. Odake, S., Sasaki, R.: Calogero-Sutherland-Moser systems, Ruijsenaars-Schneider-van Diejen systems and orthogonal polynomials. Prog. Theor. Phys. **114**, 1245–1260 (2005)
25. Perelomov, A.M.: Equilibrium configurations and small oscillations of some dynamical systems. Ann. Inst. H. Poincaré Sect. A (N.S.) **28**, 407–415 (1978)
26. Pusztai, B.G.: The hyperbolic BC_n Sutherland and the rational BC_n Ruijsenaars-Schneider-van Diejen models: Lax matrices and duality. Nuclear Phys. B **856**, 528–551 (2012)
27. Pusztai, B.G.: Scattering theory of the hyperbolic BC_n Sutherland and the rational BC_n Ruijsenaars-Schneider-van Diejen models. Nuclear Phys. B **874**, 647–662 (2013)
28. Simanek, B.: An electrostatic interpretation of the zeros of paraorthogonal polynomials on the unit circle. SIAM J. Math. Anal. **48**, 2250–2268 (2016)
29. Steinerberger, S.: Electrostatic interpretation of zeros of orthogonal polynomials. Proc. Amer. Math. Soc. **146**, 5323–5331 (2018)
30. Stieltjes, T.J.: Sur certains polynômes qui vérifient une équation différentielle linéaire du second ordre et sur la theorie des fonctions de Lamé. Acta Math. **6**, 321–326 (1885)

31. Szegö, G.: Orthogonal Polynomials, 4th edn. American Mathematical Society, Providence (1975)
32. van Diejen, J.F.: Deformations of Calogero-Moser systems and finite Toda chains. Theor. Math. Phys. **99**, 549–554 (1994)
33. van Diejen, J.F.: Difference Calogero-Moser systems and finite Toda chains. J. Math. Phys. **36**, 1299–1323 (1995)
34. van Diejen, J.F.: Multivariable continuous Hahn and Wilson polynomials related to integrable difference systems. J. Phys. A **28**, L369–L374 (1995)
35. van Diejen, J.F.: On the equilibrium configuration of the BC-type Ruijsenaars-Schneider system. J. Nonlinear Math. Phys. **12**(suppl. 1), 689–696 (2005)
36. van Diejen, J.F.: Gradient system for the roots of the Askey-Wilson polynomial. Proc. Amer. Math. Soc. **147**, 5239–5249 (2019)
37. van Diejen, J.F., Emsiz, E.: Solutions of convex Bethe Ansatz equations and the zeros of (basic) hypergeometric orthogonal polynomials. Lett. Math. Phys. **109**, 89–112 (2019)
38. Wilson, J.A.: Some hypergeometric orthogonal polynomials. SIAM J. Math. Anal. **11**, 690–701 (1980)

Convergent Non Complete Interpolatory Quadrature Rules

Ulises Fidalgo and Jacob Olson

Abstract We find a family of convergent schemes of nodes for non-complete interpolatory quadrature rules.

Keywords Interpolatory quadrature rules · Orthogonal polynomials

1 Introduction

Let $C([-1, 1])$ be the set of all continuous functions defined on $[-1, 1]$. Given an n-tuple of nodes $\mathbf{x}_n = (x_{1,n}, \ldots, x_{j,n})$ satisfying $-1 < x_{1,n} < x_{2,n} < \cdots < x_{n,n} < 1$, we consider *integration rules*

$$I_n[f] = \sum_{j=1}^{n} w_{j,n} f(x_{j,n}), \qquad f \in C[-1, 1] \tag{1}$$

associated to the integrals

$$I(f) = \int_{-1}^{1} f(x)\,\mathrm{d}\lambda_0(x), \qquad \text{where} \quad \frac{\mathrm{d}\lambda_0(x)}{\mathrm{d}x} = \frac{1}{\pi\sqrt{1-x^2}}. \tag{2}$$

The numbers $w_{j,n}$, $j = 1, \ldots, n$ are called weights.

U. Fidalgo (✉)
Department of Mathematics, Applied Mathematics, and Statistics, Case Western Reserve University, Cleveland, OH, USA
e-mail: uxf6@case.edu

J. Olson
Case Western Reserve University, Cleveland, OH, USA
e-mail: jfo27@case.edu

F. Marcellán, E. J. Huertas (eds.), *Orthogonal Polynomials: Current Trends and Applications*, SEMA SIMAI Springer Series 22,
https://doi.org/10.1007/978-3-030-56190-1_7

An integration rule $I_n[\cdot]$ is said to be interpolatory if there exists a number $m \in \{0, 1, \ldots, 2n-1\}$, such that the following equality holds for every polynomial p with degree $\leq m$ (we denote $p \in \Pi_m$):

$$I_n[p] = \int_{-1}^{1} p(x)\, \mathrm{d}\lambda_0(x). \tag{3}$$

When the equality (3) holds for certain m, and is not extendable for all polynomials with degree $m+1$, we say that $I_n[\cdot]$ is an interpolatory quadrature rule with m-degree of exactness. When $m = 2n-1$ I_n is the Gaussian quadrature rule.

Consider a sequence of interpolatory quadratures $\{I_n\}_{n\in\Lambda}$ constructed with the following schemes of nodes and weights

$$\mathbf{x} = \left\{\mathbf{x}_n = (x_{1,n}, \ldots, x_{n,n})\right\}_{n\in\Lambda} \quad \text{and} \quad \mathbf{w} = \left\{\mathbf{w}_n = (w_{1,n}, \ldots, w_{n,n})\right\}_{n\in\Lambda}, \tag{4}$$

respectively.

We say that $\{I_n\}_{n\in\Lambda}$ is convergent if

$$\lim_{n\in\Lambda} I_n[f] = \int_{-1}^{1} f(x)\, \mathrm{d}\lambda_0(x), \quad \text{for all} \quad f \in C[-1, 1]. \tag{5}$$

According to a classical result of Pólya [5, page 130], when $m(n) \to \infty$ as $n \to \infty$, the equality (5) holds true if and only if $\sup_{n\in\Lambda} \sum_{j=1}^{n} \left|w_{j,n}\right| < \infty$. This condition is satisfied if the weights $w_{j,n}$ are all positive. From (3) we observe that

$$\sum_{j=1}^{n} w_{j,n} = \int_{-1}^{1} \frac{\mathrm{d}x}{\pi\sqrt{1-x^2}} = 1 < \infty. \tag{6}$$

In the Gaussian quadrature rule (maximum degree of exactness $m(n) = 2n-1$) the weights $w_{j,n}$, $j = 1, \ldots, n$ are all positive and the convergence of the rule is guarantied. However the nodes are all fixed. For each $n \in \mathbb{N}$ the points of evaluation $x_{j,n}$, $j = 1, \ldots, n$ must be the roots of the nth orthogonal polynomial with respect to λ_0 (see for instance [16]). This is the Chebyshev polynomial with degree n. This means that if we do not have the value of f at each point $x_{j,n}$ the calculus gets stuck. It is convenient to have more flexibility in the distribution of the evaluation nodes. We study convergent interpolatory integration rules with orders of exactness $m < 2n-1$.

The authors of [3] analyze a wide class of interpolatory quadrature rules with $m(n)$ degrees of exactness behaving as follows

$$\lim_{n\to\infty} \frac{m(n)}{2n} = a \in [0,\ 1]. \tag{7}$$

They characterized all possible weak*-limit points of the sequence of counting measures associated with distribution of nodes corresponding to a convergent scheme.

A sequence $\{\nu_n\}_{n\in\mathbb{N}}$ of measures is said to converge weakly to the measure ν provided that there exists a compact set K containing the support of ν and of each ν_n, and that

$$\lim_{n\in\mathbb{N}} \int f d\nu_n = \int f d\nu$$

for each continuous function f on K. In such a case, we write $\nu_n \overset{*}{\to} \nu$. We say that ν is a weak*-limit of the sequence $\{\eta_n\}_{n\in\mathbb{N}}$ if some subsequence of $\{\nu_n\}_{n\in\Lambda\subset\mathbb{N}}$ is weakly convergent to ν.

Set two schemes of numbers as in (4) associated to an interpolatory quadrature rule $\{I_n\}_{n\in\Gamma}$ where the degree of exactness satisfies (7) for certain $a \in [0, 1]$. We also consider its corresponding sequence $\{\eta_n\}_{n\in\mathbb{N}}$ of probability counting measures

$$\eta_n := \frac{1}{n}\sum_{j=1}^{n} \delta_{x_{j,n}}, \quad n \in \mathbb{N}. \tag{8}$$

According to [2], if the rule $\{I_n\}_{n\in\Lambda}$ is convergent then every weak*-limit ν of the the sequence $\{\eta_n\}_{n\in\mathbb{N}}$ satisfies that

$$\nu \geq a\,\lambda_0. \tag{9}$$

Also from [2] we have that this necessary condition is not sufficient. Theorem 1 states conditions of convergence on the distribution of nodes.

Let us introduce some previous notation. Set $\mathcal{K}_1$ and $\mathcal{K}_2$ two compact subsets of the complex plane $\mathbb{C}$. Let $\text{dist}(\mathcal{K}_1, \mathcal{K}_2) = \min\{||x - y||| : x \in \mathcal{K}_1 \text{ and } y \in \mathcal{K}_2\}$ denote the distance between $\mathcal{K}_1$ and $\mathcal{K}_2$. Consider a compact set $K \subset \mathbb{C}\setminus[-1, 1]$, and a measure μ supported on K. A measure $\widetilde{\mu}$ supported on $[-1, 1]$ is said to be the balayage of μ if they have the same total variation $||\mu|| = ||\widetilde{\mu}||$ and their logarithmic potentials coincide on $[-1, 1]$. This is

$$V^{\widetilde{\mu}}(x) = \int \log\frac{1}{|x-t|}\mathrm{d}\widetilde{\mu}(t) = \int \log\frac{1}{|x-\zeta|}\mathrm{d}\mu(\zeta) = V^{\mu}(x), \quad x \in [-1, 1].$$

In [15, Section II.4] we can find a deep study about balayage of measures. We are now ready to state the main result of this paper:

Theorem 1 *Fix a number $\kappa \in \mathbb{N}$ and a probability discrete measure*

$$\sigma = \frac{1}{\kappa}\sum_{k=1}^{\kappa} \delta_{\zeta_k}, \quad \zeta_k \subset \mathbb{C}\setminus[-1, 1], \quad k = 1, \ldots, \kappa.$$

Assume that σ *is symmetric with respect to* $\mathbb{R}$ *with* $\operatorname{dist}(\{\zeta_1,\ldots,\zeta_\kappa\},[-1,\,1])>1$. *Denote* $\widetilde{\sigma}$ *the balayage measure associated to* σ *supported on the interval* $[-1,\,1]$. *Given a rational number* $a\in[0,\,1]$, *consider a subsequence* $\Lambda\subset\mathbb{N}$ *such that for each* $n\in\Lambda$, $2\frac{1-a}{\kappa}n\in\mathbb{N}$. *Let* $\mathbf{x}=\{\mathbf{x}_n=(x_{1,n},\ldots,x_{n,n})\}_{n\in\Lambda}$ *be a scheme of nodes. If for each* $j=1,\ldots n$, $n\in\Lambda$ *there are two constants* $A\geq 0$ *and* $\ell>0$ *satisfying*

$$\left|(1-a)\pi\int_{x_{j,n}}^{1}\mathrm{d}\widetilde{\sigma}(t)-a\arccos x_{j,n}-\frac{2j-1}{2n}\pi\right|\leq Ae^{-\ell n},\tag{10}$$

then there always exist weights $\mathbf{w}=\{\mathbf{w}_n=(w_{1,n},\ldots,w_{n,n})\}_{n\in\mathbb{N}}$, *where* $\{I_n\}_{n\in\Lambda}$ *corresponding to* $\mathbf{x}$ *and* $\mathbf{w}$ *is convergent.*

In Sect. 2 we give some explicit schemes that satisfy the relation (10). The statement of Theorem 1 is proved in Sect. 5. In such proof we use results coming from the orthogonal polynomials theory that are analyzed in Sects. 3 and 4. In Sect. 3 we study algebraic properties of families of orthogonal polynomials and their connections with convergent conditions of non-complete interpolatory quadrature rules. In Sect. 4 we describe the strong asymptotic behavior of an appropriated family of orthogonal polynomials with respect to a varying measure.

2 Some Explicit Convergent Schemes of Nodes

We consider three particular cases where the inequality (10) holds. In the three situations the measure $\sigma=\delta_\zeta$ corresponds to a Dirac delta supported on a point belonging to the real line $\zeta>2$. Hence the situations are when a takes the values 0, 1/2, and 1.

According to [15, Section II.4 equation (4.46)], the balayage measure of $\sigma=\delta_\zeta$ on $[-1,\,1]$ has the following differential form

$$\mathrm{d}\widetilde{\sigma}(t)=\frac{\sqrt{\zeta^2-1}}{\pi(\zeta-t)\sqrt{1-t^2}}\,\mathrm{d}t.\tag{11}$$

We study the function

$$I_a(x)=(1-a)\pi\int_x^1\mathrm{d}\widetilde{\sigma}(t)=(1-a)\sqrt{\zeta^2-1}\int_x^1\frac{\mathrm{d}t}{(\zeta-t)\sqrt{1-t^2}}.$$

Taking the change of variables $t=\cos\theta$ and taking into account $\zeta>2$ ($\varphi(\zeta)>2$ implies that $\arg(1-\varphi(\zeta))=\pi$), we have that

$$I_a(x)=(1-a)\left(-\arccos x+2\arg\left(e^{i\arccos x}-\varphi(\zeta)\right)-2\pi\right).$$

In this situation the condition of convergence (10) in Theorem 1 acquires the following form

$$\left| \arccos x_{j,n} + 2(1-a)\left[\pi - \arg\left(e^{i\arccos x} - \varphi(\zeta)\right)\right] + \frac{2j-1}{2n}\pi \right| \le Ae^{-\ell n}. \tag{12}$$

Then a scheme $\mathbf{x} = \left\{\mathbf{x}_n = \left(x_{1,n}, \ldots, x_{n,n}\right)\right\}_{n\in\Lambda}$ that satisfies the following relation is convergent

$$\arccos x_{j,n} + 2(1-a)\left[\pi - \arg\left(e^{i\arccos x} - \varphi(\zeta)\right)\right] = -\frac{2j-1}{2n}\pi - Ae^{-\ell n} = \kappa_{j,n},$$

with $A > 0$ and $\ell > 0$. This means that

$$\cos\left(\arccos x_{j,n} + 2(1-a)\left[\pi - \arg\left(e^{i\arccos x} - \varphi(\zeta)\right)\right]\right) = \cos\kappa_{j,n}.$$

Using the cosine addition formula we have that

$$x_{j,n}\cos\left\{2(1-a)\left[\pi - \arg\left(e^{i\arccos x} - \varphi(\zeta)\right)\right]\right\}$$
$$-\sqrt{1-x_{j,n}^2}\,\sin\left\{2(1-a)\left[\pi - \arg\left(e^{i\arccos x} - \varphi(\zeta)\right)\right]\right\} = \cos\kappa_{j,n}. \tag{13}$$

First we consider the situation $a = 1$. In this case the expressions in (13) become

$$x_{j,n} = \cos\kappa_{j,n}, \quad j = 1, \ldots, n, \quad n \in \Lambda. \tag{14}$$

The nodes are close to the zeros of the Chebyshev polynomials. That's because of the term corresponding to the σ's influence in (13) vanishes when $a = 1$.

Let us analyze now the case $a = 1/2$. We consider the following identities

$$\cos\left[\pi - \arg\left(e^{i\arccos x_{j,n}} - \varphi(\zeta)\right)\right] = \frac{\varphi(\zeta) - x_{j,n}}{\sqrt{\varphi^2(\zeta) - 2\varphi(\zeta)x_{j,n} + 1}} \tag{15}$$

and

$$\sin\left[\pi - \arg\left(e^{i\arccos x_{j,n}} - \varphi(\zeta)\right)\right] = \frac{\sqrt{1-x_{j,n}^2}}{\sqrt{\varphi^2(\zeta) - 2\varphi(\zeta)x_{j,n} + 1}}. \tag{16}$$

Substituting (15) and (16) in (13) we arrive at the quadratic equations:

$$x_{j,n}^2 - \frac{2\sin^2\kappa_{j,n}}{\varphi(\zeta)}x_{j,n} + \frac{\sin^2\kappa_{j,n}}{\varphi^2(\zeta)} - \cos^2\kappa_{j,n} = 0, \quad j = 1, \ldots, n, \quad n \in \Lambda.$$

For each $j = 1, \ldots, n$, $n \in \Lambda$ we obtained the following solutions

$$x_{j,n} = \frac{1}{\varphi(\zeta)}\left[\sin^2\kappa_{j,n} + \cos\kappa_{j,n}\sqrt{\varphi^2(\zeta) - \sin^2\kappa_{j,n}}\right]. \tag{17}$$

During the process of finding these above solutions we introduce some extra solutions that we removed. Observe that when ζ tends to ∞ the expressions in (17) reduce to (14). This is in accordance with the fact that $\widetilde{\sigma}$ approaches λ_0 as $\zeta \to \infty$, see (11), hence we only considered the positive branch of the square root in (17).

Finally take $a = 0$. From (13) we have that

$$x_{j,n}\cos\left\{2\left[\pi - \arg\left(e^{i\arccos x} - \varphi(\zeta)\right)\right]\right\}$$

$$-\sqrt{1 - x_{j,n}^2}\sin\left\{2\left[\pi - \arg\left(e^{i\arccos x} - \varphi(\zeta)\right)\right]\right\} = \cos\kappa_{j,n}.$$

We use the conditions (15) and (16), and obtain the following expression

$$x_{j,n} = \frac{2\zeta\cos\kappa_{j,n}}{\varphi(\zeta) + 2\cos\kappa_{j,n}}, \qquad j = 1, \ldots, n, \qquad n \in \Lambda.$$

Taking into account that $\varphi(\zeta) = \zeta + \sqrt{\zeta^2 + 1}$ we see that the above expression is reduced to (14) when ζ goes to infinity.

3 Connection with Orthogonal Polynomials

Let μ be a positive finite Borel measure with infinitely many points in its support $\mathrm{supp}(\mu)$. Set Δ denoting the least interval which contains $\mathrm{supp}(\mu)$. A collection of monic polynomials $\{q_{\mu,n}\}_{n\in\mathbb{Z}_+}$, $\mathbb{Z}_+ = \{0, 1, \ldots\}$ is the family orthogonal polynomials with respect to μ if its elements satisfy the following orthogonality relations

$$0 = \int x^\nu q_{\mu,n}(x)\,\mathrm{d}\mu(x), \qquad \nu = 0, 1, \ldots, n-1, \qquad n \in \mathbb{Z}_+. \tag{18}$$

Each $q_{\mu,n}$ has n single roots lying in the interior of Δ (we denote $\overset{\circ}{\Delta}$) such that it vanishes at most once in each interval of $\Delta \setminus \mathrm{supp}(\mu)$ (see [4, Theorem 5.2] or [7, Chapter 1]). We also know that $q_{\mu,n+1}$ and $q_{\mu,n}$ interlace their zeros. In [17] B. Wendroff proved that given two polynomials P_n and P_{n+1}, with $\deg P_{n+1} = \deg P_n + 1 = n + 1$, that interlace zeros, there always exist measures μ such that $P_n = q_{\mu,n}$ and $P_{n+1} = q_{\mu,n+1}$. Now we find some of these measures.

We say then a polynomial $P_n(x) = \prod_{j=1}^{n} \left(x - x_j\right)$ of degree n is admissible with respect to the measure μ, if its roots are all simple, lying in $\overset{\circ}{\Delta}$, with at most one zero into each interval of $\Delta \setminus \operatorname{supp}(\mu)$. The system of nodes $(x_1, \ldots, x_n)$ is also said to be admissible with respect to μ.

Lemma 1 *Let $P_n(x) = \prod_{j=1}^{n} \left(x - x_j\right)$ and $\widetilde{P}_n(x) = \prod_{j=1}^{n-1} \left(x - \widetilde{x}_j\right)$ be two admissible polynomials with respect to μ that satisfy $x_1 < \widetilde{x}_1 < x_2 < \cdots < \widetilde{x}_{n-1} < x_n$. Then there exists a positive integrable function ρ_n with respect to μ (ρ_n is a weight function for μ) such that for the measure μ_n which differential form $d\mu_n(x) = \rho_n(x)\, d\mu(x)$, $x \in supp(\mu)$, $P_n \equiv q_{\mu_n,n}$ and $\widetilde{P}_n \equiv q_{\mu_n,n-1}$ are the n-th and $n-1$-th monic orthogonal polynomials with respect to μ_n, respectively.*

In the proof we follow techniques used in [10].

Proof Consider Φ a set of weight functions such that for every constant $\alpha > 0$ it satisfies:

i) $\rho \in \Phi \implies \alpha\rho \in \Phi$.
ii) $(\rho, \widetilde{\rho}) \in \Phi^2 = \Phi \times \Phi \implies \alpha\rho + (1-\alpha)\widetilde{\rho} \in \Phi$, $\alpha \le 1$.
iii) If a polynomial Q satisfies $\int Q(x)\rho(x)\, \mathrm{d}\mu(x) > 0$ for all $\rho \in \Phi$, then $Q \ge 0$ in $\operatorname{supp}(\mu)$.

Two examples of sets of weight functions satisfying the above conditions are the positive polynomials and positive simple functions in [14, Definition 1.16]. In general, the positive linear combinations of a Chevyshev system (see [9, Chapter II]) conform a set as Φ. Examples of Chevyshev systems can be found in [13] (also in [8]).

Given $\rho \in \Phi$ we set

$$\mathbf{v}_\rho = \left(\int \widetilde{P}_n(x)\rho(x)\mathrm{d}\mu(x), \ldots, \int x^{n-2}\widetilde{P}_n(x)\rho(x)\mathrm{d}\mu(x),\right.$$

$$\left.\int P_n(x)\rho(x)\mathrm{d}\mu(x), \ldots, \int x^{n-1}P_n(x)\rho(x)\mathrm{d}\mu(x)\right) \in \mathbb{R}^{2n-1}.$$

Let us focus on $\mathcal{K} = \left\{\mathbf{v}_\rho : \rho \in \Phi\right\}$. Proving Lemma 1 reduces to showing that $\mathcal{K}$ contains the origin. From condition (i) we have that the origin belongs to $\mathcal{K}$'s closure, $\overline{\mathcal{K}}$. Since $\mathcal{K}$ is open we need to prove the origin is an interior point. We proceed by contradiction. Suppose that the origin belongs to the boundary of $\mathcal{K}$. This is $\mathbf{O} \in \partial\mathcal{K} = \overline{K} \setminus \mathcal{K}$. There exists a hyper-plane $\mathcal{A}$ that touches tangentially $\partial\mathcal{K}$ at $\mathbf{O}$. On the other hand we have that condition (ii) implies that $\mathcal{K}$ is convex, then there exists a vector $\mathbf{a} = \left(a_{0,n-1}, \ldots, a_{n-2,n-1}, a_{0,n}, \ldots, a_{n-1,n}\right)$ which is orthogonal with respect to $\mathcal{A}$ in the sense of the standard inner vector product ($\mathbf{a} \cdot \mathbf{u} = 0$, for all $\mathbf{u} \in \mathcal{A}$), and for each $\mathbf{v}_\rho \in \mathcal{K}$, $\mathbf{v}_\rho \cdot \mathbf{a} > 0$. So the polynomials

$$p_{n-1}(x) = a_{0,n-1} + a_{1,n-1}x + \ldots + a_{n-2,n-1}x^{n-2}$$

and

$$p_n(x) = a_{0,n} + a_{1,n}x + \ldots + a_{n-1,n}x^{n-1}$$

satisfy that

$$0 < \int \left[p_{n-1}(x)\widetilde{P}_n(x) + p_n P_n(x)\right] \rho(x)\, \mathrm{d}\mu(x), \quad \text{for all} \quad \rho \in \Phi.$$

According to condition (iii) the polynomial $\mathcal{P}(x) = p_{n-1}(x)\widetilde{P}_n(x) + p_n P_n(x)$, with real coefficients, must be non-negative in $\operatorname{supp}(\mu)$. However we shall prove that this is impossible, arriving then to a contradiction.

Assume that $\mathcal{P}(x) = p_{n-1}(x)\widetilde{P}_n(x) + p_n P_n(x)$ does not change sign in $\operatorname{supp}(\mu)$. Suppose that there is a point $t \in \operatorname{supp}(\mu)$, such that $t = x_k$ $k \in \{1, \ldots, n\}$ satisfying that $\mathcal{P}(x_k) = 0$, then taking into account that $\widetilde{P}_n$ and P_n interlace zeros, we have that $p_{n-1}(x_k) = 0$. Also, since $\mathcal{P}$ is non-negative on $\operatorname{supp}(\mu)$, we have that $t = x_k$ is a zero of multiplicity even for $\mathcal{P}$. Consider $S = \{t_1, \ldots, t_\ell\} \subset \{x_1, \ldots, x_n\}$ the set of all points where P_n and $\mathcal{P}$ vanished at same time. Then we can write

$$\mathcal{P} = q(x) \prod_{i=1}^{\ell} (x - t_i)^{2d_i}, \quad d_i \in \mathbb{N}, \quad i = 1, \ldots, \ell, \tag{19}$$

where q is a polynomial with positive values at every root of P_n. We also write

$$p_{n-1}(x) = \widetilde{p}(x) \prod_{i=1}^{\ell} (x - t_i). \tag{20}$$

The polynomial $\widetilde{p}$ has degree $\deg p_n - \ell$. Since $\widetilde{P}_n$ and P_n interlace zeros, we have that

$$\frac{\widetilde{P}_n(x)}{P_n(x)} = \sum_{j=1}^{n} \frac{\lambda_j}{x - x_j}, \quad \lambda_j > 0, \quad j = 1, \ldots, n. \tag{21}$$

Observe that

$$\frac{\prod_{i=1}^{\ell} (x - t_i)}{P_n(x)} = \frac{1}{\prod_{i=1}^{\ell} (x - t_i)^{2d_i - 1} q(x)} \left[p_{n-1}(x) \sum_{j=1}^{n} \frac{\lambda_j}{x - x_j} + p_n(x) \right].$$

This means that the above function satisfies that

$$\frac{1}{\prod_{i=1}^{\ell}(z-t_i)^{2d_i-1}\,q(z)}\left[p_{n-1}(z)\sum_{j=1}^{n}\frac{\lambda_j}{z-x_j}+p_n(z)\right]=O\left(\frac{1}{z^{n-\ell}}\right)\quad \text{as}\quad z\to\infty,$$

which is a holomorphic functions on $\overline{\mathbb{C}}\setminus(\{x_1,\ldots,x_n\}\setminus S)$. For each $\nu=0,\ldots,n-\ell-2$ we have then

$$\frac{z^{\nu}}{\prod_{i=1}^{\ell}(z-t_i)^{2d_i-1}\,q(z)}\left[p_{n-1}(z)\sum_{j=1}^{n}\frac{\lambda_j}{z-x_j}+p_n(z)\right]=O\left(\frac{1}{z^{2}}\right)\quad \text{as}\quad z\to\infty,$$

also holomorphic functions on $\overline{\mathbb{C}}\setminus(\{x_1,\ldots,x_n\}\setminus S)$. Set the elements $y_j\in\{x_1,\ldots,x_n\}\setminus S$, $j=1,\ldots,n-\ell$ with $y_1<y_2<\cdots<y_{n-\ell}$, and $\widetilde{\lambda}_j$, $j=1,\ldots,n-\ell$ the coefficients λ's defined in (21) corresponding to points y_j. Also let λ'_j denote the λ's of t_j, $j=1,\ldots,\ell$. Call F the set of the roots of the polynomial q defined in (19). Consider a closed integration path Γ with winding number 1 for all its interior points. Denote $\mathrm{Ext}(\Gamma)$ and $\mathrm{Int}(\Gamma)$ the unbounded and bounded connected components respectively of the complement of Γ. Take Γ so that $I\subset\mathrm{Int}(\Gamma)$ and $F\subset\mathrm{Ext}(\Gamma)$. From Cauchy's Theorem and the above two conditions, it follows that

$$0=\frac{1}{2\pi i}\int_{\Gamma}\frac{z^{\nu}}{\prod_{i=1}^{\ell}(z-t_i)^{2d_i-1}\,q(z)}\left[p_{n-1}(z)\sum_{j=1}^{n}\frac{\lambda_j}{z-x_j}+p_n(z)\right]\mathrm{d}z$$

$$=\frac{1}{2\pi i}\int_{\Gamma}\frac{z^{\nu}p_{n-1}(z)\sum_{j=1}^{n-\ell}\frac{\widetilde{\lambda}_j}{z-y_j}\,\mathrm{d}z}{\prod_{i=1}^{\ell}(z-t_i)^{2d_i-1}\,q(z)}+\frac{1}{2\pi i}\int_{\Gamma}\frac{z^{\nu}p_n(z)\,\mathrm{d}z}{\prod_{i=1}^{\ell}(z-t_i)^{2d_i-1}\,q(z)}.$$

Since $\dfrac{z^{\nu}p_n(z)}{\prod_{i=1}^{\ell}(z-t_i)^{2d_i-1}\,q(z)}\in\mathcal{H}(\mathrm{Int}(\Gamma))$ the second term vanishes. From (20), using the Cauchy integral formula, we obtain:

$$0 = \sum_{j=1}^{n-\ell} y_j^\nu \widetilde{p}(y_j) \frac{\widetilde{\lambda}_j}{\prod_{i=1}^{\ell} \left(y^j - t_i\right)^{2(d_i-1)} q(y_j)} = 0, \quad \nu = 0, \dots, n-\ell-1.$$

Taking into account that for each $j = 1, \dots, n-\ell$, $\dfrac{\widetilde{\lambda}_j}{\prod_{i=1}^{\ell} \left(y^j - t_i\right)^{2(d_i-1)} q(y_j)} > 0$,
we conclude that the above orthogonality relations imply that $\widetilde{p}$ must change sign at least $n-\ell$ times, hence $\deg \widetilde{p} \geq n-\ell$. Since $\deg \widetilde{p} = \deg p_{n-1} - \ell \leq n-\ell-1$ we arrive at a contradiction which completes the proof. □

Consider a monic polynomial $P_n(x) = \prod_{j=1}^{n} (x - x_j)$ with degree $n \in \mathbb{N}$ which is μ admissible. We say that a weight function ρ_n on $\mathrm{supp}(\mu)$ is orthogonal with respect to $P_n(x)$ and μ if $P_n \equiv q_{\mu_n,n}$, where $\mathrm{d}\mu_n(x) = \rho_n(x)\,\mathrm{d}\mu(x)$, $x \in \mathrm{supp}(\mu)$. We also say that ρ_n is orthogonal with respect to $\mathbf{x}_n = (x_1, \dots, x_n)$ and μ. A sequence of weight functions $\{\rho_n\}_{n\in\mathbb{N}}$ is a family of orthogonal weight functions with respect to the sequence of polynomials $\{P_n\}_{n\in\mathbb{N}}$, if for each $n \in \mathbb{N}$, $P_n \equiv q_{\mu_n,n}$.

Let $q_{m(n)}$ be an arbitrary polynomial with degree $\deg q_{m(n)}(x) = 2n - m(n) - 1$ being positive on $[-1, 1]$. Let μ_n denote the measure with differential form $\mathrm{d}\mu_n(x) = q_{m(n)}^{-1}(x)\mathrm{d}\mu(x)$, $x \in \mathrm{supp}(\mu)$. Set a system of nodes $\mathbf{x}_n = (x_1, \dots, x_n)$ such that $P_n(x) = \prod_{j=1}^{n} (x - x_j) = q_{\mu_n,n}$. This means that $\mathbf{x}_n$ is the system of n nodes corresponding to the Gaussian quadrature rule for the measure μ_n. Given an arbitrary polynomial $p \in \Pi_{m(n)}$, we have that

$$q_{m(n)}(x)p(x) - \sum_{j=1}^{n} q_{m(n)}(x_j)p(x_j)L_{j,n}(x) = q_{\mu_n,n}(x)\mathcal{P}_{n-1}(x), \tag{22}$$

where $L_{j,n}(x) := \prod_{\substack{k=1 \\ k\neq j}}^{n} \dfrac{x - x_k}{x_j - x_k}$, $\quad j = 1, \dots, n$, and $\mathcal{P}_{n-1}$ is a certain polynomial with $\deg \mathcal{P}_{n-1} = n - (m(n) - \deg p) - 1 \leq n - 1$.

Observe that $\int p(x)\,\mathrm{d}\,\mu(x) = \int q_{m(n)}(x)p(x)\mathrm{d}\mu_n(x)$. Hence from (22) we obtain

$$\int p(x)\,\mathrm{d}\,\mu(x) - \int \sum_{j=1}^{n} q_{m(n)}(x_j)p(x_j)L_{j,n}(x)\,\mathrm{d}\,\mu_n(x) = \int q_{\mu_n,n}(x)\mathcal{P}_{n-1}(x)\,\mathrm{d}\,\mu_n(x),$$

which vanishes because $q_{\mu_n,n}$ satisfies the orthogonality relations for μ_n as in (18). We conclude then

$$\int p(x)\,\mathrm{d}\,\mu(x) = \sum_{j=1}^{n} p(x_{j,n})q_{m(n)}(x_{j,n}) \int L_{j,n}(x)\frac{\mathrm{d}\,\lambda_0(x)}{q_{m(n)}(x)} = \sum_{j=1}^{n} w_{j,n}p(x_{j,n}).$$

This is an interpolatory integration rule with degree of exactness $m(n)$, where the weights can be defined via

$$w_{j,n} = q_{m(n)}(x_j) \int L_{j,n}(x)\,\mathrm{d}\mu_n(x) = q_{m(n)}(x_j)\widetilde{w}_{j,n}, \quad j = 1, \ldots, n. \tag{23}$$

The numbers $\widetilde{w}_{j,n}$, $j = 1, \ldots, n$ are the weights corresponding to a Gaussian quadrature rule, which are all positive. Since $q_{m(n)}$ is also positive the weights $w_{j,n} > 0$. According to Pólya's condition a sequence of these rules of integration is convergent.

Let us consider $\mathbf{x} = \{\mathbf{x}_n = (x_{1,n}, \ldots, x_{n,n})\}_{n\in\mathbb{N}}$ an admissible scheme of nodes for a measure μ, and take a corresponding family of orthogonal weights $\{\rho_n\}_{n\in\mathbb{N}}$. For each n, μ_n denotes the measure with differential form $\mathrm{d}\mu(x) = \rho_n(x)\mathrm{d}\mu(x)$, and introduce its family of orthonormal polynomials $\{p_{\mu_n,j}\}_{j\in\mathbb{Z}_+}$. This means that $p_{\mu_n,j} \equiv q_{\mu_n,j}/\,||q_{\mu_n,j}||_{2,\mu_n}$, $j \in \mathbb{Z}_+$ where $||\cdot||_{2,\mu_n}$ denotes the L_2 norm corresponding to the measure μ_n.

Given a function $f \in \mathrm{L}_{2,\mu_n}$ and $j \in \mathbb{Z}_+$ we consider the j-th partial sum of the Fourier series corresponding to f/ρ_n on the bases $\{p_{\mu_n,j}\}_{j\in\mathbb{Z}_+}$:

$$S_{f,\mu_n,j} = \sum_{k=0}^{j-1} f_k p_{\mu_n,k}(x), \quad f_k = \int f(x)p_{\mu_n,k}(x)\mathrm{d}\mu_n(x), \quad k = 0, \ldots, j-1.$$

Using the Christoffel-Darboux identity (see [16, Theorem 3.2.2]) we can deduce

$$S_{f,\mu_n,j}(x) = \int \frac{q_{\mu_n,j}(x)q_{\mu_n,j-1}(t) - q_{\mu_n,j}(t)q_{\mu_n,j-1}(x)}{||q_{\mu_n,j-1}||^2_{2,\mu_n}(x-t)} f(t)\mathrm{d}\mu_n(t). \tag{24}$$

The following result is an extension of [16, Theorem 15.2.4 (equality 15.2.7)]

Lemma 2 *Let* $(x_1, \ldots, x_n)$ *be an* μ *admissible system of nodes. Given a polynomial* $q_{m(n)}$ *take the system of weights* $(w_{1,n}, \ldots, w_{n,n})$ *whose elements* $w_{j,n}$, $j = 1, \ldots, n$, *are constructed using (23). Then there always exists a weight* ρ_n *such that*

$$\frac{w_{j,n}}{q_{m(n)}(x_j)} = \frac{\left|\left|q_{\tau_n,n-1}\right|\right|_{2,\tau_n}^2 S_{1/\rho_n,\tau_n,n}(x_j)}{q_{\tau_n,n-1}(x_j)q'_{\tau_n,n}(x_j)} = -\frac{\left|\left|q_{\tau_n,n}\right|\right|_{2,\tau_n}^2 S_{1/\rho_n,\tau_n,n+1}(x_j)}{q_{\tau_n,n+1}(x_j)q'_{\tau_n,n}(x_j)}, \tag{25}$$

where the measure τ_n *is such that* $\dfrac{d\tau_n}{d\mu} = \dfrac{\rho_n}{q_{m(n)}}$. *Thus* $\operatorname{sign} w_{j,n} = \operatorname{sign} S_{1/\rho_n,\tau_n,n}(x_j) = \operatorname{sign} S_{1/\rho_n,\tau_n,n+1}(x_j)$, $j = 1, \ldots, n$.

Proof Take an orthogonal weight ρ_n with respect to the system of n nodes $(x_1, \ldots, x_n)$ and the measure with differential form $d\mu(x)/q_{m(n)}(x)$. According to (23) and taking into account that $P_n \equiv q_{\mu_n,n}$ where the measure τ_n has the differential form $d\tau_n(x) = \rho_n(x)d\mu(x)/q_{m(n)}(x)$, we have the following

$$w_{j,n} = q_{m(n)}(x_j) \int \frac{q_{\tau_n,n}(x)}{q'_{\tau_n,n}(x_j)(x - x_j)} \frac{d\mu(x)}{q_{m(n)}(x)}, \qquad j = 1, \ldots, n.$$

Arranging the above formula and using the identity (24) we obtain that

$$w_{j,n} = \frac{q_{m(n)}(x_j) \left|\left|q_{\tau_n,n}\right|\right|_{2,\tau_n}^2}{q_{\tau_n,n+1}(x_j)q'_{\tau_n,n}(x_j)} \int \frac{q_{\tau_n,n+1}(x_j)q_{\tau_n,n}(x)}{\left|\left|q_{\tau_n,n}\right|\right|_{2,\tau_n}^2 (x - x_j)} \frac{1}{\rho_n(x)} \frac{\rho_n(x)\, d\mu(x)}{q_{m(n)}(x)}$$

$$= -\frac{q_{m(n)}(x_j) \left|\left|q_{\tau_n,n}\right|\right|_{2,\tau_n}^2}{q_{\tau_n,n+1}(x_j)q'_{\tau_n,n}(x_j)} S_{1/\rho_n,\tau_n,n}(x_j),$$

which proves the second identity in (25). Since $q_{\tau_n,n+1}(x_j)q'_{\tau_n,n}(x_j) < 0$, $j = 1, \ldots, n$, then $\operatorname{sign} w_{j,n} = \operatorname{sign} S_{1/\rho_n,\tau_n,n+1}(x_j)$. Following the above steps we can prove the first equality in (25) and $\operatorname{sign} w_{j,n} = \operatorname{sign} S_{1/\rho_n,\tau_n,n}(x_j)$. □

The following two results are consequences of the above Lemma 2.

Lemma 3 *An admissible scheme of nodes* $\mathbf{x} = \left\{\mathbf{x}_n = (x_{1,n}, \ldots, x_{n,n})\right\}_{n\in\mathbb{N}}$ *is convergent if there exists a family of orthogonal weights* $\{\rho_n\}_{n\in\mathbb{N}}$ *with respect to* $\mathbf{x}$ *and the sequence of measures* $\left\{d\tau_n(x) = d\mu(x)/q_{m(n)}(x)\right\}_{n\in\mathbb{N}}$ *satisfying*

$$\lim_{n\to\infty} \left|\left|1 - \rho_n(x)S_{1/\rho_n,\tau_n,n}(x)\right|\right|_{[-1,\,1],\infty} = 0, \tag{26}$$

where $||\cdot||_{[-1,1],\infty}$ *denotes the supremum norm on* $[-1,\ 1]$.

Proof Assuming the equality (26), there exists a number $N > 0$ such that for every $n \geq N$ the function $S_{1/\rho_n,\mu_n,n}(x) > 0$ on $[-1, 1]$ particularly at the nodes. According to Lemma 2, the coefficients $w_{j,n}$, $j = 1, \ldots, n$, are also positive. This completes the proof. □

Lemma 4 *Consider the varying measure* $d\mu_n(x) = d\mu(x)/q_{m(n)}(x)$ *and their orthogonal polynomials* $q_{\mu_n,n}(x) = \prod_{j=1}^{n}\left(x - x_{j,n}\right)$ *and* $q_{\mu_n,n-1}(x) = \prod_{j=1}^{n-1}\left(x - x_{j,n-1}\right)$, $n \in \mathbb{N}$. *Let* $\mathbf{y} = \left\{\mathbf{y}_n = \left(y_{1,n}, \ldots, y_{n,n}\right)\right\}_{n\in\mathbb{N}}$ *be a scheme of nodes such that for each* $n \in \mathbb{N}$

$$-1 < y_{1,n} < x_{1,n-1} < y_{2,n} < \cdots < x_{n-1,n-1} < y_{n,n} < 1. \tag{27}$$

Assume that the polynomials $P_n(x) = \prod_{j=1}^{n}\left(x - y_{j,n}\right),\quad n \in \mathbb{N}$ *satisfy*

$$\lim_{n\to\infty} \frac{q_{\mu(n)}^{-1}(t)}{||q_{\mu_n,n-1}||_{2,\mu_n}^2}\left[(q_{\mu_n,n} - P_n)q_{\mu_n,n-1}\right]'(x) = 0, \quad (x,t) \in [-1,1]^2. \tag{28}$$

Then $\mathbf{y}$ *is convergent.*

Proof From Lemma 1 we ensure the existence of a weight function ρ_n such that the polynomials $q_{\tau_n,n-1}$ and P_n belong to the family of orthogonal polynomials corresponding to the measure $\rho_n(x)\mathrm{d}\mu(x)/q_{m(n)}(x)$. Let us analyze the function

$$\left|1 - \frac{||P_n||_{2,\rho_n d\mu/q_{m(n)}}^2}{||q_{\mu_n,n-1}||_{2,\mu_n}^2} S_{1/\rho_n,\rho_n d\mu/q_{m(n)},n}(x)\right|$$

$$= \left|S_{1,\mu n,n}(x) - \frac{||\widetilde{P}_n||_{2,\rho_n d\mu/q_{m(n)}}^2}{||q_{\mu_n,n-1}||_{2,\mu_n}^2} S_{1/\rho_n,\rho_n d\mu/q_{m(n)},n}(x)\right|$$

We have used that $S_{1,\mu_n,n} \equiv 1q_{m(n)}$, hence we need to show that

$$\lim_{n\to 0}\left|S_{1,\mu_n,n} - \frac{||\widetilde{P}_n||_{2,\rho_n d\mu/q_{m(n)}}^2}{||q_{\mu_n,n-1}||_{2,\mu_n}^2} S_{1/\rho_n,\rho_n d\mu/q_{m(n)},n}(x)\right| = 0.$$

Applying (24) we observe that

$$\left(S_{1,\mu_n,n}(x) - \frac{||\widetilde{P}_n||^2_{2,\rho_n d\mu/q_{m(n)}}}{||q_{\mu_n,n-1}||^2_{2,\mu_n}} S_{1/\rho_n,\rho_n d\mu/q_{m(n)},n}(x) \right) = \frac{1}{||q_{\mu_n,n-1}||^2_{2,\mu_n}}$$

$$\times \int \left(\frac{q_{\mu_n,n}(x)q_{\mu_n,n-1}(t) - q_{\mu_n,n}(t)q_{\mu_n,n-1}(x)}{x-t} - \frac{P_n(x)q_{\mu_n,n-1}(t) - P_n(t)q_{\mu_n,n-1}(x)}{x-t} \right)$$

$$\times \frac{\mathrm{d}\mu(t)}{q_{m(n)}(t)}$$

Let us consider the kernel

$$\frac{q_{\mu_n,n}(x)q_{\mu_n,n-1}(t) - q_{\mu_n,n}(t)q_{\mu_n,n-1}(x)}{||q_{\mu_n,n-1}||^2_{2,\mu_n} q_{m(n)}(t)(x-t)} - \frac{P_n(x)q_{\mu_n,n-1}(t) - P_n(t)q_{\mu_n,n-1}(x)}{||q_{\mu_n,n-1}||^2_{2,\mu_n} q_{m(n)}(t)(x-t)}$$

$$= \frac{(q_{\mu_n,n} - P_n)(x)q_{\mu_n,n-1}(t) - (q_{\mu_n,n} - P_n)(t)q_{\mu_n,n-1}(x)}{||q_{\mu_n,n-1}||^2_{2,\mu_n} q_{m(n)}(t)(x-t)} = \mathcal{K}(x,t).$$

From Taylor's Theorem we obtain that

$$\mathcal{K}(x,t) = \frac{q_{\mu(n)}^{-1(t)}}{||q_{\mu_n,n-1}||^2_{2,\mu_n}} \left[(q_{\mu_n,n} - P_n)q_{\mu_n,n-1} \right]'(s)$$

for some s in between of x and t, so the assumption (28) completes the proof. □

4 Asymptotic Analysis

Let us consider the varying measure μ_n with $\mathrm{d}\mu_n(x)/\mathrm{d}x = (q_{m(n)}(x)\sqrt{1-x^2})^{-1}$, where $q_{m(n)}(x) = q_\kappa^{2\frac{1-a}{k}n}(x) = \left(\prod_{k=1}^{\kappa}(x-\zeta_j)\right)^{2\frac{1-a}{k}n}$, $n \in \Lambda$. Let σ be the zero counting measure of q_k. This is $\sigma = \frac{1}{\kappa}\sum_{k=1}^{\kappa}\delta_{\zeta_k}$. Set the analytic logarithmic potential corresponding to the measure σ:

$$g(z,\sigma) = -\int \log(z-\zeta)\,\mathrm{d}\sigma(\zeta). \qquad (29)$$

We take the logarithmic branch such that $g(z, \sigma)$ is analytic on a domain $D \subset K$ that contains the interval $[-1, 1]$, and also for every $x \in [-1, 1]$,

$$V^{\sigma}(x) = \int \log \frac{1}{|x-\zeta|} d\sigma(\zeta) = g(x, \sigma) = -\int \log (z-\zeta)\, d\sigma(\zeta). \tag{30}$$

Since σ is symmetric we $\displaystyle\int \arg(x-\zeta) d\sigma(\zeta) = 0$. In each compact $K \subset D$ we have that $\dfrac{1}{2n} \log \dfrac{1}{q_{m(n)}(z)} = (1-a)\, g(z, \sigma)$ on K.

Lemma 5 *Let $d\mu_n(x)/dx = (q_{m(n)}(x)\sqrt{1-x^2})^{-1}$, $n \in \mathbb{N}$ be a sequence of measures as above. Then*

$$q_{\mu_n,n} = \left(1 + O(e^{-cn})\right) \exp\left\{-nV^{\overline{\nu}}\right\} K_{1,n} + O(e^{-cn}) \exp\left\{-nV^{\overline{\nu}}\right\} K_{2,n} \tag{31}$$

and

$$\frac{d_{n,n-1}}{2^{2na}} q_{\mu_n,n-1} = \left(1 + O(e^{-cn})\right) \exp\left\{-nV^{\overline{\nu}}\right\} K_{2,n} + O(e^{-cn}) \exp\left\{-nV^{\overline{\nu}}\right\} K_{1,n} \tag{32}$$

where $d_{n,n-1} = -\left(2\pi i \left|\left|q_{\mu_n,n-1}\right|\right|^2_{\mu_n,2}\right)^{-1}$,

$$K_{1,n}(x) = 2\cos n \left((1-a)\pi \int_x^1 d\overline{\sigma}(t) - a \arccos x \right), \tag{33}$$

and

$$K_{2,n}(x) = \frac{1}{i} \cos n \left((1-a)\pi \int_x^1 d\overline{\sigma}(t) - (a - 1/n) \arccos x \right). \tag{34}$$

Proof We study a matrix Riemann-Hilbert problem like in [12, Theorem 2.4] whose solution Y is a 2×2 matrix function satisfying the following conditions:

1. $Y \in \mathcal{H}(\mathbb{C} \setminus [-1, 1])$ (all the entries of Y are analytic on $\mathbb{C} \setminus [-1, 1]$),
2. $Y_+(x) = Y_-(x) \begin{pmatrix} 1 & \left(q_{m(n)}(x)\sqrt{1-x^2}\right)^{-1} \\ 0 & 1 \end{pmatrix}$, $x \in (-1, 1)$,
3. $Y(z) \begin{pmatrix} z^{-n} & 0 \\ 0 & z^n \end{pmatrix} = \mathbb{I} + O(1/z)$ as $z \to \infty$, $\mathbb{I}$ is the 2×2 identity matrix.
4. $Y(z) = O \begin{pmatrix} 1 & |z \pm 1|^{-1/2} \\ 1 & |z \pm 1|^{-1/2} \end{pmatrix}$ as $z \to \mp 1$.

According to [12, Theorem 2.4] (see also [11]) the Y solution of above matrix Riemann-Hilbert problem (for short Y-RHP) is unique and has the form

$$Y(z) = \begin{pmatrix} q_{\mu_n,n}(z) & -\dfrac{1}{2\pi i}\displaystyle\int \dfrac{q_{\mu_n,n}(x)}{z-x}\,\mathrm{d}\mu_n(x) \\ d_{n,n-1}q_{\mu_n,n-1}(z) & -\dfrac{d_{n,n-1}}{2\pi i}\displaystyle\int \dfrac{q_{\mu_n,n-1}(x)}{z-x}\,\mathrm{d}\mu_n(x) \end{pmatrix}.$$

The key of our procedure follows the ideas introduced in [1]. We find a relationship between Y and the matrix solution $R : \mathbb{C} \setminus \gamma \to \mathbb{C}^{2\times 2}$ corresponding to another Riemann-Hilbert problem (R-RHP) for a closed Jordan curve γ positively oriented surrounding the interval $[-1,\ 1]$:

1. $R \in \mathcal{H}(\mathbb{C} \setminus \gamma)$,
2. $R_+(\zeta) = R_-(\zeta)V_n(\zeta)$, $\zeta \in \gamma$, with $V_n \in \mathcal{H}(D)$,
3. $R(z) \to \mathbb{I}$ as $z \to \infty$,

where $V_n = \mathbb{I} + \mathcal{O}(\varepsilon^n)$ with $0 \le \varepsilon < 1$, uniformly on compact subsets of K as $n \to \infty$. Those conditions imply that $R = \mathbb{I} + \mathcal{O}(\varepsilon^n)$ uniformly on $\mathbb{C}$ as $n \to \infty$. There is a chain of transformations to arrive from Y to R, which we represent $Y \to T \to S \to R$. Once we have arrived to R, we recover the entries of Y going back from R to Y.

From [3, Corollary 4] we have that the zero counting measures ν_n defined in (8) corresponding to the monic orthogonal polynomials $q_{\mu_n,n}(z) = \prod_{j=1}^{n}\left(z - x_{j,n}\right)$ with respect to the varying measures μ_n, satisfy

$$\nu_n \overset{\star}{\to} \overline{\nu} = (1-a)\widetilde{\sigma} + a\lambda_0 \quad \text{as} \quad n \to \infty, \tag{35}$$

where $\widetilde{\sigma}$ denotes the balayage of the measure σ out of $\mathbb{C} \setminus [-1,\ 1]$ onto $[-1,\ 1]$.

The measure $\overline{\nu}$ is the so called (see [15, Theorem I.1.3]) equilibrium measure under the influence of the external field $(1-a)V^{\sigma}(z)$. From (35) we have the following equilibrium condition

$$V^{\overline{\nu}}(t) - (1-a)V^{\sigma}(t) = aV^{\lambda_0}(t) = a\log 2, \quad t \in [-1,\ 1]. \tag{36}$$

Observe the conditions (3) in both Riemann Hilbert problems. Y requires a normalization at infinity to get to R's behavior at infinity. We modify Y to obtain a Riemann-Hilbert problem whose solution is defined on the same set as Y, which approaches $\mathbb{I}$ as $n \to \infty$. Let us introduce the function $g(z, \overline{\nu})$, which is the analytic potential corresponding to the measure $\overline{\nu}$ described in (35)

$$g(z,\overline{\nu}) = -\int \log(z-t)\,\mathrm{d}\overline{\nu}(t) = V^{\overline{\nu}}(z) - i\int \arg(z-t)\,\mathrm{d}\overline{\nu}(t), \tag{37}$$

with arg denoting the principal argument $g(z, \overline{\nu}) \in \mathcal{H}(K \setminus (-\infty, 1])$. Substituting $g(z, \overline{\nu})$ in (36) we obtain

$$g_+(x, \overline{\nu}) + g_-(x, \overline{\nu}) - 2a \log 2 - 2(1-a) g(x, \sigma) = 0, \quad x \in [-1, 1]. \tag{38}$$

and

$$g_-(x, \overline{\nu}) - g_+(x, \overline{\nu}) = \begin{cases} 0 & \text{if } x \geq 1 \\ 2\pi i & \text{if } x \leq 1 \\ 2\wp(x) & \text{if } x \in (-1, 1), \end{cases} \tag{39}$$

with

$$\wp(x) = \pi i \int_x^1 d\overline{\nu}(t) = \pi i \left[(1-a) \int_x^1 d\widetilde{\sigma}(t) - \frac{a}{\pi} \arccos(x) \right]. \tag{40}$$

Consider the matrices $G(z) = \begin{pmatrix} e^{ng(z,\overline{\nu})} & 0 \\ 0 & e^{-ng(z,\overline{\nu})} \end{pmatrix}$ and $L = \begin{pmatrix} 2^{na} & 0 \\ 0 & 2^{-na} \end{pmatrix}$. We define the matrix function $T = LYGL^{-1}$. So T is the unique solution of the following Riemann-Hilbert problem (T-RHP)

1. $T \in \mathcal{H}(\mathbb{C} \setminus [-1, 1])$,
2. $T_+(x) = T_-(x) M(x)$, $x \in (-1, 1)$,
3. $T(z) = \mathbb{I} + O(1/z)$ as $z \to \infty$,
4. $T(z) = O\begin{pmatrix} 1 & |z \pm 1|^{-1/2} \\ 1 & |z \pm 1|^{-1/2} \end{pmatrix}$ as $z \to \mp 1$,

where according to (38) and (39) the jump matrix $M(x) = \begin{pmatrix} e^{-2n\wp(x)} & (1-x^2)^{-1/2} \\ 0 & e^{2n\wp(x)} \end{pmatrix}$, $x \in (-1, 1)$.

According to [6, Theorem 1.34] there exists a domain D containing the interval $[-1, 1]$ where the function $\wp$ in (40) admits an analytic extension on $D \setminus [-\infty, 1]$ as

$$\mathcal{A}(z) = \pi i \int_z^1 d\overline{\nu}(\zeta) = \pi i \int_z^1 \overline{\nu}'(\zeta) d\zeta, \tag{41}$$

where $\overline{\nu}'(\zeta) = \dfrac{\psi(\zeta)}{\sqrt{1-\zeta^2}}$, with $\psi \in \mathcal{H}(D)$ and $\psi(x) > 0$, $x \in (-1, 1)$. Observe that $\mathcal{A}_+(x) = \wp(x) = -\mathcal{A}_-(x)$, then we write $M(x) = \begin{pmatrix} e^{-2n\mathcal{A}_+(x)} & (1-x^2)^{-1/2} \\ 0 & e^{-2n\mathcal{A}_-(x)} \end{pmatrix}$.

We now seek jump conditions as we have in R-RHP. Consider a closed Jordan curve $\gamma \in D$ surrounding $[-1, 1]$ as we have in R-RHP. Let Ω denote the bounded connected component of $\mathbb{C} \setminus \gamma$. We consider the function $\sqrt{z^2 - 1} \in \mathcal{H}(\mathbb{C} \setminus [-1, 1]$

with $\sqrt{x^2-1}_\pm = \pm i\sqrt{1-x^2}$, $x \in (-1,\ 1)$. We introduce the matrix function S as follows

$$S(z) = \begin{cases} T(z) & \text{when } z \in \mathbb{C} \setminus (\gamma \cup \Omega) \\ T(z)\begin{pmatrix} 1 & 0 \\ -i\sqrt{z^2-1}\,e^{-2n\mathcal{A}(z)} & 1 \end{pmatrix} & \text{when } z \in \Omega \end{cases}.$$

The matrix function S is the solution of the following Riemann-Hilbert problem (S-RHP):

1. $S \in \mathcal{H}(\mathbb{C} \setminus (\gamma \cup [-1,\ 1]))$,
2. $S_+(x) = S_-(x)\begin{pmatrix} 0 & (1-x^2)^{-1/2} \\ -(1-x^2)^{1/2} & 0 \end{pmatrix}$, when $x \in (-1,\ 1)$ and

   $S_+(\zeta) = S_-(\zeta)\begin{pmatrix} 1 & 0 \\ -i\sqrt{z^2-1}\,e^{-2n\mathcal{A}(z)} & 1 \end{pmatrix}$, when $\zeta \in \gamma$.
3. $S(z) = \mathbb{I} + O(1/z)$ as $z \to \infty$,
4. $S(z) = O\begin{pmatrix} 1 & |z \pm 1|^{-1/2} \\ 1 & |z \pm 1|^{-1/2} \end{pmatrix}$ as $z \to \mp 1$.

The jump matrix on γ approaches uniformly the identity matrix I. However it does not happen in $[-1,\ 1]$. We fix this problem in the interval following the steps in [11]. Consider the matrix

$$N(z) = \begin{pmatrix} \dfrac{a(z)+a^{-1}(z)}{2}\dfrac{D(\infty)}{D(z)} & \dfrac{a(z)-a^{-1}(z)}{2i}D(\infty)D(z) \\ \dfrac{a(z)-a^{-1}(z)}{-2i}\dfrac{1}{D(\infty)D(z)} & \dfrac{a(z)+a^{-1}(z)}{2}\dfrac{D(z)}{D(\infty)} \end{pmatrix}, \tag{42}$$

where $D(z) = \left(\dfrac{z}{\sqrt{z^2-1}} + 1\right)^{1/2}$, $D(\infty) = \sqrt{2}$ and $a(z) = \dfrac{(z-1)^{1/4}}{(z+1)^{1/4}}$. Hence N is the solution of the following Riemann-Hilbert problem

1. $N \in \mathcal{H}(\mathbb{C} \setminus [-1,\ 1])$,
2. $N_+(x) = N_-(x)\begin{pmatrix} 0 & (1-x^2)^{-1/2} \\ -(1-x^2)^{1/2} & 0 \end{pmatrix}$, $x \in (-1,\ 1)$,
3. $N(z) = \mathbb{I} + O(1/z)$ as $z \to \infty$,
4. $N(z) = O\begin{pmatrix} 1 & |z \pm 1|^{-1/2} \\ 1 & |z \pm 1|^{-1/2} \end{pmatrix}$ as $z \to \mp 1$.

Introduce the matrix function $R(z) = S(z)N^{-1}$. Taking into account that R and S satisfy the same jump conditions across $(-1,\ 1)$ we have that $R_+(x) = R_-(x)$. So $R \in \mathcal{H}(\mathbb{C} \setminus (\gamma \cup \{-1,\ 1\}))$. Since $\det N = 1$ and from (42) we have that

$$N^{-1}(z) = O\begin{pmatrix} |z \pm 1|^{-1/2} & |z \pm 1|^{-1/2} \\ 1 & 1 \end{pmatrix} \quad \text{as} \quad z \to \mp 1.$$

Thus, when $z \to \mp 1$

$$R(z) = O\begin{pmatrix} 1 & |z \pm 1|^{-1/2} \\ 1 & |z \pm 1|^{-1/2} \end{pmatrix} O\begin{pmatrix} |z \pm 1|^{-1/2} & |z \pm 1|^{-1/2} \\ 1 & 1 \end{pmatrix}.$$

This implies

$$R(z) = O\begin{pmatrix} |z \pm 1|^{-1/2} & |z \pm 1|^{-1/2} \\ |z \pm 1|^{-1/2} & |z \pm 1|^{-1/2} \end{pmatrix},$$

which means that each entry of R has isolated singularities at $z = -1$ and $z = 1$ with $R(z) = O|z \pm 1|^{-1/2}$ as $z \to \mp 1$, and they are removable. So R satisfies the following Riemann-Hilbert conditions:

1. $R \in \mathcal{H}(\mathbb{C} \setminus \gamma)$,
2. $R_+(\zeta) = R_-(\zeta)\begin{pmatrix} 1 & 0 \\ e^{-2n\mathcal{A}(\zeta)} & 1 \end{pmatrix}$, when $\zeta \in \gamma$.
3. $R(z) = \mathbb{I} + O(1/z)$ as $z \to \infty$.

From (41) $2\mathcal{A} \in \mathcal{H}(D \setminus [-\infty, 1])$ and $\mathcal{R}e(2\mathcal{A}_\pm(x)) = 0$, $x \in [-1, 1]$. Using the fact $2\mathcal{A}'_\pm(x) = \pm 2i\overline{\nu}'(x) = \mp 2\pi i \dfrac{\psi(x)}{\sqrt{1-x^2}}$, $x \in (-1, 1)$ and the Cauchy-Riemann conditions we have that $\dfrac{\partial \mathcal{R}e(2\mathcal{A}_\pm)}{\partial y}(x) > 0$, $x \in [-1, 1]$. Since $\mathcal{R}e(2\mathcal{A})$ is a harmonic function on $D\setminus[-1, 1]$ we have $\mathcal{R}e(2\mathcal{A}(z)) > 0$, $z \in D\setminus[-1, 1]$. So given an arbitrary compact set $K \subset D\setminus[-1, 1]$ there exists a constant $c(K) > 0$ and an $N \in \mathbb{N}$ large enough such that for every $n \geq N$ the function $\mathcal{R}e(2\mathcal{A}(z))(z)) > c(K)$, $z \in K$ and $n \geq N$. Note also that $\phi_n \to 0$ as $n \to \infty$. So according [1] we arrive at $R(z) = \mathbb{I} + O(e^{-cn})$ uniformly as $n \to \infty$ for each compact set $K \subset \mathbb{C} \setminus [-1, 1]$. Take $z \in \operatorname{Int}(\gamma)$. Going back now from R to Y, and considering just the first column, we have that:

$$e^{ng(z,\overline{\nu})}\begin{pmatrix} q_{\mu_n,n}(z) \\ 2^{-2na} d_{n,n-1} q_{\mu_n,n-1}(z) \end{pmatrix} = \left(\mathbb{I} + O(e^{-cn})\right)$$

$$\times \begin{pmatrix} \dfrac{a(z)+a^{-1}(z)}{2}\dfrac{D(\infty)}{D(z)} & \dfrac{a(z)-a^{-1}(z)}{2i}D(\infty)D(z) \\ \dfrac{a(z)-a^{-1}(z)}{-2i}\dfrac{1}{D(\infty)D(z)} & \dfrac{a(z)+a^{-1}(z)}{2}\dfrac{D(z)}{D(\infty)} \end{pmatrix}\begin{pmatrix} 1 \\ (1-z^2)^{1/2}e^{-2n\mathcal{A}(z)} \end{pmatrix}.$$

Take the $+$ boundary values of all quantities involved when $z \to x \in (-1, 1)$. Using the following identities from [11] or [12]

$$\frac{a_+(x) \pm a_+(x)}{2} = \frac{1}{\sqrt{2}(1-x^2)^{1/4}} \exp\left(\pm\frac{i}{2}\arccos x \mp i\frac{\pi}{4}\right),$$

we have $\exp\left\{nV^{\overline{\nu}}(x)\right\}\begin{pmatrix} q_{\mu_n,n}(x) \\ 2^{-2na}d_{n,n-1}q_{\mu_n,n-1}(x)\end{pmatrix} = \left(\mathbb{I}+O(e^{-cn})\right)\begin{pmatrix} K_{1,n}(x) \\ K_{2,n}(x)\end{pmatrix}$,

where $K_{2,n}(x) = \frac{1}{i}\cos n\left((1-a)\pi\int_x^1 \mathrm{d}\widetilde{\sigma}(t) - (a-1/n)\arccos x\right)$ and

$K_{1,n}(x) = 2\cos n\left((1-a)\pi\int_x^1 \mathrm{d}\widetilde{\sigma}(t) - a\arccos x\right)$.

Finally we obtain

$$q_{\mu_n,n}(x) = \left(1+O(e^{-cn})\right)e^{-nV^{\overline{\nu}}(x)}K_{1,n}(x) + O(e^{-cn})e^{-nV^{\overline{\nu}}(x)}K_{2,n}(x)$$

and

$$\frac{d_{n,n-1}}{2^{2na}}q_{\mu_n,n-1}(x) = \left(1+O(e^{-cn})\right)e^{-nV^{\overline{\nu}}(x)}K_{2,n}(x) + O(e^{-cn})e^{-nV^{\overline{\nu}}(x)}K_{1,n}(x),$$

which are exactly the equalities stated in (31) and (32). □

5 Proof of Theorem 1

We combine Lemma 5 and Lemma 28. First we choose a special scheme of nodes $\mathbf{y} = \left\{\mathbf{y}_n = (y_{1,n},\ldots,y_{n,n})\right\}_{n\in\Lambda}$ which satisfies (10). The corresponding polynomials have the following form

$$P_n(x) = \prod_{j=1}^{n}\left(x-y_{j,n}\right) = \Phi_n(x)\cos n\left((1-a)\pi\int_x^1 \mathrm{d}\widetilde{\sigma}(t) - a\arccos x\right), \tag{43}$$

Where Φ_n is a real valued function on $[-1, 1]$ that never vanishes. Let us rewrite the relation (31) as follows

$$q_{\mu_n,n}(x) = 2\exp\left\{-nV^{\overline{\nu}}(x)\right\}\left(\cos n\left((1-a)\pi\int_x^1 \mathrm{d}\widetilde{\sigma}(t) - a\arccos x\right) + O(e^{-cn})\right)$$

$$= 2\exp\left\{-nV^{\overline{\nu}}(x)\right\}\cos n\left((1-a)\pi\int_x^1 \mathrm{d}\widetilde{\sigma}(t) - a\arccos x + O(e^{-cn})\right).$$

Combining the above equality with (43) we obtain that $\int_{x_{j,n}}^{y_{j,n}} \mathrm{d}\overline{\nu}(t) = O(e^{-cn})$ and $x_{j,n} - y_{j,n} = O(e^{-cn})$. This implies that $\limsup_{n\to\infty} |q_{\mu_n,n} - P_n|^{1/n}(x) = \exp(-c -$

$V^{\overline{\nu}}(x))$ on $[-1,\ 1]$. Hence

$$\limsup_{n\to\infty}\left(\frac{1}{||q_{\mu_n,n-1}||^2_{2,\mu_n}}\left[\frac{(q_{\mu_n,n}-P_n)q_{\mu_n,n-1}}{q^2_{m(n)}}\right]'\right)^{1/n}(x)$$
$$=\exp(-c+2(1-9)V^{\sigma}(x)+2(1-9)V^{\sigma}(t).$$

Here we have taken into account that dist $(\{\zeta_1,\ldots,\zeta_\kappa\},[-1,\ 1])>1$, which yields $V^{\sigma}(x)<0$, $x\in[-1,\ 1]$. Then we see that condition (28) in Lemma 4 is satisfied. We now prove that condition (27) holds.

Taking into account the equality (32) we have that the zeros of the polynomials $q_{\mu_n,n-1}$ satisfy that for each $j=1,\ldots,n-1$, $n\in\mathbb{N}$

$$(1-a)\pi\int_{x_{j,n-1}}^{1} d\widetilde{\sigma}(t)-(a-1/n)\arccos x+O(e^{-cn})=\frac{2j-1}{2n}\pi. \tag{44}$$

For each $j=1,\ldots,n-1$, we subtract the above equality (44) to (10), and we obtain that

$$\int_{y_{j,n}}^{x_{j,n-1}} d\overline{\nu}(t)=\frac{1}{n}(1+o(1))\quad\text{as}\quad n\to\infty.$$

This means that for n large enough $y_{j,n}<x_{j,n-1}$, $j=1,\ldots,n-1$. Considering now the jth equality in (10) and the $j+1$th in (44) we have that

$$\int_{x_{j,n-1}}^{y_{j+1,n}} d\overline{\nu}(t)=\frac{1}{n}(\pi-1+o(1))\quad\text{as}\quad n\to\infty,$$

which implies that $x_{j,n-1}<y_{j+1,n}$. So condition (27) holds. This proves that the scheme $\mathbf{y}$ is convergent.

Once we know that $\mathbf{y}$ is convergent, we can construct another convergent scheme $\mathbf{x}=\{\mathbf{x}_n=(x_{1,n},\ldots,x_{n,n})\}_{n\in\Lambda}$ taking

$$x_{j,n}-y_{j,n}\le Ae^{-\ell n},\quad j=1,\ldots,n,\quad n\in\Lambda,$$

and follow the previous process. This completes the proof.

Acknowledgments This author's research was supported in part by the research grant MTM2012-36372-C03-01 funded by "Ministerio de Economía y Competitividad", Spain.

References

1. Aptekarev, A.I., Van Assche, W.: Scalar and matrix Riemann-Hilbert approach to the strong asymptotics of Padé approximants and complex orthogonal polynomials with vaying weight. J. Approx. Theory **129**, 129–166 (2004)
2. Bloom, T., Lubinsky, D.S., Stahl, H.: What distribution of points are possible for convergent sequences of interpolatory integration rules. Constr. Appr. Theo. **9**, 41–58 (1993)
3. Bloom, T., Lubinsky, D.S., Stahl, H.: Interpolatory integration rules and orthogonal polynomials with varying weights. Numer. Algorithms **3**, 55–66 (1992)
4. Chihara, T.S.: An Introduction to Orthogonal Polynomials. Gordon and Breach, Science Publishers, New York, London, Paris (1978)
5. Davis, P.J., Rabinowitz, P.: Methods of Numerical Integration, 2nd edn. Academic Press, San Diego, CA (1984)
6. Deift, P., Kriecherbauer, T., McLaughlin, K.T.-R.: New results on the equilibrium measures for logarithmic potentials in presence of an external field. J. Approx. Theory. **95**, 388–475 (1998)
7. Freud, G.: Orthogonal Polynomials. Pergamon Press, London, Toronto, New York (1971)
8. Fidalgo, U., Medina Peralta, S., Mínguez Cenicero, J.: Mixed type multiple orthogonal polynomials: Perfectness and interlacing properties. Linear Algebra Appl. **438**, 1229-1239 (2013)
9. Krein, M.G., Nudel'man, A.A.: The Markov Moment Problem and Extremal Problems. Transl. Math. Monogr., Vol. 50. Amer. Math. Soc., Providence, RI (1977)
10. Kroó, A., Schmidt, D., Sommer, M.: On some properties of A-spaces and their relation to Hobby-Rice theorem. J. Approx. Theory **68**, 136–141 (1999)
11. Kuijlaars, A.B.J.: Riemann-Hilbert analysis for orthogonal polynomials. In: Koelink, E., Van Assche, W. (eds.), Orthogonal Polynomials and Special Functions. Lect. Notes Math., vol. 1817, pp. 167–210. Springer, Berlin (2003)
12. Kuijlaars, A.B.J., Mc Laughlin, K.T.-R., Van Assche, W., Vanlessen, M.: The Riemann-Hilbert approach to strong asymptotics for orthogonal polynomials on $[-1, 1]$. Adv. Math. **188**, 337–398 (2004)
13. Nikishin, E.M.: On simultaneous Padé approximants. Matem. Sb. **113**, 499–519 (1980) (Russian); English translation in Math. USSR Sb. **41**, 409–425 (1982)
14. Rudin, W.: Real and Complex Analysis, 3rd edn. McGraw-Hill, New York (1986)
15. Saff, E.B., Totik, V.: Logarithmic Potentials with External Fields. Series of Comprehensive Studies in Mathematics, Vol. 316. Springer, New York, NY (1997)
16. Szegő, G.: Orthogonal Polynomials, Vol. XXIII. Coll. Pub. Amer. Math. Soc., New York, NY (1939)
17. Wendroff, B.: On orthogonal polynomials. Proc. Amer. Math. Soc. **12**, 554–555 (1961)

Asymptotic Computation of Classical Orthogonal Polynomials

Amparo Gil, Javier Segura, and Nico M. Temme

Abstract The classical orthogonal polynomials (Hermite, Laguerre and Jacobi) are involved in a vast number of applications in physics and engineering. When large degrees n are needed, the use of recursion to compute the polynomials is not a good strategy for computation and a more efficient approach, such as the use of asymptotic expansions, is recommended. In this paper, we give an overview of the asymptotic expansions considered in Gil et al. (Comput. Phys. Commun. **210**, 124–131 (2017)) for computing Laguerre polynomials $L_n^{(\alpha)}(x)$ for bounded values of the parameter α. Additionally, we show examples of the computational performance of an asymptotic expansion for $L_n^{(\alpha)}(x)$ valid for large values of α and n. This expansion was used in Gil et al. (Stud. Appl. Math. **140**(3), (2018)) as starting point for obtaining asymptotic approximations to the zeros. Finally, we analyze the expansions considered in recent publications of the authors for computing the Jacobi polynomials for large degrees, summarizing some of the results from Gil et al. (SIGMA 14(73):9, 2018; SIAM J. Sci. Comput. 41(1):A668-A693, 2019; J. Math. Anal. Appl. 494(2):124642, 2021).

A. Gil
Departamento de Matemática Aplicada y CC. de la Computación. ETSI Caminos, Universidad de Cantabria, Santander, Spain
e-mail: amparo.gil@unican.es

J. Segura (✉)
Departamento de Matemáticas, Estadística y Computación, Universidad de Cantabria, Santander, Spain
e-mail: segurajj@unican.es

N. M. Temme
IAA, Alkmaar, The Netherlands

CWI, Amsterdam, The Netherlands
e-mail: nicot@cwi.nl

F. Marcellán, E. J. Huertas (eds.), *Orthogonal Polynomials: Current Trends and Applications*, SEMA SIMAI Springer Series 22,
https://doi.org/10.1007/978-3-030-56190-1_8

Keywords Classical orthogonal polynomials · Asymptotic expansions · Numerical computation

1 Introduction

As it is well known, Laguerre $L_n^{(\alpha)}(x)$ and Jacobi $P_n^{(\alpha,\beta)}(x)$ polynomials present monotonic and oscillatory regimes depending on the parameter values. Sharp bounds limiting the oscillatory region for these polynomials are given in [3]. In the particular case of Jacobi polynomials, these bounds are

$$\frac{B - 4(n-1)\sqrt{C}}{A} \le x_{n,k}(\alpha,\beta) \le \frac{B + 4(n-1)\sqrt{C}}{A}, \tag{1}$$

where $x_{n,k}(\alpha,\beta)$ are the zeros of the Jacobi polynomials and

$$\begin{aligned} A &= (2n+\alpha+\beta)(n(2n+\alpha+\beta)+2(\alpha+\beta+2)),\\ B &= (\beta-\alpha)((\alpha+\beta+6)n+2(\alpha+\beta)),\\ C &= n^2(n+\alpha+\beta+1)^2+(\alpha+1)(\beta+1)(n^2+(\alpha+\beta+4)n+2(\alpha+\beta)). \end{aligned} \tag{2}$$

An example of the oscillatory behaviour of the Jacobi polynomials for two different values of n is given in Fig. 1.

On the other hand, the classical orthogonal polynomials satisfy (also well-known) three-term recurrence relations

$$p_{n+1}(x) = (A_n x + B_n) p_n(x) - C_n p_{n-1}, \tag{3}$$

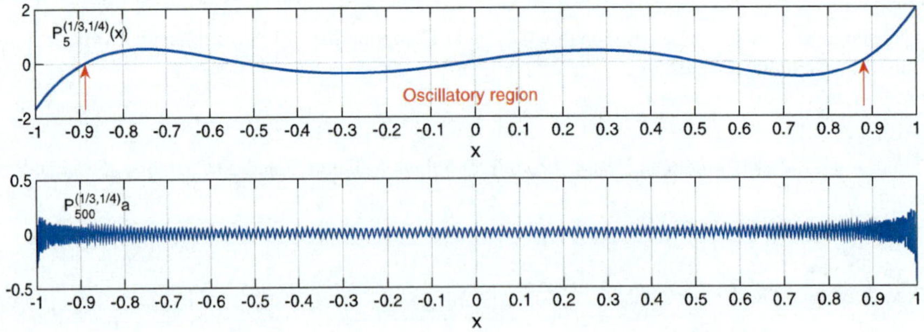

Fig. 1 Two plots of Jacobi polynomials ($P_5^{(1/3,1/4)}(x)$ and $P_{500}^{(1/3,1/4)}(x)$) showing the oscillatory behaviour of the polynomials

where, for Laguerre polynomials $L_n^{(\alpha)}(x)$, the coefficients are given by

$$A_n = \frac{-1}{n+1}, \quad B_n = \frac{2n+\alpha+1}{n+1}, \quad C_n = \frac{n+\alpha}{n+1}, \tag{4}$$

and for Jacobi polynomials $P_n^{(\alpha,\beta)}(x)$,

$$\begin{aligned} A_n &= \frac{(2n+\alpha+\beta+1)(2n+\alpha+\beta+2)}{2(n+1)(n+\alpha+\beta+1)}, \\ B_n &= \frac{(\alpha^2-\beta^2)(2n+\alpha+\beta+1)}{2(n+1)(n+\alpha+\beta+1)(2n+\alpha+\beta)}, \\ C_n &= \frac{(n+\alpha)(n+\beta)(2n+\alpha+\beta+2)}{(n+1)(n+\alpha+\beta+1)(2n+\alpha+\beta)}. \end{aligned} \tag{5}$$

By using the Perron-Kreuser theorem [15] it is easy to show that the three-term recurrence relations for the classical orthogonal polynomials are not badly conditioned in the sense that there is no solution which dominates exponentially over the rest. Therefore the recurrence relations can be used in the forward direction, for example, with starting values $L_0^{(\alpha)}(x) = 1$ and $L_1^{(\alpha)}(x) = 1 + \alpha - x$ for Laguerre polynomials, or $P_0^{(\alpha,\beta)}(x) = 1$ and $P_1^{(\alpha,\beta)}(x) = \frac{1}{2}(\alpha - \beta) + \frac{1}{2}(\alpha + \beta + 2)x$, to compute the Jacobi polynomials when n is small/moderate. However, when large degrees n are needed, the use of recursion to compute the polynomials is not a good strategy for computation and a more efficient approach, such as the use of asymptotic expansions, is recommended.

The use of asymptotic expansions in the computation of Hermite polynomials has already been discussed in [7], since Hermite polynomials are a particular case of the parabolic cylinder function $U(a, x)$. In this paper, we consider the asymptotic computation of Laguerre and Jacobi polynomials. Other recent references dealing with the numerical evaluation of these polynomials using asymptotic methods are, for example, [2] and [13].

We first give an overview of the asymptotic expansions considered in [8] for computing Laguerre polynomials $L_n^{(\alpha)}(x)$ for bounded values of the parameter α. Additionally, we show examples of the computational performance of an asymptotic expansion for $L_n^{(\alpha)}(x)$ given in [6] valid for large values of α and n. We also analyze the expansions considered in [9, 10] and [11] to compute the Jacobi polynomials for large degrees n. These expansions have been used as starting point to obtain asymptotic approximations to the nodes and/or weights of the Gauss-Jacobi (G-J) quadrature. In the asymptotic expansions for large degree n described in [11], we assume that α and β should be of order $O(1)$ for large values of n. It is interesting to note that these approximations can be used as standalone methods for the non-iterative computation (15 − 16 digits accuracy) of the nodes and weights of G-J quadratures of high degree ($n \geq 100$). Alternatively, they can be used as accurate starting points of methods for solving nonlinear equations. This is the approach followed, for example, in [12] in which simple asymptotic approximations for the nodes are iteratively refined by Newton's method.

2 Laguerre Polynomials

The computation of Laguerre polynomials $L_n^{(\alpha)}(z)$ for bounded values of the parameter α was considered in [8]. In that paper, we analyzed the computational performance of three types of asymptotic expansions valid for large values of n: an expansion in terms of Airy functions and two expansions in terms of Bessel functions (one for small values of the variable z and other one in which larger values are allowed). It is possible to obtain asymptotic expansions in terms of elementary functions starting from Bessel expansions, as done in [1]; however, the Bessel expansions appear to have a larger range of validity.

We next summarize results given in [8] and [6]; more details about the expansions can be found in [18].

a) An expansion in terms of Airy functions

The representation used is

$$L_n^{(\alpha)}(\nu x) = (-1)^n \frac{e^{\frac{1}{2}\nu x}\chi(\zeta)}{2^\alpha \nu^{\frac{1}{3}}} \left(\mathrm{Ai}\left(\nu^{2/3}\zeta\right) A(\zeta) + \nu^{-\frac{4}{3}} \mathrm{Ai}'\left(\nu^{2/3}\zeta\right) B(\zeta)\right) \tag{6}$$

with expansions

$$A(\zeta) \sim \sum_{j=0}^{\infty} \frac{\alpha_{2j}}{\nu^{2j}}, \quad B(\zeta) \sim \sum_{j=0}^{\infty} \frac{\beta_{2j+1}}{\nu^{2j}}, \quad \nu \to \infty. \tag{7}$$

Here

$$\nu = 4\kappa, \quad \kappa = n + \tfrac{1}{2}(\alpha+1), \quad \chi(\zeta) = 2^{\frac{1}{2}} x^{-\frac{1}{4}-\frac{1}{2}\alpha} \left(\frac{\zeta}{x-1}\right)^{\frac{1}{4}}, \tag{8}$$

and

$$\begin{aligned} \tfrac{2}{3}(-\zeta)^{\frac{3}{2}} &= \tfrac{1}{2}\left(\arccos\sqrt{x} - \sqrt{x-x^2}\right) \ \text{if } 0 < x \le 1, \\ \tfrac{2}{3}\zeta^{\frac{3}{2}} &= \tfrac{1}{2}\left(\sqrt{x^2-x} - \operatorname{arccosh}\sqrt{x}\right) \quad \text{if } x \ge 1. \end{aligned} \tag{9}$$

The expansions in (7) hold uniformly for bounded α and $x \in (x_0, \infty)$, where $x_0 \in (0, 1)$, a fixed number. We observe that $x = 1$ is a turning point, where the argument of the Airy functions vanishes (see (9)). Left of $x = 1$, where $\zeta < 0$, the zeros of $L_n^{(\alpha)}(\nu x)$ occur.

The first few coefficients in the expansions (7) are $\alpha_0 = 1$ and

For $0 < x \leq 1$

$$\begin{aligned}\alpha_2 = &-\frac{1}{1152}\left(4608u^8v^6\mu^3 + 2016v^6u^8\mu^2 + 2304u^8v^6\mu^4\right.\\ &-672v^6u^4\mu^2 - 192v^6\mu u^6 + 2880v^6u^6\mu^2 + 2688v^3u^7\mu^2\\ &-288v^6\mu u^8 + 3072u^6v^6\mu^3 - 672v^6\mu u^4 + 2688v^3\mu u^7 +\\ &924v^6u^2 + 558v^6u^4 - 180u^6v^6 - 135v^6u^8 + 504u^7v^3 + 336v^3u^5\\ &\left.+280u^3v^3 - 7280u^6 + 385v^6\right)/(v^6u^6),\\ \beta_1 = &\frac{1}{12}\left(48v^3u^4\mu^2 + 9v^3u^4 + 48v^3\mu u^4 - 20u^3 + 6v^3u^2 + 5v^3\right)/(u^3v^4),\end{aligned}$$

where $u = \sqrt{1/x - 1}$, $v = 2\sqrt{-\zeta}$ and $\mu = (\alpha - 1)/2$.

For $x > 1$

$$\begin{aligned}\alpha_2 = &\frac{1}{1152}\left(4608u^8v^6\mu^3 + 2016v^6u^8\mu^2 +\right.\\ &2304u^8v^6\mu^4 - 672v^6u^4\mu^2 + 192v^6\mu u^6\\ &-2880v^6u^6\mu^2 - 336v^3u^5 - 288v^6\mu u^8 - 3072u^6v^6\mu^3\\ &-672v^6\mu u^4 + 280u^3v^3 - 924v^6u^2 + 558v^6u^4 + 180u^6v^6\\ &-135v^6u^8 + 2688v^3u^7\mu^2 + 504u^7v^3 + 2688v^3\mu u^7\\ &\left.-7280u^6 + 385v^6\right)/(u^6v^6),\\ \beta_1 = &\frac{1}{12}\left(48v^3u^4\mu^2 + 9v^3u^4 + 48v^3\mu u^4 - 20u^3 - 6v^3u^2 + 5v^3\right)/(u^3v^4),\end{aligned}$$

where $u = \sqrt{1 - 1/x}$, $v = 2\sqrt{\zeta}$ and $\mu = (\alpha - 1)/2$.

A detailed explanation of the method used to obtain the coefficients of Airy expansions can be found in [18, §23.3]. The expansion in terms of Airy functions was derived in [4].

b) A simple Bessel-type expansion

The starting point to obtain this expansion is the well-known relation between Laguerre polynomials and the Kummer function ${}_1F_1(a; c; x)$

$$L_n^{(\alpha)}(z) = \binom{n+\alpha}{n} {}_1F_1\left(\begin{matrix}-n\\ \alpha+1\end{matrix}; z\right). \tag{10}$$

Then, we use an expansion for ${}_1F_1(a; c; x)$ for large negative values of a; see [18, §10.3.4]

$$\begin{aligned}\frac{1}{\Gamma(c)} {}_1F_1\left(\begin{matrix}-a\\ c\end{matrix}; z\right) \sim \left(\frac{z}{a}\right)^{\frac{1}{2}(1-c)} \frac{\Gamma(1+a)e^{\frac{1}{2}z}}{\Gamma(a+c)} \times\\ \left(J_{c-1}\left(2\sqrt{az}\right) \sum_{k=0}^{\infty} \frac{a_k(z)}{(-a)^k} - \sqrt{\frac{z}{a}} J_c\left(2\sqrt{az}\right) \sum_{k=0}^{\infty} \frac{b_k(z)}{(-a)^k}\right).\end{aligned} \tag{11}$$

This expansion of ${}_1F_1(-a;c;z)$ is valid for bounded values of z and c, with $a\to\infty$ inside the sector $-\pi+\delta\le \operatorname{ph} a\le \pi-\delta$. Using this, we obtain

$$\begin{aligned}L_n^{(\alpha)}(x)\sim \left(\frac{x}{n}\right)^{-\frac12\alpha} e^{\frac12 x}\times \\ \left(J_\alpha\left(2\sqrt{nx}\right)\sum_{k=0}^{\infty}(-1)^k\frac{a_k(x)}{n^k}-\sqrt{\frac{x}{n}}J_{\alpha+1}\left(2\sqrt{nx}\right)\sum_{k=0}^{\infty}(-1)^k\frac{b_k(x)}{n^k}\right),\end{aligned} \tag{12}$$

as $n\to\infty$.

The coefficients $a_k(x)$ and $b_k(x)$ follow from the expansion of the function

$$f(x,s)=e^{xg(s)}\left(\frac{s}{1-e^{-s}}\right)^{\alpha+1},\quad g(s)=\frac{1}{s}-\frac{1}{e^s-1}-\frac12. \tag{13}$$

The function f is analytic in the strip $|\Im s|<2\pi$ and it can be expanded for $|s|<2\pi$ into

$$f(x,s)=\sum_{k=0}^{\infty}c_k(x)s^k. \tag{14}$$

The coefficients $a_k(x)$ and $b_k(x)$ are given in terms of the $c_k(x)$ coefficients:

$$\begin{aligned}a_k(x)&=\sum_{m=0}^{k}\binom{k}{m}(m+1-c)_{k-m}x^m c_{k+m}(x),\\ b_k(x)&=\sum_{m=0}^{k}\binom{k}{m}(m+2-c)_{k-m}x^m c_{k+m+1}(x),\end{aligned} \tag{15}$$

$k=0,1,2,\ldots,c=\alpha+1$.

The coefficients $c_k(x)$ are combinations of Bernoulli numbers and Bernoulli polynomials, the first ones being

$$\begin{aligned}&c_0(x)=1,\quad c_1(x)=\tfrac{1}{12}\left(6c-x\right),\\ &c_2(x)=\tfrac{1}{288}\left(-12c+36c^2-12xc+x^2\right),\\ &c_3(x)=\tfrac{1}{51840}\left(-5x^3+90x^2c+(-540c^2+180c+72)x+1080c^2(c-1)\right),\end{aligned} \tag{16}$$

and the first relations are

$$
\begin{aligned}
&a_0(x) = c_0(x) = 1, \quad b_0(x) = c_1(x),\\
&a_1(x) = (1-c)c_1(x) + xc_2(x), \quad b_1(x) = (2-c)c_2(x) + xc_3(x),\\
&a_2(x) = (c^2 - 3c + 2)c_2(x) + (4x - 2xc)c_3(x) + x^2 c_4(x),\\
&b_2(x) = (c^2 - 5c + 6)c_3(x) + (6x - 2xc)c_4(x) + x^2 c_5(x),
\end{aligned}
\tag{17}
$$

again with $c = \alpha + 1$.

c) **A not so simple expansion in terms of Bessel functions**
To understand how this representation and the coefficients of the expansion are obtained, we start with the standard form

$$
F_\zeta(\nu) = \frac{1}{2\pi i}\int_C e^{\nu(u-\zeta/u)} f(u)\,\frac{du}{u^{\alpha+1}}, \tag{18}
$$

where the contour C starts at $-\infty$ with $\operatorname{ph} u = -\pi$, encircles the origin anti-clockwise, and returns to $-\infty$ with $\operatorname{ph} u = \pi$. The function $f(u)$ is assumed to be analytic in a neighborhood of C, and in particular in a domain that contains the saddle points $\pm ib$, where $b = \sqrt{\zeta}$.

When f is replaced by unity, we obtain the Bessel function:

$$
F_\zeta(\nu) = \zeta^{-\frac{1}{2}\alpha} J_\alpha\left(2\nu\sqrt{\zeta}\right). \tag{19}
$$

For Laguerre polynomials, we have the integral representation

$$
L_n^{(\alpha)}(z) = \frac{1}{2\pi i}\int_{\mathcal{L}} (1-t)^{-\alpha-1} e^{-tz/(1-t)}\,\frac{dt}{t^{n+1}}, \tag{20}
$$

where $\mathcal{L}$ is a circle around the origin with radius less than unity.
After substituting $t = e^{-s}$, we obtain

$$
e^{-\nu x} L_n^{(\alpha)}(2\nu x) = \frac{2^{-\alpha}}{2\pi i}\int_{-\infty}^{(0+)} e^{\nu h(s,x)} \left(\frac{\sinh s}{s}\right)^{-\alpha-1} \frac{ds}{s^{\alpha+1}}, \tag{21}
$$

where $\nu = 2n + \alpha + 1$ and $h(s,x) = s - x\coth s$. The contour is the same as in (18). Using the transformation $h(s,x) = u - \zeta/u$, we obtain the integral representation in the standard form of (18)

$$
2^\alpha e^{-\nu x} L_n^{(\alpha)}(2\nu x) = \frac{1}{2\pi i}\int_{-\infty}^{(0+)} e^{\nu(u-\zeta/u)} f(u)\,\frac{du}{u^{\alpha+1}}, \tag{22}
$$

where

$$f(u) = \left(\frac{u}{\sinh s}\right)^{\alpha+1} \frac{ds}{du}. \tag{23}$$

We now consider the following recursive scheme

$$\begin{aligned} f_j(u) &= A_j(\zeta) + B_j(\zeta)/u + \left(1 + b^2/u^2\right) g_j(u), \\ f_{j+1}(u) &= g_j'(u) - \frac{\alpha+1}{u} g_j(u), \\ A_j(\zeta) &= \frac{f_j(ib) + f_j(-ib)}{2}, \quad B_j(\zeta) = i\frac{f_j(ib) - f_j(-ib)}{2b}, \end{aligned} \tag{24}$$

with $f_0(u) = f(u)$, the coefficient function.
Using this scheme and integration by parts in (22), we obtain the following representation

$$L_n^{(\alpha)}(2\nu x) = \frac{e^{\nu x}\chi(\zeta)}{2^\alpha \zeta^{\frac{1}{2}\alpha}} \left(J_\alpha(2\nu\sqrt{\zeta})A(\zeta) - \frac{1}{\sqrt{\zeta}} J_{\alpha+1}(2\nu\sqrt{\zeta})B(\zeta)\right), \tag{25}$$

with expansions

$$A(\zeta) \sim \sum_{j=0}^{\infty} \frac{A_{2j}(\zeta)}{\nu^{2j}}, \quad B(\zeta) \sim \sum_{j=0}^{\infty} \frac{B_{2j+1}(\zeta)}{\nu^{2j+1}}, \quad \nu \to \infty. \tag{26}$$

Here,

$$\nu = 2n + \alpha + 1, \quad \chi(\zeta) = (1-x)^{-\frac{1}{4}} \left(\frac{\zeta}{x}\right)^{\frac{1}{2}\alpha + \frac{1}{4}}, \quad x < 1, \tag{27}$$

with ζ given by

$$\begin{aligned} \sqrt{-\zeta} &= \tfrac{1}{2}\left(\sqrt{x^2 - x} + \operatorname{arcsinh}\sqrt{-x}\right), && \text{if} \quad x \le 0, \\ \sqrt{\zeta} &= \tfrac{1}{2}\left(\sqrt{x - x^2} + \arcsin\sqrt{x}\right), && \text{if} \quad 0 \le x < 1. \end{aligned} \tag{28}$$

This expansion in terms of Bessel functions was derived in [4]. It is uniformly valid for $x \le 1 - \delta$, where $\delta \in (0, 1)$ is a fixed number. Recall that the expansion in terms of Airy functions is valid around the turning point $x = 1$ and up to infinity, but not for small positive x. As shown in [4], the Bessel-type expansion (left of $x = 1$) and the Airy-type expansion are valid in overlapping x-domains of $\mathbb{R}$. For $x < 0$ the J-Bessel functions can be written in terms of modified I-Bessel functions.

The coefficients $A_j(\zeta)$ and $B_j(\zeta)$ in (26) can all be expressed in terms of the derivatives $f^{(k)}(\pm ib)$ of $f(u)$ (see (23)) at the saddle points $\pm ib$. We have to proceed as follows:
We expand the functions $f_j(u)$ (24) in two-point Taylor expansions

$$f_j(u) = \sum_{k=0}^{\infty} C_k^{(j)}(u^2 - b^2)^k + u \sum_{k=0}^{\infty} D_k^{(j)}(u^2 - b^2)^k. \tag{29}$$

Using the recursive scheme for the functions $f_j(u)$ given in (24), we derive the following scheme for the coefficients

$$\begin{aligned} C_k^{(j+1)} &= (2k - \alpha)D_k^{(j)} + b^2(\alpha - 4k - 2)D_{k+1}^{(j)} + 2(k+1)b^4 D_{k+2}^{(j)}, \\ D_k^{(j+1)} &= (2k + 1 - \alpha)C_{k+1}^{(j)} - 2(k+1)b^2 C_{k+2}^{(j)}, \end{aligned} \tag{30}$$

for $j, k = 0, 1, 2, \ldots$, and the coefficients A_j and B_j follow from

$$A_j(\zeta) = C_0^{(j)}, \quad B_j(\zeta) = -b^2 D_0^{(j)}, \quad j \geq 0. \tag{31}$$

In the present case of the Laguerre polynomials the functions f_{2j} are even and f_{2j+1} are odd, and we have $A_{2j+1}(\zeta) = 0$ and $B_{2j}(\zeta) = 0$. A few non–vanishing coefficients are

$$\begin{aligned} A_0(\zeta) &= f(ib), \\ B_1(\zeta) &= -\tfrac{1}{4}b\big((2\alpha - 1)if^{(1)}(ib) + bf^{(2)}(ib)\big), \\ A_2(\zeta) &= -\tfrac{1}{32b}\big(3i(4^2\alpha - 1)f^{(1)}(ib) - (3 - 16\alpha + 4\alpha^2)bf^{(2)}(ib) \\ &\quad +2i(2\alpha - 3)b^2 f^{(3)}(ib)b^2 + b^3 f^{(4)}(ib)\big), \\ B_3(\zeta) &= -\tfrac{1}{384b}\big(3(4\alpha^2 - 1)(2\alpha - 3)(if^{(1)}(ib) + bf^{(2)}(ib)) \\ &\quad +2i(\alpha - 7)(2\alpha - 1)(2\alpha - 3)b^2 f^{(3)}(ib) \\ &\quad +3(19 - 20\alpha + 4\alpha^2))b^3 f^{(4)}(ib) \\ &\quad -3i(2\alpha - 5)b^4 f^{(5)}(ib) - b^5 f^{(6)}(ib)\big). \end{aligned} \tag{32}$$

To have $A_0(\zeta) = 1$ we have to scale all A and B-coefficients with respect to $A_0(\zeta) = \chi(\zeta)$. As an example, using this normalization and the explicit expression for the first two derivatives $f^{(k)}(ib)$ of $f(u)$, we obtain for $B_1(\zeta)$ the following expression

$$B_1(\zeta) = \frac{1}{48\xi}\left(5\xi^4 b + 6\xi^2 b + 3\xi + 12\alpha^2(b - \xi) - 3b\right), \tag{33}$$

where $\xi = \sqrt{\dfrac{x}{1-x}}$.

In the resulting algorithm for Laguerre polynomials described in [8], the two Bessel-type expansions were used in the oscillatory region of the functions. The use of the simple Bessel-type expansion was restricted, as expected, to values of the argument of the Laguerre polynomials very close to the origin. The second Bessel-type expansion was used in the part of the oscillatory region where the Airy expansion failed (computational schemes for Airy functions are more efficient than the corresponding schemes for Bessel functions). Therefore, the Airy expansion was used in the monotonic region and in part of the oscillatory region of the Laguerre polynomials $L_n^{(\alpha)}(x)$, with α small.

Recall that we have observed (see below (28)) that the Airy-type and Bessel-type expansions in (6) and (25), respectively, are valid in overlapping x-intervals of $\mathbb{R}$.

d) An expansion in terms of Bessel functions for large α and n

When α is large, an expansion suitable for the oscillatory region of the Laguerre polynomials and valid for large values of n, is the expansion given in [6]

$$\begin{aligned} L_n^{(\alpha)}(4\kappa x) = e^{-\kappa A}\chi(b)\left(\frac{b}{2\kappa x}\right)^{\alpha}\frac{\Gamma(n+\alpha+1)}{n!}\times \\ \left(J_\alpha(4\kappa b)A(b) - 2bJ'_\alpha(4\kappa x)B(b)\right), \end{aligned} \tag{34}$$

with expansions

$$A(b) \sim \sum_{k=0}^{\infty}\frac{A_k(b)}{\kappa^k}, \quad B(b) \sim \sum_{k=0}^{\infty}\frac{B_k(b)}{\kappa^k}, \quad \kappa\to\infty, \tag{35}$$

where

$$\kappa = n + \tfrac{1}{2}(\alpha+1), \quad \chi(b) = \left(\frac{4b^2-\tau^2}{4x-4x^2-\tau^2}\right)^{\frac{1}{4}}, \quad \tau = \frac{\alpha}{2\kappa}. \tag{36}$$

We have $\tau < 1$. The quantity b is a function of x and follows from the relation

$$\begin{aligned} 2W - 2\tau\arctan\frac{W}{\tau} = \\ 2R - \arcsin\frac{1-2x}{\sqrt{1-\tau^2}} - \tau\arcsin\frac{x-\frac{1}{2}\tau^2}{x\sqrt{1-\tau^2}} + \tfrac{1}{2}\pi(1-\tau), \end{aligned} \tag{37}$$

where

$$R = \tfrac{1}{2}\sqrt{4x-4x^2-\tau^2} = \sqrt{(x_2-x)(x-x_1)}, \quad W = \sqrt{4b^2-\tau^2}, \tag{38}$$

and

$$x_1 = \tfrac{1}{2}\left(1-\sqrt{1-\tau^2}\right), \quad x_2 = \tfrac{1}{2}\left(1+\sqrt{1-\tau^2}\right). \tag{39}$$

The relation in (37) can be used for $x \in [x_1, x_2]$, in which case $b \geq \frac{1}{2}\tau$. In this interval the Laguerre polynomial $L_n^{(\alpha)}(4\kappa x)$ oscillates. The endpoints of this interval can be seen as two turning points, where in fact Airy functions can be used. We assume that $\tau \leq \tau_0$, where τ_0 is a fixed number in the interval (0, 1). When $\tau \to 1$, the points x_1 and x_2, which can be considered as turning points, become the same, and we expect expansions in terms of parabolic cylinder function (in the present case these are Hermite polynomials).

The expansion is valid for $x \in [0, (1-\delta)x_2]$, with δ a small positive number. For $x < x_1$, the relation in (37) should be modified, because the arguments of the square roots of W and R become negative.

At the turning point $x = x_1$ we have $b = \frac{1}{2}\tau$, and the argument of the Bessel functions in (34) becomes $4\kappa b = 2\kappa\tau = \alpha$, where we have used the definition of τ in (36). We see that argument and order of the Bessel functions become equal, and because we assume that α is large, this explains that at the turning point x_1 we can expect Airy-type behaviour. The Airy-type approximation around x_1 cannot be valid near $x = 0$, but the Bessel function can handle this.

The first coefficients of the expansions in (35) are

$$\begin{aligned} A_0(b) &= 1, \quad B_0(b) = 0, \\ A_1 &= \frac{\tau}{24(\tau^2-1)}, \quad B_1 = \frac{PR^3+QW^3}{192R^3W^4(\tau^2-1)}, \\ &\quad P = 4(2\tau^2+12b^2)(1-\tau^2), \\ &\quad Q = 2\tau^4 - 12x^2\tau^2 - \tau^2 - 8x^3 + 24x^2 - 6x, \\ B_2(b) &= A_1(b)B_1(b). \end{aligned} \tag{40}$$

An example of the computational performance of the expansion (34) is given in Fig. 2. For comparison, we show the results for two different values of α ($\alpha = 198.5, \alpha = 20.6$) and a fixed value of n ($n = 340$). The coefficients $A_k,\ B_k,\ k = 0, 1, 2$ in (35) have been considered in the computations. The expansion has been tested by considering the relation given in Eq. (18.9.13) of [14], written in the form

$$\epsilon = \left| \frac{L_{n-1}^{(\alpha+1)}(z) + L_n^{(\alpha)}(z)}{L_n^{(\alpha+1)}(z)} - 1 \right|, \tag{41}$$

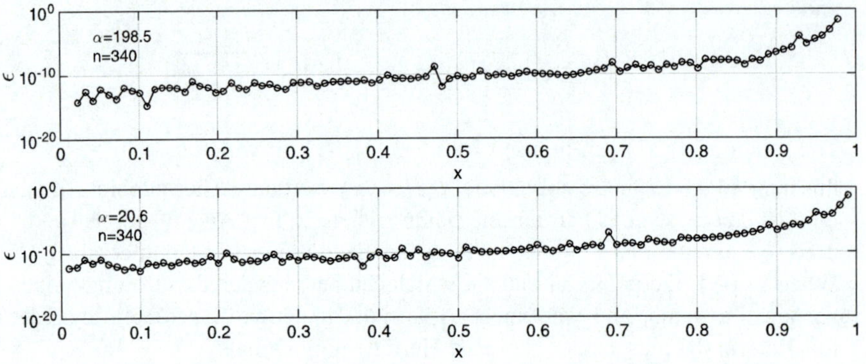

Fig. 2 Test of the performance of the expansion (34) for large values of n and α. The ϵ-value in Eq. (41) is plotted

where $z = 4\kappa x$ is the argument of the Laguerre polynomial in the expansion (34). As can be seen, an accuracy better than 10^{-10} can be obtained in the first half of the x-interval for the two values of α tested.

3 Jacobi Polynomials

Next we review of some useful asymptotic expansions for computing Jacobi polynomials.

3.1 An Expansion in Terms of Elementary Functions

In [11] we considered the following representation

$$P_n^{(\alpha,\beta)}(\cos\theta) = \frac{G_\kappa(\alpha,\beta)}{\sqrt{\pi\kappa}} \frac{\cos\chi\, U(x) - \sin\chi\, V(x)}{\sin^{\alpha+\frac{1}{2}}\frac{1}{2}\theta\, \cos^{\beta+\frac{1}{2}}\frac{1}{2}\theta}, \tag{42}$$

where

$$x = \cos\theta, \quad \chi = \kappa\theta - \left(\tfrac{1}{2}\alpha + \tfrac{1}{4}\right)\pi, \quad \kappa = n + \tfrac{1}{2}(\alpha+\beta+1). \tag{43}$$

The expansion is valid for $x \in [-1+\delta, 1-\delta]$, where δ is a small positive number.

The functions $U(x)$ and $V(x)$ have the expansions

$$U(x) \sim \sum_{k=0}^{\infty} \frac{u_{2k}(x)}{\kappa^{2k}}, \quad V(x) \sim \sum_{k=0}^{\infty} \frac{v_{2k+1}(x)}{\kappa^{2k+1}}. \tag{44}$$

The first coefficients are

$$\begin{aligned} u_0(x) &= 1, \quad v_1(x) = \frac{2\alpha^2 - 2\beta^2 + (2\alpha^2 + 2\beta^2 - 1)x}{8\sin\theta}, \\ u_2(x) &= \frac{1}{384\sin^2\theta}\Big(12(5 - 2\alpha^2 - 2\beta^2)(\alpha^2 - \beta^2)x + \\ &\quad 4(-3(\alpha^2 - \beta^2)^2 + 3(\alpha^2 + \beta^2) - 6 + 4\alpha(\alpha^2 - 1 + 3\beta^2) + \\ &\quad (-12(\alpha^2 + \beta^2)(\alpha^2 + \beta^2 - 1) - 16\alpha(\alpha^2 - 1 + 3\beta^2) - 3)x^2\Big). \end{aligned} \tag{45}$$

The function $G_\kappa(\alpha, \beta)$ is given by

$$G_\kappa(\alpha, \beta) = \frac{\Gamma(n + \alpha + 1)}{n!\,\kappa^\alpha} = \frac{\Gamma\left(\kappa + \frac{1}{2}(\alpha - \beta + 1)\right)}{\Gamma\left(\kappa - \frac{1}{2}(\alpha + \beta - 1)\right)\kappa^\alpha}, \tag{46}$$

with expansion in negative powers of $\left(\kappa - \frac{1}{2}\beta\right)^2$

$$G_\kappa(\alpha, \beta) \sim (w/\kappa)^\alpha \sum_{m=0}^{\infty} \frac{C_m(\rho)(-\alpha)_{2m}}{w^{2m}}, \tag{47}$$

where

$$w = \kappa - \tfrac{1}{2}\beta, \quad \rho = \tfrac{1}{2}(\alpha + 1), \tag{48}$$

and the first few $C_m(\rho)$ coefficients are

$$\begin{aligned} C_0(\rho) &= 1, \quad C_1(\rho) = -\tfrac{1}{12}\rho, \quad C_2(\rho) = \tfrac{1}{1440}(5\rho + 1), \\ C_3(\rho) &= -\rho\frac{4 + 21\rho + 35\rho^2}{362880}, \\ C_4(\rho) &= \rho\frac{18 + 101\rho + 210\rho^2 + 175\rho^3}{87091200}. \end{aligned} \tag{49}$$

3.2 An Expansion in Terms of Bessel Functions

Another expansion, this time in terms of Bessel functions, considered in [11] is

$$\begin{aligned} P_n^{(\alpha,\beta)}(\cos\theta) &= \frac{G_\kappa(\alpha,\beta)}{\sin^\alpha \frac{1}{2}\theta\, \cos^\beta \frac{1}{2}\theta}\sqrt{\frac{\theta}{\sin\theta}}\, W(\theta), \\ W(\theta) &= J_\alpha(\kappa\theta)\, S(\theta) + \frac{1}{\kappa} J_{\alpha+1}(\kappa\theta)\, T(\theta), \end{aligned} \tag{50}$$

where $G_\kappa(\alpha,\beta)$ is defined in (46), and

$$\kappa = n + \tfrac{1}{2}(\alpha+\beta+1). \tag{51}$$

This expansion holds uniformly with respect to $\theta \in [0, \pi-\delta]$, where δ is a small positive number. When an asymptotic is needed for θ near π (or for x near -1), we can use the symmetry relation

$$P_n^{(\alpha,\beta)}(-x) = (-1)^n P_n^{(\beta,\alpha)}(x). \tag{52}$$

The functions $S(\theta)$ and $T(\theta)$ have the expansions

$$S(\theta) \sim \sum_{k=0}^{\infty} \frac{S_k(\theta)}{\kappa^{2k}}, \quad T(\theta) \sim \sum_{k=0}^{\infty} \frac{T_k(\theta)}{\kappa^{2k}}, \quad \kappa \to \infty, \tag{53}$$

with $S_0(\theta) = A_0(\theta) = 1$, $T_0(\theta) = A_1(\theta)$, and for $k = 1, 2, 3, \ldots$

$$\begin{aligned} S_k(\theta) &= -\frac{1}{\theta^{k-1}} \sum_{j=0}^{k-1} \binom{k-1}{j} A_{j+k+1}(\theta)(-\theta)^j 2^{k-1-j} (\alpha+2+j)_{k-j-1}, \\ T_k(\theta) &= \frac{1}{\theta^k} \sum_{j=0}^{k} \binom{k}{j} A_{j+k+1}(\theta)(-\theta)^j 2^{k-j} (\alpha+1+j)_{k-j}. \end{aligned} \tag{54}$$

The coefficients $A_k(\theta)$ are analytic functions for $0 \le \theta < \pi$. The first ones are $A_0(\theta) = 1$ and

$$A_1(\theta) = \frac{\left(4\alpha^2-1\right)(\sin\theta - \theta\cos\theta) + 2\theta\left(\alpha^2-\beta^2\right)(\cos\theta - 1)}{8\theta\sin\theta}. \tag{55}$$

For small values of θ, it is convenient to consider expansions of the coefficients $A_k(\theta)$:

$$A_k(\theta) = \chi^k \theta^k \sum_{j=0}^{\infty} A_{jk}\theta^{2j}, \quad \chi = \frac{\theta}{\sin\theta}, \tag{56}$$

where the series represent entire functions of θ. The first few A_{jk} are

$$\begin{aligned}
A_{0,1} &= \tfrac{1}{24}\left(\alpha^2 + 3\beta^2 - 1\right),\\
A_{1,1} &= \tfrac{1}{480}\left(-3\alpha^2 - 5\beta^2 + 2\right),\\
A_{0,2} &= \tfrac{1}{5760}\left(-16\alpha - 14\alpha^2 - 90\beta^2 + 5\alpha^4 + 4\alpha^3 + 45\beta^4 + 30\beta^2\alpha^2 + 60\beta^2\alpha + 21\right).
\end{aligned} \tag{57}$$

More details about the coefficients $A_k(\theta)$ are given in [11].

As in the case of the Bessel-type expansions for Laguerre polynomials, for computing the representation (50), an algorithm for evaluating the Bessel functions $J_\nu(z)$ is needed. We use a computational scheme based on the methods of approximation described in the Appendix.

3.3 An Expansion for Large β

In [9] we discussed two asymptotic approximations of Jacobi polynomials for large values of the β-parameter. The expansions are in terms of Laguerre polynomials. One of the two expansions given in [9] is the following:

$$\begin{aligned}
P_n^{(\alpha,\beta)}\left(1 - \frac{2z}{b}\right) &= (1 - z/b)^n W(n, \alpha, z),\\
W(n, \alpha, z) &= L_n^{(\alpha)}(z) R(n, \alpha, \beta, z) + L_{n-1}^{(\alpha)}(z) S(n, \alpha, \beta, z),
\end{aligned} \tag{58}$$

where $0 < z < \beta$, $b = \beta + n$ and $L_n^{(\alpha)}(z)$, $L_{n-1}^{(\alpha)}(z)$ are Laguerre polynomials. $R(n, \alpha, \beta, z)$, $S(n, \alpha, \beta, z)$ have the expansions

$$R(n, \alpha, \beta, z) \sim \sum_{k=0}^{\infty} \frac{r_k(n, \alpha, z)}{b^k}, \quad S(n, \alpha, \beta, z) \sim \sum_{k=0}^{\infty} \frac{s_k(n, \alpha, z)}{b^k}. \tag{59}$$

The first few coefficients are

$$
\begin{aligned}
r_0(n,\alpha,z) &= 1, \quad s_0(n,\alpha,z) = 0,\\
r_1(n,\alpha,z) &= \tfrac{1}{2}n(2z+\alpha+1),\\
s_1(n,\alpha,z) &= -\tfrac{1}{2}(n+\alpha)(\alpha+z+1),\\
r_2(n,\alpha,z) &= -\tfrac{1}{24}\left(3\alpha^3+6\alpha^2z-6\alpha nz+3\alpha z^2\right.\\
&\quad \left.-9nz^2+10\alpha^2+8\alpha z-4nz-12z^2+9\alpha+2\right),\\
s_2(n,\alpha,z) &= \tfrac{1}{24}\left(n+\alpha)(3\alpha^3+3\alpha^2z-6\alpha nz-3\alpha z^2-6nz^2\right.\\
&\quad \left.-3z^3+10\alpha^2+\alpha z-4nz-11z^2+9\alpha-2z+2\right).
\end{aligned}
\tag{60}
$$

In this expansion for the Jacobi polynomials $P_n^{(\alpha,\beta)}(x)$, the range of validity in the x-interval is very limited near $x = 1$ (to reach the whole x-interval $(-1, 1)$, z should be of order $\mathcal{O}(\beta)$). For large values of the parameter α, we can obtain a similar expansion valid near $x = -1$ by using the symmetry relation (52) of the Jacobi polynomials.

3.4 An Expansion for Large α and β

In [10] an expansion for Jacobi polynomials in terms of elementary functions valid for large values of the parameters α and β has been given:

$$
P_n^{(\alpha,\beta)}(x) = \frac{2^{\frac{1}{2}(\alpha+\beta+1)}e^{-\kappa\psi}}{\sqrt{\pi\kappa w(x)U(x)}}\left(\cos\left(\kappa\chi+\tfrac{1}{4}\pi\right)P+\sin\left(\kappa\chi+\tfrac{1}{4}\pi\right)Q\right), \tag{61}
$$

where

$$
\begin{aligned}
w(x) &= (1-x)^\alpha(1+x)^\beta,\\
\psi &= -\tfrac{1}{2}(1-\tau)\ln(1-\tau)-\tfrac{1}{2}(1+\tau)\ln(1+\tau)+\\
&\quad \tfrac{1}{2}(1+\sigma)\ln(1+\sigma)+\tfrac{1}{2}(1-\sigma)\ln(1-\sigma),\\
\chi &= (\tau+1)\arctan\frac{U(x)}{1-x+\sigma+\tau}+\\
&\quad (\tau-1)\arctan\frac{U(x)}{1+x+\sigma-\tau}+(1-\sigma)\,\mathrm{atan2}(-U(x),\tau+x\sigma).\\
U(x) &= \sqrt{1-2\sigma\tau x-\tau^2-\sigma^2-x^2},
\end{aligned}
\tag{62}
$$

and

$$\sigma = \frac{\alpha+\beta}{2\kappa}, \qquad \tau = \frac{\alpha-\beta}{2\kappa}, \qquad \kappa = n + \tfrac{1}{2}(\alpha+\beta+1). \tag{63}$$

The function $\operatorname{atan2}(y, x)$ in the third term of $\chi(x)$ in (62) denotes the phase $\in (-\pi, \pi]$ of the complex number $x + iy$.

P and Q have expansions

$$P \sim \sum_{j=0}^{\infty} \frac{p_j}{\kappa^j}, \qquad Q \sim \sum_{j=0}^{\infty} \frac{q_j}{\kappa^j}, \tag{64}$$

and the first coefficients are $p_0 = 1$, $q_0 = 0$. These expansions are valid for $x \in [x_-(1+\delta), x_+(1-\delta)]$, where $x_\pm$ are the turning points

$$x_\pm = -\sigma\tau \pm \sqrt{(1-\sigma^2)(1-\tau^2)},$$

and δ is a fixed positive small number. For more details about the coefficients p_j and q_j, see [10].

3.5 *Computational Performance*

The performance of the asymptotic expansions for the Jacobi polynomials has been tested by considering the relation given in Eq. (18.9.3) of [14] written in one of the two following forms

$$\frac{P_n^{(\alpha,\beta-1)}(x) - P_n^{(\alpha-1,\beta)}(x)}{P_{n-1}^{(\alpha,\beta)}(x)} = 1, \tag{65}$$

$$\frac{P_n^{(\alpha-1,\beta)}(x) + P_{n-1}^{(\alpha,\beta)}(x)}{P_n^{(\alpha,\beta-1)}(x)} = 1. \tag{66}$$

Note that the test (65) fails close to the zeros of $P_{n-1}^{(\alpha,\beta)}(x)$; in this case, we can consider the alternative test (66): the zeros of $P_{n-1}^{(\alpha,\beta)}(x)$ and $P_n^{(\alpha,\beta-1)}(x)$ interlace according to Theorem 2 of [17], and therefore they can not vanish simultaneously.

First, we test the performance of the asymptotic expansion (42). Figure 3 shows the points where the error in the computation of the test is greater than 10^{-12} for two different sets of values of the parameters (α, β). The first four nonzero coefficients of the expansions (44) have been considered in the computations. Random values have been generated in the parameter region $(n, \theta) \in (10, 1000) \times (0, \pi)$. As

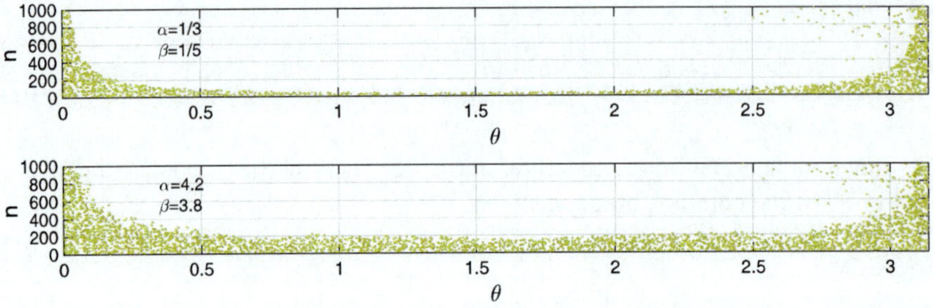

Fig. 3 Performance of the asymptotic expansion (42). The points where the error in the computation is greater than 10^{-12}. Two different sets of values of the parameters α and β have been considered in the tests

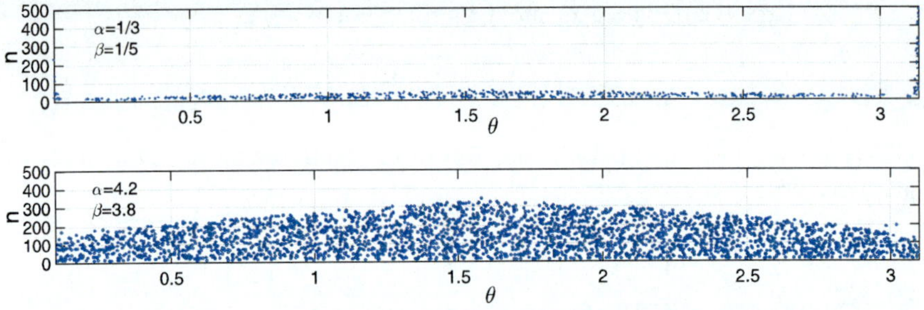

Fig. 4 Tests of the performance of the asymptotic expansion (50) for two different sets of values of the parameters α and β . The points where the error in the computation is greater than 10^{-12} are shown

expected, the expansion works well for values of $x = \cos(\theta)$ not close to ± 1. Also, the parameters $\alpha,\ \beta$ should be of order $O(1)$, and smaller values of these parameters give better results, as can be seen when comparing the results for $P_n^{(1/3,\,1/5)}(\cos\theta)$ and $P_n^{(4.2,\,3.8)}(\cos\theta)$.

The performance of the asymptotic expansion in terms of Bessel functions (50) is illustrated in Fig. 4. As before, the points where the error in the computation is greater than 10^{-12} are shown. The asymptotic expansion (50) has been computed by using the first three coefficients $S_k(\theta),\ T_k(\theta),\ k = 0, 1, 2$ in (53). The results are as expected: better accuracy is obtained near $x = \pm 1$, although the expansion performs well even for small values of x (θ-values close to $\pi/2$) when α is small. Also, for a fixed number of terms used in the expansion, the accuracy obtained with the expansion decreases as the parameter values $(\alpha,\ \beta)$ increase.

Regarding CPU times, the Bessel expansions, as expected, are slower than the elementary expansions. Therefore, when both expansions are valid in the same parameter region, it is always convenient to use the expansion in terms of elementary functions.

When the parameter values α and/or β are large, it is interesting to consider the expansions given in Sects. 3.3 and 3.4. An example of the performance of the expansion for $P_n^{(\alpha,\beta)}(x)$ given in (58) is shown in Fig. 5 for a fixed value of α and two different values of the parameter β ($\beta = 250$, $\beta = 500$). The first five coefficients of the expansions (59) have been considered in the computations and the algorithm given in [8] is used to compute the Laguerre polynomials. The variable x in the abscissas axis is the argument $1-2z/b$ in the expansion. Random points have been generated in the parameter region $(n, x) \in (10, 110) \times (0.994, 1)$. The points where the error in the computation is greater than 10^{-8} are plotted in the figure. The results show that the range of validity of the expansion is quite narrow; therefore, it is of limited interest as a tool to compute the polynomials. The expansion is, however, very useful for obtaining expansions for the zeros of Jacobi polynomials, as discussed in [9].

More interesting for computational purposes is the expansion in terms of elementary functions given in (61). In Fig. 6 we show an example of the performance of this expansion. The first four coefficients of the expansions (64) have been considered in the computations. The points where the error in the computation is greater than

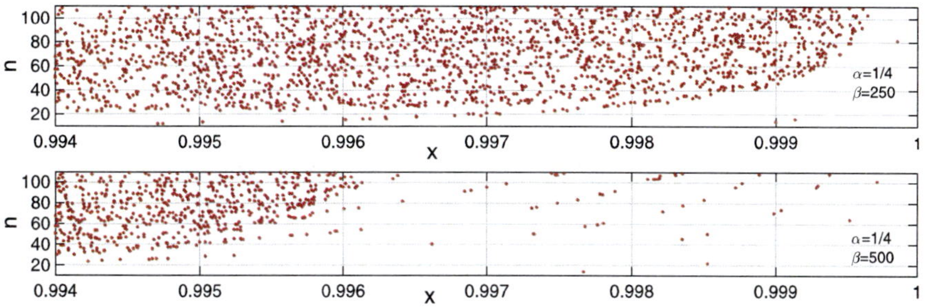

Fig. 5 Performance of the asymptotic expansion (58) for a fixed value of α and two different (large) values of the parameter β. The variable x in the abscissas axis is the argument $1 - 2z/b$ in the expansion. The points where the error in the computation is greater than 10^{-8} are plotted

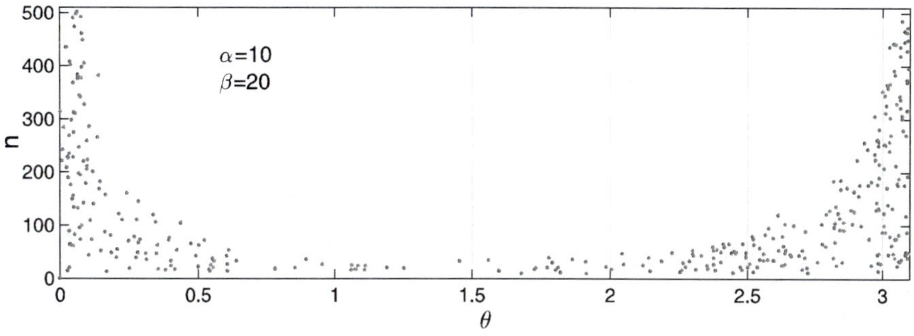

Fig. 6 Example of the performance of the asymptotic expansion (61). The points where the error in the computation is greater than 10^{-8} are plotted

10^{-8} are plotted in the figure. We have chosen θ ($x = \cos(\theta)$) as the variable in the abscissas axis.

As can be seen, the expansion performs well even for not too large values of α and β ($\alpha = 10$, $\beta = 20$ in the example shown.) Also, as expected, the values of $x = \cos(\theta)$ should not be close to ± 1.

Appendix

The following methods of approximation were used in our algorithm for computing the Bessel functions $J_\nu(x)$:

Power Series
When z is small, we use the power series given in Eq. (10.2.2) of [16, §10.19(ii)]:

$$J_\nu(z) = (\tfrac{1}{2}z)^\nu \sum_{k=0}^{\infty} (-1)^k \frac{(\frac{1}{4}z^2)^k}{k!\,\Gamma(\nu+k+1)}.$$

Debye's Asymptotic Expansions
The expressions are given in Eq. (10.19.3) and Eq. (10.19.6) of [16, §10.19(ii)]: When $\nu < z$, we use

$$J_\nu(\nu \operatorname{sech}\alpha) \sim \frac{e^{\nu(\tanh\alpha - \alpha)}}{(2\pi\nu\tanh\alpha)^{\frac{1}{2}}} \sum_{k=0}^{\infty} \frac{U_k(\coth\alpha)}{\nu^k},$$

and for $\nu > z$

$$J_\nu(\nu \sec\beta) \sim \left(\frac{2}{\pi\nu\tan\beta}\right)^{\frac{1}{2}} \left(\cos\xi \sum_{k=0}^{\infty} \frac{U_{2k}(i\cot\beta)}{\nu^{2k}} - i\sin\xi \sum_{k=0}^{\infty} \frac{U_{2k+1}(i\cot\beta)}{\nu^{2k+1}}\right).$$

The coefficients $U_k(p)$ are polynomials in p of degree $3k$ given by $U_0(p) = 1$ and

$$U_{k+1}(p) = \tfrac{1}{2}p^2(1-p^2)U_k'(p) + \frac{1}{8}\int_0^p (1-5t^2)U_k(t)dt.$$

Hankel's Expansion
For large values of the argument z, we use Hankel's expansion given in [16, §10.17(i)]:

$$J_\nu(z) \sim \left(\frac{2}{\pi z}\right)^{\frac{1}{2}} \left(\cos\omega \sum_{k=0}^{\infty} (-1)^k \frac{a_{2k}(\nu)}{z^{2k}} - \sin\omega \sum_{k=0}^{\infty} (-1)^k \frac{a_{2k+1}(\nu)}{z^{2k+1}}\right),$$

where

$$\omega = z - \tfrac{1}{2}\nu\pi - \tfrac{1}{4}\pi.$$

The coefficients $a_k(\nu)$ are given by $a_0(\nu) = 1$ and

$$a_k(\nu) = \frac{(4\nu^2 - 1^2)(4\nu^2 - 3^2)\cdots(4\nu^2 - (2k-1)^2)}{k!8^k}, \quad k = 1, 2, 3, \ldots.$$

Airy-Type Expansions

We use the representation given in [5, Chapter 8]

$$J_\nu(\nu x) = \frac{\phi(\zeta)}{\nu^{1/3}}\left[\mathrm{Ai}(\nu^{2/3}\zeta)\, A_\nu(\zeta) + \nu^{-4/3}\mathrm{Ai}'(\nu^{2/3}\zeta)\, B_\nu(\zeta)\right],$$

where

$$\phi(\zeta) = \left(\frac{4\zeta}{1-x^2}\right)^{\frac{1}{4}}, \quad \phi(0) = 2^{\frac{1}{3}}.$$

The variable ζ is written in terms of the variable x as

$$\begin{aligned} \tfrac{2}{3}\zeta^{3/2} &= \ln\frac{1+\sqrt{1-x^2}}{x} - \sqrt{1-x^2}, && 0 < x \le 1, \\ \tfrac{2}{3}(-\zeta)^{3/2} &= \sqrt{x^2-1} - \arccos\frac{1}{x}, && x \ge 1. \end{aligned}$$

Three-Term Recurrence Relation Using Miller's Algorithm

The standard three-term recurrence relation for the cylinder functions

$$J_{\nu-1}(z) + J_{\nu+1}(z) = (2\nu/z)\, J_\nu(z),$$

is computed backwards (starting from large values of ν) using Miller's algorithm; see [5, §4.6].

Acknowledgments We thank the referee for valuable comments. We acknowledge financial support from Ministerio de Ciencia e Innovación, Spain, projects MTM2015-67142-P (MINECO/FEDER, UE) and PGC2018-098279-B-I00 (MCIU/AEI/FEDER, UE). NMT thanks CWI, Amsterdam, for scientific support.

References

1. Deaño, A., Huertas, E.J., Marcellán, F.: Strong and ratio asymptotics for Laguerre polynomials revisited. J. Math. Anal. Appl. **403**(2), 477–486 (2013)
2. Deaño, A., Huybrechs, D., Opsomer, P.: Construction and implementation of asymptotic expansions for Jacobi-type orthogonal polynomials. Adv. Comput. Math. **42**(4), 791–822 (2016)
3. Dimitrov, D.K., Nikolov, G.P.: Sharp bounds for the extreme zeros of classical orthogonal polynomials. J. Approx. Theor. **162**(10), 1793–1804 (2010)
4. Frenzen, C.L., Wong, R.: Uniform asymptotic expansions of Laguerre polynomials. SIAM J. Math. Anal. **19**(5), 1232–1248 (1988)
5. Gil, A., Segura, J., Temme, N.M.: Numerical Methods for Special Functions. Society for Industrial and Applied Mathematics (SIAM), Philadelphia, PA (2007)
6. Gil, A., Segura, J., Temme, N.M.: Asymptotic approximations to the nodes and weights of Gauss–Hermite and Gauss–Laguerre quadratures. Stud. Appl. Math. **140**(3), 298–332 (2018)
7. Gil, A., Segura, J., Temme, N.M.: Computing the real parabolic cylinder functions $U(a,x)$, $V(a,x)$. ACM Trans. Math. Softw. **32**(1), 70–101 (2006)
8. Gil, A., Segura, J., Temme, N.M.: Efficient computation of Laguerre polynomials. Comput. Phys. Commun. **210**, 124–131 (2017)
9. Gil, A., Segura, J., Temme, N.M.: Expansions of Jacobi polynomials for large values of beta and of their zeros. SIGMA **14**(73), 9 p. (2018)
10. Gil, A., Segura, J., Temme, N.M.: Asymptotic expansions of Jacobi polynomials and of the nodes and weights of Gauss-Jacobi quadrature for large degree and parameters in terms of elementary functions. J. Math. Anal. Appl. **494**(2), 124642 (2021)
11. Gil, A., Segura, J., Temme, N.M.: Noniterative computation of Jacobi polynomials. SIAM J. Sci. Comput. **41**(1), A668–A693 (2019)
12. Hale, N., Townsend, A.: Fast and accurate computation of Gauss–Legendre and Gauss–Jacobi quadrature nodes and weights. SIAM. J. Sci. Comput. **35**(2), A652–A674 (2013)
13. Huybrechs, D., Opsomer, P.: Construction and implementation of asymptotic expansions for Laguerre-type orthogonal polynomials. IMA J. Numer. Anal. **38**(3), 1085–1118 (2018)
14. Koornwinder, T.H., Wong, R., Koekoek, R., Swarttouw, R.F.: Chapter 18, Orthogonal polynomials. In: NIST Handbook of Mathematical Functions, pp. 435–484. U.S. Dept. Commerce, Washington, DC (2010). http://dlmf.nist.gov/18
15. Kreuser, P.: Über das Verhalten der Integrale homogener linearer Differenzengleichungen im Unendlichen. Diss. Tübingen, 48 S (1914)
16. Olver, F.W.J., Maximon, L.C.: Chapter 10, Bessel functions. In: NIST Handbook of Mathematical Functions, pp. 215–286. U.S. Dept. Commerce, Washington, DC (2010b). http://dlmf.nist.gov/10
17. Segura, J.: Interlacing of the zeros of contiguous hypergeometric functions. Numer. Algorithms **49**(1–4), 387–407 (2008)
18. Temme, N.M.: Asymptotic Methods for Integrals, vol. 6. Series in Analysis. World Scientific Publishing, Hackensack, NJ (2015)

An Introduction to Multiple Orthogonal Polynomials and Hermite-Padé Approximation

Guillermo López-Lagomasino

Abstract We present a brief introduction to the theory of multiple orthogonal polynomials on the basis of known results for an important class of measures known as Nikishin systems. For type I and type II multiple orthogonal polynomials with respect to such systems of measures, we describe some of their most relevant properties regarding location and distribution of zeros as well as their weak and ratio asymptotic behavior.

Keywords Perfect systems · Nikishin systems · Multiple orthogonal polynomials · Hermite-Padé approximation · Logarithmic asymptotics · Ratio asymptotics · Asymptotic properties

1 Introduction

The object of this paper is to provide an introduction to the study of Hermite-Padé approximation, multiple orthogonal polynomials, and some of their asymptotic properties. For the most part, the attention is restricted to the case of multiple orthogonality with respect to an important class of measures introduced by E.M. Nikishin in [40]. For this reason, they are referred to in the specialized literature as Nikishin systems.

Throughout the years, with the assistance of colleagues and former Ph.D. students, I have dedicated a great part of my research to this subject area and I wish to express here my gratitude to all those involved. This material is not intended to present new results, but a mere account of the experience I have accumulated for the benefit of future research from the younger generations.

G. López Lagomasino (✉)
Departamento de Matemáticas, Universidad Carlos III de Madrid, Leganés, Spain
e-mail: lago@math.uc3m.es

F. Marcellán, E. J. Huertas (eds.), *Orthogonal Polynomials: Current Trends and Applications*, SEMA SIMAI Springer Series 22,
https://doi.org/10.1007/978-3-030-56190-1_9

1.1 Some Historical Background

In 1873, Charles Hermite publishes in [28] his proof of the transcendence of e making use of simultaneous rational approximation of systems of exponentials. That paper marked the beginning of the modern analytic theory of numbers.

The formal theory of simultaneous rational approximation for general systems of analytic functions was initiated by K. Mahler in lectures delivered at the University of Groningen in 1934–35. These lectures were published years later in [37]. Important contributions in this respect are also due to his students J. Coates and H. Jager, see [13] and [29]. K. Mahler's approach to the simultaneous approximation of finite systems of analytic functions may be reformulated in the following terms.

Let $\mathbf{f} = (f_1, \dots, f_m)$ be a family of analytic functions in some domain D of the extended complex plane containing ∞. Fix a non-zero multi-index $\mathbf{n} = (n_1, \dots, n_m) \in \mathbb{Z}_+^m$, $|\mathbf{n}| = n_1 + \dots, n_m$. There exist polynomials $a_{\mathbf{n},1}, \dots, a_{\mathbf{n},m}$, not all identically equal to zero, such that

(i) $\deg a_{\mathbf{n},j} \leq n_j - 1$, $j = 1, \dots, m$ ($\deg a_{\mathbf{n},j} \leq -1$ means that $a_{\mathbf{n},j} \equiv 0$),
(ii) $a_{\mathbf{n},0} + \sum_{j=1}^{m} a_{\mathbf{n},j}(z) f_j(z) = \mathcal{O}(1/z^{|\mathbf{n}|})$, $z \to \infty$.

for some polynomial $a_{\mathbf{n},0}$. Analogously, there exists $Q_{\mathbf{n}}$, not identically equal to zero, such that

(i) $\deg Q_{\mathbf{n}} \leq |\mathbf{n}|$,
(ii) $Q_{\mathbf{n}}(z) f_j(z) - P_{\mathbf{n},j}(z) = \mathcal{O}(1/z^{n_j+1})$, $z \to \infty$, $j = 1, \dots, m$,

for some polynomials $P_{\mathbf{n},j}$, $j = 1, \dots, m$.

The polynomials $a_{\mathbf{n},0}$ and $P_{\mathbf{n},j}$, $j = 0, \dots, m$, are uniquely determined from (ii) once their partners are found. The two constructions are called type I and type II polynomials (approximants) of the system $(f_1, \dots, f_m)$. Algebraically, they are closely related. This is clearly exposed in [13, 29], and [37]. When $m = 1$ both definitions coincide with that of the well-known Padé approximation in its linear presentation.

Apart from Hermite's result, type I, type II, and a combination of the two (called mixed type), have been employed in the proof of the irrationality of other numbers. For example, in [11] F. Beukers shows that Apery's proof (see [1]) of the irrationality of $\zeta(3)$ can be placed in the context of mixed type Hermite-Padé approximation. See [46] for a brief introduction and survey on the subject. More recently, mixed type approximation has appeared in random matrix and non-intersecting brownian motion theories (see, for example, [9, 14, 31], and [32]), and the Degasperi-Procesi equation [10].

In applications in the areas of number theory, convergence of simultaneous rational approximation, and asymptotic properties of type I and type II polynomials, a central question is if these polynomials have no defect; that is, if they attain the maximal degree possible.

Definition 1 A multi-index $\mathbf{n}$ is said to be **normal** for the system $\mathbf{f}$ for type I approximation (respectively, for type II,) if $\deg a_{\mathbf{n},j} = n_j - 1$, $j = 1, \dots, m$

(respectively, $\deg Q_{\mathbf{n}} = |\mathbf{n}|$). A system of functions $\mathbf{f}$ is said to be **perfect** if all multi-indices are normal.

It is easy to verify that $(a_{\mathbf{n},0}, \ldots, a_{\mathbf{n},m})$ and $Q_{\mathbf{n}}$ are uniquely determined to within a constant factor when $\mathbf{n}$ is normal. Moreover, if a system is perfect, the order of approximation in parts ii) above is exact for all $\mathbf{n}$. The convenience of these properties is quite clear.

Considering the construction at the origin (instead of $z = \infty$ which we chose for convenience), the system of exponentials considered by Hermite, $(e^{w_1 z}, \ldots, e^{w_m z})$, $w_i \neq w_j, i \neq j, i, j = 1, \ldots, m$, is known to be perfect for type I and type II. A second example of a perfect system for both types is that given by the binomial functions $(1-z)^{w_1}, \ldots, (1-z)^{w_m}$, $w_i - w_j \notin \mathbb{Z}$. When normality occurs for multi-indices with decreasing components the system is said to be weakly perfect. Basically, these are the only examples known of perfect systems, except for certain ones formed by Cauchy transforms of measures.

1.2 Markov Systems and Orthogonality

Let s be a finite Borel measure with constant sign whose compact support consists of infinitely many points and is contained in the real line. In the sequel, we only consider such measures. By Δ we denote the smallest interval which contains the support, $\operatorname{supp} s$, of s. We denote this class of measures by $\mathcal{M}(\Delta)$. Let

$$\widehat{s}(z) = \int \frac{ds(x)}{z - x}$$

denote the Cauchy transform of s. Obviously, $\widehat{s} \in \mathcal{H}(\overline{\mathbb{C}} \setminus \Delta)$; that is, it is analytic in $\overline{\mathbb{C}} \setminus \Delta$.

If we apply the construction above to the system formed by $\widehat{s}$ ($m = 1$), it is easy to verify that $Q_{\mathbf{n}}$ turns out to be orthogonal to all polynomials of degree less than $n \in \mathbb{Z}_+$. Consequently, $\deg Q_{\mathbf{n}} = n$, all its zeros are simple and lie in the open convex hull $\operatorname{Co}(\operatorname{supp} s)$ of $\operatorname{supp} s$. Therefore, such systems of one function are perfect. These properties allow to deduce Markov's theorem on the convergence of (diagonal) Padé approximations of $\widehat{s}$ published in [38]. For this reason, $\widehat{s}$ is also called a Markov function.

Markov functions are quite relevant in several respects. Many elementary functions can be expressed as such. The resolvent function of self-adjoint operators admits that type of representation. If one allows complex weights, any reasonable analytic function in the extended complex plane with a finite number of algebraic singularities adopts that form. This fact, and the use of Padé approximation, has played a central role in some of the most relevant achievements in the last decades concerning the exact rate of convergence of the best rational approximation: namely, A.A. Gonchar and E.A. Rakhmanov's result, see [24, 26], and [5], on the best rational approximation of e^{-x} on $[0, +\infty)$; and H. Stahl's theorem, see [44], on the best rational approximation of x^α on $[0, 1]$.

Let us see two examples of general systems of Markov functions which play a central role in the theory of multiple orthogonal polynomials.

1.2.1 Angelesco Systems

In [2], A. Angelesco considered the following systems of functions. Let $\Delta_j, j = 1, \ldots, m$, be pairwise disjoint bounded intervals contained in the real line and $s_j, j = 1, \ldots, m$, a system of measures such that $\mathrm{Co}(\operatorname{supp} s_j) = \Delta_j$.

Fix $\mathbf{n} \in \mathbb{Z}_+^m$ and consider the type II approximant of the so called Angelesco system of functions $(\widehat{s}_1, \ldots, \widehat{s}_m)$ relative to $\mathbf{n}$. It turns out that

$$\int x^\nu Q_{\mathbf{n}}(x) ds_j(x) = 0, \quad \nu = 0, \ldots, n_j - 1, \quad j = 1, \ldots, m.$$

Therefore, $Q_{\mathbf{n}}$ has n_j simple zeros in the interior (with respect to the euclidean topology of $\mathbb{R}$) of Δ_j. In consequence, since the intervals Δ_j are pairwise disjoint, $\deg Q_{\mathbf{n}} = |\mathbf{n}|$ and Angelesco systems are type II perfect.

Unfortunately, Angelesco's paper received little attention and such systems reappear many years later in [39] where E.M. Nikishin deduces some of their formal properties.

In [25] and [3], their logarithmic and strong asymptotic behavior, respectively, are given. These multiple orthogonal polynomials and the rational approximations associated have nice asymptotic formulas but not so good convergence properties. In this respect, a different system of Markov functions turns out to be more interesting and foundational from the geometric and analytic points of view.

1.2.2 Nikishin Systems

In an attempt to construct general classes of functions for which normality takes place, in [40] E.M. Nikishin introduced the concept of MT-system (now called Nikishin system). Let $\Delta_\alpha, \Delta_\beta$ be two non intersecting bounded intervals contained in the real line and $\sigma_\alpha \in \mathcal{M}(\Delta_\alpha), \sigma_\beta \in \mathcal{M}(\Delta_\beta)$. With these two measures we define a third one as follows (using the differential notation)

$$d\langle \sigma_\alpha, \sigma_\beta \rangle(x) = \widehat{\sigma}_\beta(x) d\sigma_\alpha(x);$$

that is, one multiplies the first measure by a weight formed by the Cauchy transform of the second measure. Certainly, this product of measures is non commutative. Above, $\widehat{\sigma}_\beta$ denotes the Cauchy transform of the measure σ_β.

Definition 2 Take a collection Δ_j, $j = 1, \ldots, m$, of intervals such that

$$\Delta_j \cap \Delta_{j+1} = \emptyset, \qquad j = 1, \ldots, m-1.$$

Let $(\sigma_1, \ldots, \sigma_m)$ be a system of measures such that $\mathrm{Co}(\operatorname{supp} \sigma_j) = \Delta_j, \sigma_j \in \mathcal{M}(\Delta_j)$, $j = 1, \ldots, m$. We say that $(s_{1,1}, \ldots, s_{1,m}) = \mathcal{N}(\sigma_1, \ldots, \sigma_m)$, where

$$s_{1,1} = \sigma_1, \quad s_{1,2} = \langle \sigma_1, \sigma_2 \rangle, \ldots \quad , s_{1,m} = \langle \sigma_1, \langle \sigma_2, \ldots, \sigma_m \rangle \rangle$$

is the Nikishin system of measures generated by $(\sigma_1, \ldots, \sigma_m)$.

Fix $\mathbf{n} \in \mathbb{Z}_+^m$ and consider the type II approximant of the Nikishin system of functions $(\widehat{s}_{1,1}, \ldots, \widehat{s}_{1,m})$ relative to $\mathbf{n}$. It is easy to prove that

$$\int x^\nu Q_{\mathbf{n}}(x) ds_{1,j}(x) = 0, \quad \nu = 0, \ldots, n_j - 1, \quad j = 1, \ldots, m.$$

All the measures $s_{1,j}$ have the same support; therefore, it is not immediate to conclude that $\deg Q_{\mathbf{n}} = |\mathbf{n}|$. Nevertheless, if we denote

$$s_{j,k} = \langle \sigma_j, \sigma_{j+1}, \ldots, \sigma_k \rangle, \qquad j < k, \quad s_{j,j} = \langle \sigma_j \rangle = \sigma_j,$$

the previous orthogonality relations may be rewritten as follows

$$\int (p_1(x) + \sum_{k=2}^{m} p_k(x) \widehat{s}_{2,k}(x)) Q_{\mathbf{n}}(x) d\sigma_1(x) = 0, \tag{1}$$

where $p_1, \ldots, p_m$ are arbitrary polynomials such that $\deg p_k \leq n_k - 1, k = 1, \ldots, m$.

Definition 3 A system of real continuous functions $u_1, \ldots, u_m$ defined on an interval Δ is called an AT-system on Δ for the multi-index $\mathbf{n} \in \mathbb{Z}_+^m$ if for any choice of real polynomials (that is, with real coefficients) $p_1, \ldots, p_m$, $\deg p_k \leq n_k - 1$, the function

$$\sum_{k=1}^{m} p_k(x) u_k(x)$$

has at most $|\mathbf{n}| - 1$ zeros on Δ. If this is true for all $\mathbf{n} \in \mathbb{Z}_+^m$ we have an AT system on Δ.

In other words, $u_1, \ldots, u_m$ forms an AT-system for $\mathbf{n}$ on Δ when the system of functions

$$(u_1, \ldots, x^{n_1 - 1} u_1, u_2, \ldots, x^{n_m - 1} u_m)$$

is a Chebyshev system on Δ of order $|\mathbf{n}| - 1$. From the properties of Chebyshev systems (see [30, Theorem 1.1]), it follows that given $x_1, \ldots, x_N$, $N < |\mathbf{n}|$, points in the interior of Δ one can find polynomials $h_1, \ldots, h_m$, conveniently, with $\deg h_k \leq n_k - 1$, such that $\sum_{k=1}^{m} h_k(x)u_k(x)$ changes sign at $x_1, \ldots, x_N$, and has no other points where it changes sign on Δ.

In [40], Nikishin stated without proof that the system of functions $(1, \widehat{s}_{2,2}, \ldots, \widehat{s}_{2,m})$ forms an AT-system for all multi-indices $\mathbf{n} \in \mathbb{Z}_+^m$ such that $n_1 \geq n_1 \geq \cdots \geq n_m$ (he proved it when additionally $n_1 - n_m \leq 1$). Due to (1) this implies that Nikishin systems are type II weakly perfect. Ever since the appearance of [40], a subject of major interest for those involved in simultaneous approximation was to determine whether or not Nikishin systems are perfect. This problem was settled positively in [20] (see also [21] where Nikishin systems with unbounded and/or touching supports are considered).

In the last two decades, a general theory of multiple orthogonal polynomials and Hermite-Padé approximation has emerged which to a great extent matches what is known to occur for standard orthogonal polynomials and Padé approximation. From the approximation point of view, Markov and Stieltjes type theorems have been obtained (see, for example, [12, 16–19, 25, 27, 35, 36, 40]). From the point of view of the asymptotic properties of multiple orthogonal polynomials there are results concerning their weak, ratio, and strong asymptotic behavior (see, for example, [3, 4, 6, 22, 33, 35]). This is specially so for Nikishin systems of measures on which we will focus in this brief introduction.

2 On the Perfectness of Nikishin Systems

Let us begin with the following result which was established in [20]. It constitutes the key to the proof of many interesting properties of Nikishin systems; in particular, the fact that they are perfect. From the definition it is obvious that if $(\sigma_1, \ldots, \sigma_m)$ generates a Nikishin system so does $(\sigma_j, \ldots, \sigma_k)$ where $1 \leq j < k \leq m$.

Theorem 1 *Let* $(s_{1,1}, \ldots, s_{1,m}) = \mathcal{N}(\sigma_1, \ldots, \sigma_m)$ *be given. Then,* $(1, \widehat{s}_{1,1}, \ldots, \widehat{s}_{1,m})$ *forms an AT-system on any interval* Δ *disjoint from* $\Delta_1 = \mathrm{Co}(\operatorname{supp} \sigma_1)$. *Moreover, for each* $\mathbf{n} \in \mathbb{Z}_+^{m+1}$, *and arbitrary polynomials with real coefficients* p_k, $\deg p_k \leq n_k - 1$, $k = 0, \ldots, m$, *the linear form* $p_0 + \sum_{k=1}^{m} p_k \widehat{s}_{1,k}$, *has at most* $|\mathbf{n}| - 1$ *zeros in* $\mathbb{C} \setminus \Delta_1$.

For arbitrary multi-indices the proof is quite complicated and based on intricate transformations which allow to reduce the problem to the case of multi-indices with decreasing components. For such multi-indices the proof is pretty straightforward and we will limit ourselves to that situation. Let us first present two auxiliary lemmas.

Lemma 1 *Let* $(s_{1,1}, \ldots, s_{1,m}) = \mathcal{N}(\sigma_1, \ldots, \sigma_m)$ *be given. Assume that there exist polynomials with real coefficients* $\ell_0, \ldots, \ell_m$ *and a polynomial* w *with real* *coefficients whose zeros lie in* $\mathbb{C} \setminus \Delta_1$ *such that*

$$\frac{\mathcal{L}_0(z)}{w(z)} \in \mathcal{H}(\mathbb{C} \setminus \Delta_1) \qquad \text{and} \qquad \frac{\mathcal{L}_0(z)}{w(z)} = O\left(\frac{1}{z^N}\right), \qquad z \to \infty,$$

where $\mathcal{L}_0 := \ell_0 + \sum_{k=1}^m \ell_k \widehat{s}_{1,k}$ *and* $N \geq 1$. *Let* $\mathcal{L}_1 := \ell_1 + \sum_{k=2}^m \ell_k \widehat{s}_{2,k}$. *Then*

$$\frac{\mathcal{L}_0(z)}{w(z)} = \int \frac{\mathcal{L}_1(x)}{(z-x)} \frac{d\sigma_1(x)}{w(x)}. \tag{2}$$

If $N \geq 2$, *we also have*

$$\int x^\nu \mathcal{L}_1(x) \frac{d\sigma_1(x)}{w(x)} = 0, \qquad \nu = 0, \ldots, N-2. \tag{3}$$

In particular, $\mathcal{L}_1$ *has at least* $N-1$ *sign changes in the interior of* Δ_1 *(with respect to the Euclidean topology of* $\mathbb{R}$*).*

Proof Let Γ be a positively oriented closed smooth Jordan curve that surrounds Δ_1 sufficiently close to Δ_1. Since $\frac{\mathcal{L}_0(z)}{w(z)} = O(1/z), z \to \infty$, if z and the zeros of $w(z)$ are in the unbounded connected component of the complement of Γ, Cauchy's integral formula and Fubini's theorem render

$$\frac{\mathcal{L}_0(z)}{w(z)} = \frac{1}{2\pi i} \int_\Gamma \frac{\mathcal{L}_0(\zeta)}{w(\zeta)} \frac{d\zeta}{z-\zeta} = \frac{1}{2\pi i} \sum_{k=1}^m \int_\Gamma \frac{\ell_k(\zeta)\widehat{s}_{1,k}(\zeta) d\zeta}{w(\zeta)(z-\zeta)} =$$

$$\sum_{k=1}^m \int \frac{1}{2\pi i} \int_\Gamma \frac{\ell_k(\zeta) d\zeta}{w(\zeta)(z-\zeta)(\zeta-x)} ds_{1,k}(x) = \sum_{k=1}^m \int \frac{\ell_k(x) ds_{1,k}(x)}{w(x)(z-x)} = \int \frac{\mathcal{L}_1(x)}{(z-x)} \frac{d\sigma_1(x)}{w(x)}$$

which is (2).

When $N \geq 2$, it follows that $\frac{z^\nu \mathcal{L}_0(z)}{w(z)} = O(1/z^2), z \to \infty$, for $\nu = 0, \ldots, N-2$. Then, using Cauchy's theorem, Fubini's theorem and Cauchy's integral formula, it follows that

$$0 = \int_\Gamma \frac{z^\nu \mathcal{L}_0(z)}{w(z)} dz = \sum_{k=1}^m \int_\Gamma \frac{z^\nu \ell_k(z) \widehat{s}_{1,k}(z)}{w(z)} dz = \sum_{k=1}^m \int \int_\Gamma \frac{z^\nu \ell_k(z) dz}{w_k(z)(z-x)} ds_{1,k}(x) =$$

$$2\pi i \sum_{k=1}^m \int \frac{x^\nu \ell_k(x)}{w_k(x)} ds_{1,k}(x) = 2\pi i \int x^\nu \mathcal{L}_1(x) \frac{d\sigma_1(x)}{w(x)},$$

and we obtain (3). □

Lemma 2 *Let* $(s_{1,1},\ldots,s_{1,m}) = \mathcal{N}(\sigma_1,\ldots,\sigma_m)$ *and* $\mathbf{n} = (n_0,\ldots,n_m) \in \mathbb{Z}_+^{m+1}$ *be given. Consider the linear form*

$$\mathcal{L}_{\mathbf{n}} = p_0 + \sum_{k=1}^{m} p_k \widehat{s}_{1,k}, \quad \deg p_k \le n_k - 1, \quad k = 0,\ldots,m,$$

where the polynomials p_k *have real coefficients. Assume that* $n_0 = \max\{n_0, n_1 - 1,\ldots,n_m - 1\}$. *If* $\mathcal{L}_{\mathbf{n}}$ *had at least* $|\mathbf{n}|$ *zeros in* $\mathbb{C} \setminus \Delta_1$ *the reduced form* $p_1 + \sum_{k=2}^{m} p_k \widehat{s}_{2,k}$ *would have at least* $|\mathbf{n}| - n_0$ *zeros in* $\mathbb{C} \setminus \Delta_2$.

Proof The function $\mathcal{L}_{\mathbf{n}}$ is symmetric with respect to the real line $\mathcal{L}_{\mathbf{n}}(\overline{z}) = \overline{\mathcal{L}_{\mathbf{n}}(z)}$; therefore, its zeros come in conjugate pairs. Thus, if $\mathcal{L}_{\mathbf{n}}$ has at least $|\mathbf{n}|$ zeros in $\mathbb{C} \setminus \Delta_1$, there exists a polynomial $w_{\mathbf{n}}$, $\deg w_{\mathbf{n}} \ge |\mathbf{n}|$, with real coefficients and zeros contained in $\mathbb{C} \setminus \Delta_1$ such that $\mathcal{L}_{\mathbf{n}}/w_{\mathbf{n}} \in \mathcal{H}(\mathbb{C} \setminus \Delta_1)$. This function has a zero of order $\ge |\mathbf{n}| - n_0 + 1$ at ∞. Consequently, for all $\nu = 0,\ldots,|\mathbf{n}| - n_0 - 1$,

$$\frac{z^\nu \mathcal{L}_{\mathbf{n}}}{w_{\mathbf{n}}} = \mathcal{O}(1/z^2) \in \mathcal{H}(\overline{\mathbb{C}} \setminus \Delta_1), \qquad z \to \infty,$$

and

$$\frac{z^\nu \mathcal{L}_{\mathbf{n}}}{w_{\mathbf{n}}} = \frac{z^\nu p_0}{w_{\mathbf{n}}} + \sum_{k=1}^{m} \frac{z^\nu p_k}{w_{\mathbf{n}}} \widehat{s}_{1,k} \,.$$

From (3), it follows that

$$0 = \int x^\nu (p_1 + \sum_{k=2}^{m} p_k \widehat{s}_{2,k})(x) \frac{d\sigma_1(x)}{w_{\mathbf{n}}(x)}, \qquad \nu = 0,\ldots,|\mathbf{n}| - n_0 - 1,$$

taking into consideration that $s_{1,1} = \sigma_1$ and $ds_{1,k}(x) = \widehat{s}_{2,k}(x) d\sigma_1(x)$, $k = 2,\ldots,m$.

These orthogonality relations imply that $p_1 + \sum_{k=2}^{m} p_k \widehat{s}_{2,k}$ has at least $|\mathbf{n}| - n_0$ sign changes in the interior of Δ_1. In fact, if there were at most $|\mathbf{n}| - n_0 - 1$ sign changes one can easily construct a polynomial p of degree $\le |\mathbf{n}| - n_0 - 1$ such that $p(p_1 + \sum_{k=2}^{m} p_k \widehat{s}_{2,k})$ does not change sign on Δ_1 which contradicts the orthogonality relations. Therefore, already in the interior of $\Delta_1 \subset \mathbb{C} \setminus \Delta_2$, the reduced form would have the number of zeros claimed. □

Proof of Theorem 1 When $n_0 \ge n_1 \ge \cdots \ge n_m$ In this situation assume that the linear form $p_0 + \sum_{k=1}^{m} p_k \widehat{s}_{1,k}$, has at least $|\mathbf{n}|$ zeros in $\overline{\mathbb{C}} \setminus \Delta_1$. Applying Lemma 2 consecutively m times we would arrive to the conclusion that p_m has at least n_m zeros in $\mathbb{C}$ but this is impossible since its degree is $\le n_m - 1$. □

From Theorem 1 the following result readily follows.

Theorem 2 *Nikishin systems are type I and type II perfect.*

Proof Consider a Nikishin system $\mathcal{N}(\sigma_1, \ldots, \sigma_m)$. Let us prove that it is type I perfect. Given a multi-index $\mathbf{n} \in \mathbb{Z}_+^m$ condition ii) for the system of functions $(\widehat{s}_{1,1}, \ldots, \widehat{s}_{1,m})$ implies that

$$\int x^\nu \left(a_{\mathbf{n},1} + a_{\mathbf{n},2}\widehat{s}_{2,2} + \cdots + a_{\mathbf{n},m}\widehat{s}_{2,m}\right)(x)d\sigma_1(x) = 0, \qquad \nu = 0, \ldots, |\mathbf{n}| - 2.$$

These orthogonality relations imply that $\mathcal{A}_{\mathbf{n},1} := a_{\mathbf{n},1} + a_{\mathbf{n},2}\widehat{s}_{2,2} + \cdots + a_{\mathbf{n},m}\widehat{s}_{2,m}$ has at least $|\mathbf{n}|-1$ sign changes in the interior of Δ_1 (with the Euclidean topology of $\mathbb{R}$). Suppose that $\mathbf{n}$ is not normal; that is, $\deg a_{\mathbf{n},j} \leq n_j - 2$ for some $j = 1, \ldots, m$. Then, according to Theorem 1, $\mathcal{A}_{\mathbf{n},1}$ can have at most $|\mathbf{n}| - 2$ zeros in $\mathbb{C} \setminus \Delta_2$. This contradicts the previous assertion, so such a component j cannot exist in $\mathbf{n}$.

The proof of type II perfectness is analogous. Suppose that there exists an $\mathbf{n}$ such that $Q_\mathbf{n}$ has less than $|\mathbf{n}|$ sign changes in the interior of Δ_1. Let $x_k, k = 1, \ldots x_N, N \leq |\mathbf{n}| - 1$ be the points where it changes sign. Construct a linear form

$$p_1 + p_2\widehat{s}_{2,2} + \cdots + p_m\widehat{s}_{2,m}, \qquad \deg p_k \leq n_k - 1, \qquad k = 1, \ldots, m.$$

with a simple zero at each of the points x_k and a zero of multiplicity $|\mathbf{n}| - N - 1$ at one of the end points of Δ_1. This is possible because there are sufficient free parameters in the coefficients of the polynomials p_k and the form is analytic on a neighborhood of Δ_1. By Theorem 1 this linear form cannot have any more zeros in the complement of Δ_2 than those that have been assigned. However, using (1), we have

$$\int (p_1(x) + \sum_{k=2}^{m} p_k(x)\widehat{s}_{2,k}(x))Q_\mathbf{n}(x)d\sigma_1(x) = 0,$$

which is not possible since the function under the integral sign has constant sign on Δ_1 and it is not identically equal to zero. □

3 On the Interlacing Property of Zeros

In the sequel, we restrict our attention to multi-indices in

$$\mathbb{Z}_+^m(\bullet) := \{\mathbf{n} \in \mathbb{Z}_+^m : n_1 \geq \cdots \geq n_m\}.$$

3.1 Interlacing for Type I

Fix $\mathbf{n} \in \mathbb{Z}_+^m(\bullet)$. Consider the type I Hermite-Padé approximant $(a_{\mathbf{n},0}, \ldots, a_{\mathbf{n},m})$ of $(\widehat{s}_{1,1}, \ldots, \widehat{s}_{1,m})$ for the multi-index $\mathbf{n} = (n_1, \ldots, n_m)$. Set

$$\mathcal{A}_{\mathbf{n},k} = a_{\mathbf{n},k} + \sum_{j=k+1}^{m} a_{\mathbf{n},j}\widehat{s}_{k+1,j}, \qquad k = 0, \ldots, m-1.$$

We take $\mathcal{A}_{\mathbf{n},m} = a_{\mathbf{n},m}$.

Proposition 1 *For each $k = 1, \ldots, m$ the linear form $\mathcal{A}_{\mathbf{n},k}$ has exactly $n_k + \cdots + n_m - 1$ zeros in $\mathbb{C} \setminus \Delta_{k+1}$, where $\Delta_{m+1} = \emptyset$, they are all simple and lie in the interior of Δ_k. Let $A_{\mathbf{n},k}$ be the monic polynomial whose roots are the zeros of $\mathcal{A}_{\mathbf{n},k}$ on Δ_k. Then*

$$\int x^\nu \mathcal{A}_{\mathbf{n},k}(x) \frac{d\sigma_k(x)}{A_{\mathbf{n},k-1}(x)} = 0, \qquad \nu = 0, \ldots, n_k + \cdots + n_m - 2, \qquad k = 1, \ldots, m, \tag{4}$$

where $A_{\mathbf{n},0} \equiv 1$.

Proof According to Theorem 1 applied to the Nikishin system $\mathcal{N}(\sigma_{k+1}, \ldots, \sigma_m)$, the linear form $\mathcal{A}_{\mathbf{n},k}$ cannot have more than $n_k + \cdots, n_m - 1$ zeros in $\mathbb{C} \setminus \Delta_{k+1}$. So it suffices to show that it has at least $n_k + \cdots + n_m - 1$ sign changes on Δ_k to prove that it has exactly $n_k + \cdots + n_m - 1$ simple roots in $\mathbb{C} \setminus \Delta_{k+1}$ which lie in the interior of Δ_k. We do this producing consecutively the orthogonality relations (4).

From the definition of type I Hermite-Padé approximation it follows that $z^\nu \mathcal{A}_{\mathbf{n},0} = \mathcal{O}(1/z^2)$, $\nu = 0, \ldots, |\mathbf{n}| - 2$. From (3) we get that

$$\int x^\nu \mathcal{A}_{\mathbf{n},1}(x) d\sigma_1(x) = 0, \qquad \nu = 0, \ldots, n_1 + \cdots + n_m - 2,$$

which is (4) when $k = 1$. Therefore, $\mathcal{A}_{\mathbf{n},1}$ has at least $n_1 + \cdots + n_m - 1$ sign changes on Δ_1 as we needed to prove. Let $A_{\mathbf{n},1}$ be the monic polynomial whose roots are the zeros of $\mathcal{A}_{\mathbf{n},1}$ on Δ_1. Since $z^\nu \mathcal{A}_{\mathbf{n},1}/A_{\mathbf{n},1} = \mathcal{O}(1/z^2)$, $\nu = n_2 + \cdots + n_m - 2$ (recall that the multi-index has decreasing components) from (3) we get (4) for $k = 2$ which implies that $\mathcal{A}_{\mathbf{n},2}$ has at least $n_2 + \cdots + n_m - 1$ sign changes on Δ_2 as needed. We repeat the process until we arrive to $\mathcal{A}_{\mathbf{n},m} = a_{\mathbf{n},m} = A_{\mathbf{n},m}$. □

Fix $\ell \in \{1, \ldots, m\}$. Given $\mathbf{n} \in \mathbb{Z}_+^m(\bullet)$, we denote

$$\mathbf{n}^\ell = \mathbf{n} + \mathbf{e}_\ell,$$

where $\mathbf{e}_\ell$ is the m dimensional canonical vector with 1 in its ℓ-th component and 0 everywhere else. the multi-index which is obtained adding 1 to the ℓ-th component of $\mathbf{n}$. Notice that $\mathbf{n}^\ell$ need not belong to $\mathbb{Z}_+^m(\bullet)$; however, we can construct the linear forms $\mathcal{A}_{\mathbf{n}^\ell,k}$ corresponding to the type I Hermite-Padé approximation with respect to $\mathbf{n}^\ell$.

Theorem 3 *Assume that* $\mathbf{n} \in \mathbb{Z}_+^m(\bullet)$ *and* $\ell \in \{1,\ldots,m\}$. *For each* $k = 1,\ldots,m$, $\mathcal{A}_{\mathbf{n}^\ell,k}$ *has at most* $n_k + \cdots + n_m$ *zeros in* $\mathbb{C} \setminus \Delta_{k+1}$ *and at least* $n_k + \cdots + n_m - 1$ *sign changes in* Δ_k. *Therefore, all its zeros are real and simple. The zeros of* $\mathcal{A}_{\mathbf{n},k}$ *and* $\mathcal{A}_{\mathbf{n}^\ell,k}$ *in* $\mathbb{C} \setminus \Delta_{k+1}$ *interlace.*

Proof Let A, B be real constants such that $|A| + |B| > 0$. For $k = 1,\ldots,m$, consider the forms $G_{\mathbf{n},k} := A\mathcal{A}_{\mathbf{n},k} + B\mathcal{A}_{\mathbf{n}^\ell,k}$. From Theorem 1 it follows that $G_{\mathbf{n},k}$ has at most $n_k + \cdots + n_m$ zeros in $\mathbb{C} \setminus \Delta_{k+1}$. Let us prove that it has at least $n_k + \cdots + n_m - 1$ sign changes on Δ_k. Once this is achieved we know that all the zeros of $G_{\mathbf{n},k}$ are simple and lie on the real line. In particular, this would be true for $\mathcal{A}_{\mathbf{n}^\ell,k}$.

Let us start with $k = 1$. Notice that

$$\int x^\nu G_{\mathbf{n},1}(x) d\sigma_1(x) = 0, \qquad \nu = 0,\ldots,n_1 + \cdots + n_m - 2.$$

Consequently, $G_{\mathbf{n},1}$ has at least $n_1 + \cdots + n_m - 1$ sign changes on Δ_1 as claimed. Therefore, its zeros in $\mathbb{C} \setminus \Delta_2$ are real and simple. Let $W_{\mathbf{n},1}$ be the monic polynomial whose roots are the simple zeros of $G_{\mathbf{n},1}$ in $\mathbb{C} \setminus \Delta_2$.

Observe that $z^\nu G_{\mathbf{n},1}/W_{\mathbf{n},1} = \mathcal{O}(1/z^2)$, $\nu = 0,\ldots,n_2 + \cdots + n_m - 2$. Consequently, from (3)

$$\int x^\nu G_{\mathbf{n},2}(x) \frac{d\sigma_2(x)}{W_{\mathbf{n},1}(x)} = 0, \qquad \nu = 0,\ldots,n_2 + \cdots + n_m - 2,$$

which implies that $G_{\mathbf{n},2}$ has at least $n_2 + \cdots + n_m - 1$ sign changes on Δ_2. So all its zeros are real and simple. Repeating the same arguments we obtain the claim about the number of sign changes of $G_{\mathbf{n},k}$ on Δ_k and that all its zeros are real and simple for each $k = 1,\ldots,m$.

Let us check that $\mathcal{A}_{\mathbf{n},k}$ and $\mathcal{A}_{\mathbf{n}^\ell,k}$ do not have common zeros in $\mathbb{C} \setminus \Delta_{k+1}$. To the contrary, assume that x_0 is such a common zero. We have $\mathcal{A}'_{\mathbf{n},k}(x_0) \neq 0 \neq \mathcal{A}'_{\mathbf{n}^\ell,k}(x_0)$ because the zeros of these forms are simple. Then

$$\mathcal{A}'_{\mathbf{n}^\ell,k}(x_0)\mathcal{A}_{\mathbf{n},k} - \mathcal{A}'_{\mathbf{n},k}(x_0)\mathcal{A}_{\mathbf{n}^\ell,k}$$

has a double zero at x_0 against what was proved above.

Fix $y \in \mathbb{R} \setminus \Delta_{k+1}$ and set

$$G^y_{\mathbf{n},k}(z) = \mathcal{A}_{\mathbf{n}^l,k}(z)\mathcal{A}_{\mathbf{n},k}(y) - \mathcal{A}_{\mathbf{n}^l,k}(y)\mathcal{A}_{\mathbf{n},k}(z).$$

Let $x_1, x_2, x_1 < x_2$, be two consecutive zeros of $\mathcal{A}_{\mathbf{n}^l,k}$ in $\mathbb{R} \setminus \Delta_{k+1}$ and let $y \in (x_1, x_2)$. The function $G^y_{\mathbf{n},k}(z)$ is a real valued function when restricted to $\mathbb{R} \setminus \Delta_{k+1}$ and analytic in $\mathbb{C} \setminus \Delta_{k+1}$. We have $(G^y_{\mathbf{n},k})'(z) = \mathcal{A}'_{\mathbf{n}^\ell,k}(z)\mathcal{A}_{\mathbf{n},k}(y) - \mathcal{A}_{\mathbf{n}^\ell,k}(y)\mathcal{A}'_{\mathbf{n},k}(z)$. Assume that $(G^{y_0}_{\mathbf{n},k})'(y_0) = 0$ for some $y_0 \in (x_1, x_2)$. Since $G^y_{\mathbf{n},k}(y) = 0$ for all $y \in (x_1, x_2)$ we obtain that $G^{y_0}_{\mathbf{n},k}(z)$ has a zero of order ≥ 2 (with respect to z) at y_0 which contradicts what was proved above. Consequently,

$$(G^y_{\mathbf{n},k})'(y) = \mathcal{A}'_{\mathbf{n}^l,k}(y)\mathcal{A}_{\mathbf{n},k}(y) - \mathcal{A}_{\mathbf{n}^l,k}(y)\mathcal{A}'_{\mathbf{n},k}(y)$$

takes values with constant sign for all $y \in (x_1, x_2)$. At the end points x_1, x_2, this function cannot be equal to zero because $\mathcal{A}_{\mathbf{n},k}, \mathcal{A}_{\mathbf{n}^l,k}$ do not have common zeros. By continuity, $(G^y_{\mathbf{n},k})'$ preserves the same sign on all $[x_\nu, x_{\nu+1}]$ (and, consequently, on each side of the interval Δ_{k+1}). Thus

$$\mathrm{sign}(G^{x_1}_{\mathbf{n},k})'(x_1) = \mathrm{sign}((\mathcal{A}_{\mathbf{n}^l,k})'(x_1)\mathcal{A}_{\mathbf{n},k}(x_1)) =$$

$$\mathrm{sign}((\mathcal{A}_{\mathbf{n}^l,k})'(x_2)\mathcal{A}_{\mathbf{n},k}(x_2)) = \mathrm{sign}(G^{x_2}_{\mathbf{n},k})'(x_2)\,.$$

Since

$$\mathrm{sign}(\mathcal{A}_{\mathbf{n}^l,k})'(x_1) \neq \mathrm{sign}(\mathcal{A}_{\mathbf{n}^l,k})'(x_2)\,,$$

we obtain that

$$\mathrm{sign}\mathcal{A}_{\mathbf{n},k}(x_1) \neq \mathrm{sign}\mathcal{A}_{\mathbf{n},k}(x_2)\,;$$

consequently, there must be an intermediate zero of $\mathcal{A}_{\mathbf{n},k}$ between x_1 and x_2. □

The linear forms $\mathcal{A}_{\mathbf{n},k}, k = 0, \ldots, m$ satisfy nice iterative integral representations. Let us introduce the following functions

$$H_{\mathbf{n},k}(z) := \frac{A_{\mathbf{n},k+1}(z)\mathcal{A}_{\mathbf{n},k}(z)}{A_{\mathbf{n},k}(z)}, \qquad k = 0, \ldots, m, \tag{5}$$

where the $A_{\mathbf{n},k}$ are the polynomials introduced in the statement of Proposition 1. We take $A_{\mathbf{n},0} \equiv 1 \equiv A_{\mathbf{n},m+1}$. We have:

Proposition 2 *For each* $k = 0, \ldots, m-1$

$$H_{\mathbf{n},k}(z) = \int \frac{A^2_{\mathbf{n},k+1}(x)}{z-x} \frac{H_{\mathbf{n},k+1}(x)d\sigma_{k+1}(x)}{A_{\mathbf{n},k}(x)A_{\mathbf{n},k+2}(x)} \tag{6}$$

and

$$\int x^\nu A_{\mathbf{n},k+1}(x)\frac{H_{\mathbf{n},k+1}(x)d\sigma_{k+1}(x)}{A_{\mathbf{n},k}(x)A_{\mathbf{n},k+2}(x)} = 0, \qquad \nu = 0,\ldots,n_{k+1}+\cdots+n_m-2. \tag{7}$$

Proof Relation (7) is (4) (with the index k shifted by one) using the notation introduced for the functions $H_{\mathbf{n},k}(z)$. From (2) applied to the function $\mathcal{A}_{\mathbf{n},k}/A_{\mathbf{n},k}$, $k = 0,\ldots,m-1$, we obtain

$$\frac{\mathcal{A}_{\mathbf{n},k}(z)}{A_{\mathbf{n},k}(z)} = \int \frac{\mathcal{A}_{\mathbf{n},k+1}(x)}{z-x}\frac{d\sigma_{k+1}(x)}{A_{\mathbf{n},k}(x)}.$$

Using (4) with $k+1$ replacing k it follows that

$$\int \frac{A_{\mathbf{n},k+1}(z) - A_{\mathbf{n},k+1}(x)}{z-x}\mathcal{A}_{\mathbf{n},k+1}(x)\frac{d\sigma_{k+1}(x)}{A_{\mathbf{n},k}(x)} = 0.$$

Combining these two integral formulas we get

$$\frac{A_{\mathbf{n},k+1}(z)\mathcal{A}_{\mathbf{n},k}(z)}{A_{\mathbf{n},k}(z)} = \int \frac{A_{\mathbf{n},k+1}(x)}{z-x}\mathcal{A}_{\mathbf{n},k+1}(x)\frac{d\sigma_{k+1}(x)}{A_{\mathbf{n},k}(x)} =$$

$$\int \frac{A^2_{\mathbf{n},k+1}(x)}{z-x}\frac{A_{\mathbf{n},k+2}(x)\mathcal{A}_{\mathbf{n},k+1}(x)}{A_{\mathbf{n},k+1}(x)}\frac{d\sigma_{k+1}(x)}{A_{\mathbf{n},k}(x)A_{\mathbf{n},k+2}(x)}$$

which is (6). □

Notice that the varying measure appearing in (6) has constant sign on Δ_{k+1}.

3.2 *Interlacing for Type II*

Now, let us see what happens with the type II Hermite Padé approximants. Let us introduce the following functions.

$$\Psi_{\mathbf{n},0}(z) = Q_{\mathbf{n}}(z), \qquad \Psi_{\mathbf{n},k}(z) = \int \frac{\Psi_{\mathbf{n},k-1}(x)}{z-x}d\sigma_k(x), \qquad k = 1,\ldots,m.$$

where $Q_{\mathbf{n}}$ is the type II multiple orthogonal polynomial. The next result is [27, Proposition 1].

Proposition 3 *Let* $\mathbf{n} \in \mathbb{Z}^m_+(\bullet)$. *For each* $k = 1,\ldots,m$

$$\int \Psi_{\mathbf{n},k-1}(x)\left(p_k(x) + \sum_{j=k+1}^{m} p_j(x)\widehat{s}_{k+1,j}(x)\right)d\sigma_k(x) = 0, \tag{8}$$

where $\deg p_j \leq n_j - 1$, $j = k, \ldots, m$. *For* $k = 0, \ldots, m-1$, *the function* $\Psi_{\mathbf{n},k}$ *has exactly* $n_{k+1} + \cdots + n_m$ *zeros in* $\mathbb{C} \setminus \Delta_k$, *where* $\Delta_0 = \emptyset$), *they are all simple and lie in the interior of* Δ_{k+1}. *The function* $\Psi_{\mathbf{n},m}$ *has no roots in* $\mathbb{C} \setminus \Delta_m$.

Proof The statement about the zeros of $Q_{\mathbf{n}} = \Psi_{\mathbf{n},0}$ was proved above. The proof of (8) is carried out by induction on k. The statement is equivalent to showing that for each $k = 1, \ldots, m$, we have

$$\int x^\nu \Psi_{\mathbf{n},k-1}(x) ds_{k,j}(x) = 0, \qquad \nu = 0, \ldots, n_j - 1, \qquad j = k, \ldots, m. \tag{9}$$

When $k = 1$ we have $\Psi_{\mathbf{n},0} = Q_{\mathbf{n}}$ and (9) reduces to the orthogonality relations which define $Q_{\mathbf{n}}$. So the basis of induction is settled.

Assume that (9) is true for $k \leq m - 1$ and let us show that it also holds for $k + 1$. Take $k + 1 \leq j \leq m$ and $\nu \leq n_j - 1$. Then

$$\int x^\nu \Psi_{\mathbf{n},k}(x) ds_{k+1,j}(x) = \int x^\nu \int \frac{\Psi_{\mathbf{n},k-1}(t) d\sigma_k(t)}{x - t} ds_{k+1,j}(x) =$$

$$\int \Psi_{\mathbf{n},k-1}(t) \int \frac{x^\nu - t^\nu + t^\nu}{x - t} ds_{k+1,j}(x) d\sigma_k(t) =$$

$$\int \Psi_{\mathbf{n},k-1}(t) p_{\nu-1}(t) d\sigma_k(t) + \int t^\nu \Psi_{\mathbf{n},k-1}(t) \widehat{s}_{k+1,j}(t) d\sigma_k(t),$$

where $p_{\nu-1}$ is a polynomial of degree $\leq \nu - 1 \leq n_j - 2 < n_k - 1$ (because the indices have decreasing components). Since $\widehat{s}_{k+1,j}(t) d\sigma_k(t) = ds_{k,j}(t)$ the induction hypothesis renders that both integrals on the last line equal zero and we obtain what we need.

Applying (9) with $j = k$, we have

$$\int \frac{z^{n_k} - x^{n_k}}{z - x} \Psi_{\mathbf{n},k-1}(x) d\sigma_k(x) = 0.$$

Therefore, using the definition of $\Psi_{\mathbf{n},k}$, it follows that

$$z^{n_k} \Psi_{\mathbf{n},k}(z) = \int \frac{x^{n_k}}{z - x} \Psi_{\mathbf{n},k-1}(x) d\sigma_k(x) = \mathcal{O}(1/z), \qquad z \to \infty.$$

In other words

$$\Psi_{\mathbf{n},k}(z) = \mathcal{O}(1/z^{n_k+1}), \qquad z \to \infty.$$

Relations (8) applied with k replaced with $k + 1 \leq m$, together with Theorem 1, imply that $\Psi_{\mathbf{n},k}$ has at least $n_{k+1} + \cdots + n_m$ sign changes on the interval Δ_{k+1}. Let

$Q_{\mathbf{n},k+1}$ be the monic polynomial whose roots are the zeros of $\Psi_{\mathbf{n},k}$ in $\mathbb{C} \setminus \Delta_k$. Take $Q_{\mathbf{n},m+1} \equiv 1$. Obviously $\deg Q_{\mathbf{n},k+1} \geq N_{k+1} = n_{k+1} + \cdots, n_m, N_{m+1} = 0$.

Notice that $\Psi_{\mathbf{n},k}/Q_{\mathbf{n},k+1} \in \mathcal{H}(\mathbb{C} \setminus \Delta_k)$ and $\Psi_{\mathbf{n},k}/Q_{\mathbf{n},k+1} = O(1/z^{n_k+N_{k+1}+1})$, $z \to \infty$, and the order is $> N_k + 1$ if $\deg Q_{\mathbf{n},k+1} > N_{k+1}$. Similar to the way in which (2) and (3) were proved, it follows that

$$\frac{\Psi_{\mathbf{n},k}(z)}{Q_{\mathbf{n},k+1}(z)} = \int \frac{\Psi_{\mathbf{n},k-1}(x)}{z-x} \frac{d\sigma_k(x)}{Q_{\mathbf{n},k+1}(x)} \tag{10}$$

and

$$\int x^\nu \Psi_{\mathbf{n},k-1}(x) \frac{d\sigma_k(x)}{Q_{\mathbf{n},k+1}(x)} = 0, \qquad \nu = 0, \ldots, N_k - 1. \tag{11}$$

The second one of these relations also implies that $\Psi_{\mathbf{n},k-1}$ has at least N_k sign changes on Δ_k. Notice that should $\Psi_{\mathbf{n},k}$ have more than N_{k+1} zeros in $\mathbb{C} \setminus \Delta_k$ we get at least one more orthogonality relation in (11) and then $\Psi_{\mathbf{n},k-1}$ would have more than N_k zeros in $\mathbb{C} \setminus \Delta_{k-1}$, $\Delta_0 = \emptyset$. Applying this argument for decreasing values of k we obtain that if for some $k = 1, \ldots, m$, $\Psi_{\mathbf{n},k}$ has more than N_{k+1} zeros in $\mathbb{C}\setminus\Delta_k$ then $\Psi_{\mathbf{n},0} = Q_{\mathbf{n}}$ would have more that $|\mathbf{n}| = N_1 = n_1 + \cdots, n_m$ zeros in $\mathbb{C}$ which is not possible. Consequently, for $k = 1, \ldots, m$ the function $\Psi_{\mathbf{n},k}$ has exactly N_{k+1} zeros in $\mathbb{C} \setminus \Delta_k$ they are all simple and lie in the interior of Δ_{k+1}, $\Delta_{m+1} = \emptyset$ as stated. □

Set

$$\mathcal{H}_{\mathbf{n},k} := \frac{Q_{\mathbf{n},k-1}\Psi_{\mathbf{n},k-1}}{Q_{\mathbf{n},k}}.$$

Proposition 4 *Fix* $\mathbf{n} \in \mathbb{Z}_+^m(\bullet)$. *For each* $k = 1, \ldots, m$

$$\int x^\nu Q_{\mathbf{n},k}(x) \frac{\mathcal{H}_{\mathbf{n},k}(x) d\sigma_k(x)}{Q_{\mathbf{n},k-1}(x) Q_{\mathbf{n},k+1}(x)} = 0, \qquad \nu = 0, \ldots, n_k + \cdots + n_m - 1, \tag{12}$$

and

$$\mathcal{H}_{\mathbf{n},k+1}(z) = \int \frac{Q^2_{\mathbf{n},k}(x)}{z-x} \frac{\mathcal{H}_{\mathbf{n},k}(x) d\sigma_k(x)}{Q_{\mathbf{n},k-1}(x) Q_{\mathbf{n},k+1}(x)}, \tag{13}$$

where $Q_{\mathbf{n},0} = Q_{\mathbf{n},m+1} \equiv 1$.

Proof Using the notation introduced for the functions $\mathcal{H}_{\mathbf{n},k}$, (11) adopts the form (12). In turn, this implies that

$$\int \frac{\mathcal{Q}_{\mathbf{n},k}(z) - \mathcal{Q}_{\mathbf{n},k}(x)}{z - x} \mathcal{Q}_{\mathbf{n},k}(x) \frac{\mathcal{H}_{\mathbf{n},k}(x) d\sigma_k(x)}{\mathcal{Q}_{\mathbf{n},k-1}(x)\mathcal{Q}_{\mathbf{n},k+1}(x)} = 0.$$

Separating this integral in two and using (10) we obtain (13). □

For $\mathbf{n} \in \mathbb{Z}_+^m(\bullet)$ and $\ell \in \{1, \ldots, m\}$ we define $\mathbf{n}^\ell$ as was done above. Though $\mathbf{n}^\ell$ need not belong to $\mathbb{Z}_+^m(\bullet)$, we can define the corresponding functions $\Psi_{\mathbf{n}^\ell,k}$.

Theorem 4 *Assume that* $\mathbf{n} \in \mathbb{Z}_+^m(\bullet)$ *and* $\ell \in \{1, \ldots, m\}$. *For each* $k = 0, \ldots, m-1$, $\Psi_{\mathbf{n}^\ell,k}$ *has at most* $n_{k+1} + \cdots + n_m + 1$ *zeros in* $\mathbb{C} \setminus \Delta_k$ *and at least* $n_{k+1} + \cdots + n_m$ *sign changes in* Δ_{k+1}. *Therefore, all its zeros are real and simple. The zeros of* $\Psi_{\mathbf{n},k}$ *and* $\Psi_{\mathbf{n}^\ell,k}$ *in* $\mathbb{C} \setminus \Delta_k$ *interlace.*

Proof The proof is similar to that of Theorem 3 so we will not dwell into details. For each $k = 0, \ldots, m-1$ and $A, B \in \mathbb{R}$, $|A| + |B| > 0$ define

$$G_{\mathbf{n},k} = A\Psi_{\mathbf{n},k} + B\Psi_{\mathbf{n}^\ell,k}.$$

Then

$$\int G_{\mathbf{n},k-1}(x) \left(p_k(x) + \sum_{j=k+1}^{m} p_j(x)\widehat{s}_{k+1,j}(x) \right) d\sigma_k(x) = 0,$$

$$\deg p_j \leq n_j - 1 \qquad j = k, \ldots, m.$$

From here, it follows that there exists a monic polynomial $W_{\mathbf{n},k+1}$, $\deg W_{\mathbf{n},k+1} \geq N_{k+1} = n_{k+1} + \cdots, n_m$, $N_{m+1} = 0$, whose roots are the zeros of $G_{\mathbf{n},k}$ in $\mathbb{C} \setminus \Delta_k$ such that

$$\int x^\nu G_{\mathbf{n},k-1}(x) \frac{d\sigma_k(x)}{W_{\mathbf{n},k+1}(x)} = 0, \qquad \nu = 0, \ldots, n_k + \deg W_{\mathbf{n},k+1} - 1.$$

These relations imply that $G_{\mathbf{n},k}$ has at most $n_{k+1} + \cdots + n_m + 1$ zeros in $\mathbb{C} \setminus \Delta_k$ and at least $n_{k+1} + \cdots + n_m$ sign changes in Δ_{k+1}. In particular, this is true for $\Psi_{\mathbf{n}^\ell,k}$. So, all the zeros of $G_{\mathbf{n},k}$ (and $\Psi_{\mathbf{n}^\ell,k}$) in $\mathbb{C} \setminus \Delta_k$ are real and simple. The interlacing is proved following the same arguments as in Theorem 3. The details are left to the reader. □

4 Weak Asymptotic

Following standard techniques, the weak asymptotic for type I and type II Hermite-Padé polynomials is derived using arguments from potential theory. We will briefly summarize what is needed.

4.1 Preliminaries from Potential Theory

Let E_k, $k = 1\ldots, m$, be (not necessarily distinct) compact subsets of the real line and

$$C = (c_{j,k}), \qquad 1 \leq j, k \leq m,$$

a real, positive definite, symmetric matrix of order m. C will be called the interaction matrix. Let $\mathcal{M}_1(E_k)$ be the subclass of probability measures in $\mathcal{M}(E_k)$. Set

$$\mathcal{M}_1 = \mathcal{M}_1(E_1) \times \cdots \times \mathcal{M}_1(E_m) .$$

Given a vector measure $\boldsymbol{\mu} = (\mu_1, \ldots, \mu_m) \in \mathcal{M}_1$ and $j = 1, \ldots, m$, we define the combined potential

$$W_j^{\boldsymbol{\mu}}(x) = \sum_{k=1}^{m} c_{j,k} V^{\mu_k}(x) ,$$

where

$$V^{\mu_k}(x) := \int \log \frac{1}{|x-t|} d\mu_k(t) ,$$

denotes the standard logarithmic potential of μ_k. Set

$$\omega_j^{\boldsymbol{\mu}} := \inf\{W_j^{\boldsymbol{\mu}}(x) : x \in E_j\}, \quad j = 1, \ldots, m .$$

It is said that $\sigma \in \mathcal{M}(\Delta)$ is regular, and we write $\sigma \in \textbf{Reg}$, if

$$\lim \gamma_n^{1/n} = \frac{1}{\text{cap}(\text{supp}(\sigma))},$$

where cap(supp(σ)) denotes the logarithmic capacity of supp(σ) and γ_n is the leading coefficient of the (standard) n-th orthonormal polynomial with respect to σ. See [45, Theorems 3.1.1, 3.2.1] for different equivalent forms of defining regular measures and its basic properties. In connection with regular measures it is frequently convenient that the support of the measure be regular. A compact set E is said to be regular when the Green's function, corresponding to the unbounded connected component of $\mathbb{C} \setminus E$, with singularity at ∞ can be extended continuously to E.

In Chapter 5 of [42] the authors prove (we state the result in a form convenient for our purpose).

Lemma 3 *Assume that the compact sets $E_k, k = 1, \ldots, m$, are regular. Let C be a real, positive definite, symmetric matrix of order m. If there exists $\boldsymbol{\lambda}$ =*

$(\lambda_1, \ldots, \lambda_m) \in \mathcal{M}_1$ *such that for each* $j = 1, \ldots, m$

$$W_j^{\boldsymbol{\lambda}}(x) = \omega_j^{\boldsymbol{\lambda}}, \qquad x \in \operatorname{supp} \lambda_j,$$

then $\boldsymbol{\lambda}$ *is unique. Moreover, if* $c_{j,k} \geq 0$ *when* $E_j \cap E_k \neq \emptyset$*, then* $\boldsymbol{\lambda}$ *exists.*

For details on how Lemma 3 is derived from [42, Chapter 5] see [8, Section 4]. The vector measure $\boldsymbol{\lambda}$ is called the equilibrium solution for the vector potential problem determined by the interaction matrix C on the system of compact sets $E_j, j = 1, \ldots, m$ and $\omega^{\boldsymbol{\lambda}} := (\omega_1^{\boldsymbol{\lambda}}, \ldots, \omega_m^{\boldsymbol{\lambda}})$ is the vector equilibrium constant. There are other characterizations of the equilibrium measure and constant but we will not dwell into that because they will not be used and their formulation requires introducing additional notions and notation.

We also need

Lemma 4 *Let* $E \subset \mathbb{R}$ *be a regular compact set and* ϕ *a continuous function on* E*. Then, there exists a unique* $\lambda \in \mathcal{M}_1(E)$ *and a constant* w *such that*

$$V^{\lambda}(z) + \phi(z) \begin{cases} \leq w, & z \in \operatorname{supp} \lambda, \\ \geq w, & z \in E. \end{cases}$$

In particular, equality takes place on all $\operatorname{supp} \lambda$. If the compact set E is not regular with respect to the Dirichlet problem, the second part of the statement is true except on a set e such that $\operatorname{cap}(e) = 0$. Theorem I.1.3 in [43] contains a proof of this lemma in this context. When E is regular, it is well known that this inequality except on a set of capacity zero implies the inequality for all points in the set (cf. Theorem I.4.8 from [43]). λ is called the equilibrium measure in the presence of the external field ϕ on E and w is the equilibrium constant.

As usual, a sequence of measures (μ_n) supported on a compact set E is said to converge to a measure μ in the weak star topology if for every continuous function f on E we have

$$\lim_n \int f d\mu_n = \int f d\mu.$$

We write $* \lim_n \mu_n = \mu$. Given a polynomial Q of degree n we denote

$$\mu_Q = \frac{1}{n} \sum_{Q(x)=0} \delta_x,$$

where δ_x is the Dirac measure with mass 1 at point x. In the previous sum, each zero of Q is repeated taking account of its multiplicity. The measure μ_Q is usually called the normalized zero counting measure of Q.

One last ingredient needed is a result which relates the asymptotic zero distribution of polynomials orthogonal with respect to varying measures with the solution of a vector equilibrium problem in the presence of an external field contained in

Lemma 4. Different versions of it appear in [23], and [45]. In [23], it was proved assuming that $\operatorname{supp}\sigma$ is an interval on which $\sigma' > 0$ a.e.. Theorem 3.3.3 in [45] does not cover the type of external field we need to consider. As stated here, the proof appears in [22, Lemma 4.2].

Lemma 5 *Assume that* $\sigma \in \mathbf{Reg}$ *and* $\operatorname{supp}\sigma \subset \mathbb{R}$ *is regular. Let* $\{\phi_n\}, n \in \Lambda \subset \mathbb{Z}_+$, *be a sequence of positive continuous functions on* $\operatorname{supp}\sigma$ *such that*

$$\lim_{n\in\Lambda} \frac{1}{2n} \log \frac{1}{\phi_n(x)} = \phi(x) > -\infty, \tag{14}$$

uniformly on $\operatorname{supp}\sigma$. *Let* $(q_n), n \in \Lambda$, *be a sequence of monic polynomials such that* $\deg q_n = n$ *and*

$$\int x^k q_n(x)\phi_n(x)d\sigma(x) = 0, \qquad k = 0, \ldots, n-1.$$

Then

$$*\lim_{n\in\Lambda} \mu_{q_n} = \lambda, \tag{15}$$

and

$$\lim_{n\in\Lambda} \left(\int |q_n(x)|^2 \phi_n(x) d\sigma(x) \right)^{1/2n} = e^{-w}, \tag{16}$$

where λ *and* w *are the equilibrium measure and equilibrium constant in the presence of the external field* ϕ *on* $\operatorname{supp}\sigma$ *given by Lemma 4. We also have*

$$\lim_{n\in\Lambda} \left(\frac{|q_n(z)|}{\|q_n\phi_n^{1/2}\|_E} \right)^{1/n} = \exp\left(w - V^{\lambda}(z)\right), \qquad \mathcal{K} \subset \mathbb{C} \setminus \Delta, \tag{17}$$

where $\|\cdot\|_E$ *denotes the uniform norm on* E *and* Δ *is the smallest interval containing* $\operatorname{supp}\sigma$.

4.2 Weak Asymptotic Behavior for Type II

In the proof of the asymptotic zero distribution of the polynomials $Q_{\mathbf{n},j}$ we take $E_j = \operatorname{supp}\sigma_j$. We need to specify the sequence of multi-indices for which the result takes place and the relevant interaction matrix for the vector equilibrium problem which arises.

Let $\Lambda = \Lambda(p_1, \dots, p_n) \subset \mathbb{Z}_+^m(\bullet)$ be an infinite sequence of distinct multi-indices such that

$$\lim_{\mathbf{n}\in\Lambda} \frac{n_j}{|\mathbf{n}|} = p_j \in (0, 1), \qquad j = 1, \dots, m. \tag{18}$$

Obviously, $p_1 \geq \cdots \geq p_m$ and $\sum_{j=1}^m p_j = 1$. Set

$$P_j = \sum_{k=j}^m p_k, \qquad j = 1, \dots, m.$$

Let us define the interaction matrix $C_{\mathcal{N}}$ which is relevant in the next result. Set

$$C_{\mathcal{N}} := \begin{pmatrix} P_1^2 & -\frac{P_1P_2}{2} & 0 & \cdots & 0 \\ -\frac{P_1P_2}{2} & P_2^2 & -\frac{P_2P_3}{2} & \cdots & 0 \\ 0 & -\frac{P_2P_3}{2} & P_3^2 & \cdots & 0 \\ \vdots & \vdots & \vdots & \ddots & \vdots \\ 0 & 0 & 0 & \cdots & P_m^2 \end{pmatrix}. \tag{19}$$

This matrix satisfies all the assumptions of Lemma 3 on the compact sets $E_j = \operatorname{supp}(\sigma_j)$, $j = 1, \dots, m$, including $c_{j,k} \geq 0$ when $E_j \cap E_k \neq \emptyset$ and it is positive definite because the principal section $(C_{\mathcal{N}})_r$, $r = 1, \dots, m$ of $C_{\mathcal{N}}$ satisfies

$$\det(C_{\mathcal{N}})_r = P_1^2 \cdots P_r^2 \det \begin{pmatrix} 1 & -\frac{1}{2} & 0 & \cdots & 0 & 0 \\ -\frac{1}{2} & 1 & -\frac{1}{2} & \cdots & 0 & 0 \\ 0 & -\frac{1}{2} & 1 & \cdots & 0 & 0 \\ \vdots & \vdots & \vdots & \ddots & \vdots & \vdots \\ 0 & 0 & 0 & \cdots & 1 & -\frac{1}{2} \\ 0 & 0 & 0 & \cdots & -\frac{1}{2} & 1 \end{pmatrix}_{r\times r} > 0.$$

Let $\boldsymbol{\lambda}(C_{\mathcal{N}}) = (\lambda_1, \dots, \lambda_m)$ be the solution of the corresponding vector equilibrium problem stated in Lemma 3.

The next result, under more restrictive conditions on the measures but in the framework of so called Nikishin systems on a graph tree is contained in [27]. We have practically reproduced their arguments which incidentally can also be adapted to the study of so called mixed type Hermite-Padé approximation in which the definition contains a mixture of type I and type II interpolation conditions. For details see [22].

Theorem 5 *Let Λ be a sequence of multi-indices verifying* (18). *Assume that $\sigma_j \in$* **Reg** *and* $\operatorname{supp} \sigma_j = E_j$ *is regular for each $j = 1, \dots, m$. Then,*

$$* \lim_{\mathbf{n}\in\Lambda} \mu_{Q_{\mathbf{n},j}} = \lambda_j, \qquad j = 1, \dots, m. \tag{20}$$

where $\boldsymbol{\lambda} = (\lambda_1, \ldots, \lambda_m) \in \mathcal{M}_1$ *is the vector equilibrium measure determined by the matrix* $C_{\mathcal{N}}$ *on the system of compact sets* E_j, $j = 1, \ldots, m$. *Moreover,*

$$\lim_{\mathbf{n}\in\Lambda} \left| \int Q_{\mathbf{n},j}^2(x) \frac{\mathcal{H}_{\mathbf{n},j}(x)\,\mathrm{d}\sigma_j(x)}{Q_{\mathbf{n},j-1}(x)Q_{\mathbf{n},j+1}(x)} \right|^{1/2|\mathbf{n}|} = \exp\left(-\sum_{k=1}^{j} \omega_k^{\lambda}/P_k\right), \tag{21}$$

where $\omega^{\lambda} = (\omega_1^{\lambda}, \ldots, \omega_m^{\lambda})$ *is the vector equilibrium constant. For* $j = 1, \ldots, m$

$$\lim_{\mathbf{n}\in\Lambda} |\Psi_{\mathbf{n},j}(z)|^{1/|\mathbf{n}|} = \exp\left(P_j V^{\lambda_j}(z) - P_{j+1} V^{\lambda_{j+1}}(z) - 2\sum_{k=1}^{j} \omega_k^{\lambda}/P_k\right) \tag{22}$$

uniformly on compact subsets of $\mathbb{C} \setminus (\Delta_j \cup \Delta_{j+1})$ *where* $\Delta_{m+1} = \emptyset$ *and the term with* P_{m+1} *is dropped when* $j = m$.

Proof The unit ball in the cone of positive Borel measures is weak star compact; therefore, it is sufficient to show that each one of the sequences of measures $(\mu_{Q_{\mathbf{n},j}})$, $\mathbf{n} \in \Lambda$, $j = 1, \ldots, m$, has only one accumulation point which coincides with the corresponding component of the vector equilibrium measure $\boldsymbol{\lambda}$ determined by the matrix $C_{\mathcal{N}}$ on the system of compact sets E_j, $j = 1, \ldots, m$.

Let $\Lambda' \subset \Lambda$ be such that for each $j = 1, \ldots, m$

$$* \lim_{\mathbf{n}\in\Lambda'} \mu_{Q_{\mathbf{n},j}} = \mu_j.$$

Notice that $\mu_j \in \mathcal{M}_1(E_j)$, $j = 1, \ldots, m$. Taking into account that all the zeros of $Q_{\mathbf{n},j}$ lie in Δ_j, it follows that

$$\lim_{\mathbf{n}\in\Lambda'} |Q_{\mathbf{n},j}(z)|^{1/|\mathbf{n}|} = \exp(-P_j V^{\mu_j}(z)), \tag{23}$$

uniformly on compact subsets of $\mathbb{C} \setminus \Delta_j$.

When $k = 1$, (12) reduces to

$$\int x^{\nu} Q_{\mathbf{n},1}(x) \frac{\mathrm{d}\sigma_1(x)}{|Q_{\mathbf{n},2}(x)|} = 0, \qquad \nu = 0, \ldots, |\mathbf{n}| - 1.$$

According to (23)

$$\lim_{\mathbf{n}\in\Lambda'} \frac{1}{2|\mathbf{n}|} \log |Q_{\mathbf{n},2}(x)| = -\frac{P_2}{2} V^{\mu_2}(x),$$

uniformly on Δ_2. Using Lemma 5, it follows that μ_1 is the unique solution of the extremal problem

$$V^{\mu_1}(x) - \frac{P_2}{2} V^{\mu_2}(x) \begin{cases} = \omega_1, & x \in \operatorname{supp} \mu_1 , \\ \geq \omega_1, & x \in E_1 , \end{cases} \tag{24}$$

and

$$\lim_{\mathbf{n}\in\Lambda'} \left| \int \frac{Q^2_{\mathbf{n},1}(x)}{|Q_{\mathbf{n},2}(x)|} \mathrm{d}\sigma_1(x) \right|^{1/2|\mathbf{n}|} = e^{-\omega_1} . \tag{25}$$

Using induction on increasing values of j, let us show that for all $j = 1, \ldots, m$

$$V^{\mu_j}(x) - \frac{P_{j-1}}{2P_j} V^{\mu_{j-1}}(x) - \frac{P_{j+1}}{2P_j} V^{\mu_{j+1}}(x) + \frac{P_{j-1}}{P_j}\omega_{j-1} \begin{cases} = \omega_j, & x \in \operatorname{supp} \mu_j , \\ \geq \omega_j, & x \in E_j , \end{cases} , \tag{26}$$

(when $j = 1$ or $j = m$ the terms with P_0 and P_{m+1} do not appear,) and

$$\lim_{n\in\Lambda'} \left| \int Q^2_{\mathbf{n},j}(x) \frac{|\mathcal{H}_{\mathbf{n},j}(x)| \mathrm{d}\sigma_j(x)}{|Q_{\mathbf{n},j-1}(x) Q_{\mathbf{n},j+1}(x)|} \right|^{1/2N_{\mathbf{n},j}} = e^{-\omega_j} , \tag{27}$$

where $Q_{n,0} \equiv Q_{n,m+1} \equiv 1$ and $N_{\mathbf{n},j} = n_j + \cdots + n_m$. For $j = 1$ these relations are non other than (24)–(25) and the initial induction step is settled. Let us assume that the statement is true for $j - 1 \in \{1, \ldots, m-1\}$ and let us prove it for j.

Taking account of the fact that $Q_{\mathbf{n},j-1}$, $Q_{\mathbf{n},j+1}$ and $\mathcal{H}_{\mathbf{n},j}$ have constant sign on Δ_j, for $j = 1, \ldots, m$, the orthogonality relations (12) can be expressed as

$$\int x^\nu Q_{\mathbf{n},j}(x) \frac{|\mathcal{H}_{\mathbf{n},j}(x)| \mathrm{d}\sigma_j(x)}{|Q_{\mathbf{n},j-1}(x) Q_{\mathbf{n},j+1}(x)|} = 0 , \qquad \nu = 0, \ldots, N_{\mathbf{n},j} - 1 ,$$

and using (13) it follows that

$$\int x^\nu Q_{\mathbf{n},j}(x) \left| \int \frac{Q^2_{\mathbf{n},j-1}(t)}{|x-t|} \frac{|\mathcal{H}_{\mathbf{n},j-1}(t)| \mathrm{d}\sigma_{j-1}(t)}{|Q_{\mathbf{n},j-2}(t) Q_{\mathbf{n},j}(t)|} \right| \frac{\mathrm{d}\sigma_j(x)}{|Q_{\mathbf{n},j-1}(x) Q_{\mathbf{n},j+1}(x)|} = 0 ,$$

for $\nu = 0, \ldots, N_{\mathbf{n},j} - 1$.

Relation (23) implies that

$$\lim_{n\in\Lambda'} \frac{1}{2N_{\mathbf{n},j}} \log |Q_{\mathbf{n},j-1}(x) Q_{\mathbf{n},j+1}(x)| = -\frac{P_{j-1}}{2P_j} V^{\mu_{j-1}}(x) - \frac{P_{j+1}}{2P_j} V^{\mu_{j+1}}(x) , \tag{28}$$

uniformly on Δ_j. (Since $Q_{n,0} \equiv 1$, when $j = 1$ we only get the second term on the right hand side of this limit.)

Set

$$K_{n,j-1} := \left| \int Q^2_{\mathbf{n},j-1}(t) \frac{|\mathcal{H}_{\mathbf{n},j-1}(t)| d\sigma_{j-1}(t)}{|Q_{\mathbf{n},j-2}(t) Q_{\mathbf{n},j}(t)|} \right|^{-1/2}. \tag{29}$$

It follows that for $x \in \Delta_j$

$$\frac{1}{\delta^*_{j-1} K^2_{\mathbf{n},j-1}} \leq \left| \int \frac{Q^2_{\mathbf{n},j-1}(t)}{|x-t|} \frac{|\mathcal{H}_{\mathbf{n},j-1}(t)| d\sigma_{j-1}(t)}{|Q_{\mathbf{n},j-2}(t) Q_{\mathbf{n},j}(t)|} \right| \leq \frac{1}{\delta_{j-1} K^2_{n,j-1}},$$

where $0 < \delta_{j-1} = \min\{|x-t| : t \in \Delta_{j-1}, x \in \Delta_j\} \leq \max\{|x-t| : t \in \Delta_{j-1}, x \in \Delta_j\} = \delta^*_{j-1} < \infty$. Taking into consideration these inequalities, from the induction hypothesis, we obtain that

$$\lim_{n \in \Lambda'} \left| \int \frac{Q^2_{\mathbf{n},j-1}(t)}{|x-t|} \frac{|\mathcal{H}_{\mathbf{n},j-1}(t)| d\sigma_{j-1}(t)}{|Q_{\mathbf{n},j-2}(t) Q_{\mathbf{n},j}(t)|} \right|^{1/2N_{\mathbf{n},j}} = e^{-P_{j-1}\omega_{j-1}/P_j}. \tag{30}$$

Taking (28) and (30) into account, Lemma 5 yields that μ_j is the unique solution of the extremal problem (26) and

$$\lim_{n \in \Lambda'} \left| \int \int \frac{Q^2_{\mathbf{n},j-1}(t)}{|x-t|} \frac{|\mathcal{H}_{\mathbf{n},j-1}(t)| d\sigma_{j-1}(t)}{|Q_{\mathbf{n},j-2}(t) Q_{\mathbf{n},j}(t)|} \frac{Q^2_{\mathbf{n},j}(x) d\sigma_j(x)}{|Q_{\mathbf{n},j-1}(x) Q_{\mathbf{n},j+1}(x)|} \right|^{1/2N_{\mathbf{n},j}} = e^{-\omega_j}.$$

According to (13) the previous formula reduces to (27). We have concluded the induction.

Now, we can rewrite (26) as

$$P_j^2 V^{\mu_j}(x) - \frac{P_j P_{j-1}}{2} V^{\mu_{j-1}}(x) - \frac{P_j P_{j+1}}{2} V^{\mu_{j+1}}(x) \begin{cases} = \omega'_j, & x \in \operatorname{supp} \mu_j, \\ \geq \omega'_j, & x \in E_j, \end{cases} \tag{31}$$

for $j = 1, \ldots, m$, where

$$\omega'_j = P_j^2 \omega_j - P_j P_{j-1} \omega_{j-1}, \qquad (\omega_0 = 0). \tag{32}$$

(Recall that the terms with V^{μ_0} and $V^{\mu_{m+1}}$ do not appear when $j = 0$ and $j = m$, respectively.) By Lemma 3, $\boldsymbol{\lambda} = (\mu_1, \ldots, \mu_m)$ is the solution of the equilibrium problem determined by the interaction matrix $C_\mathcal{N}$ on the system of compact sets E_j, $j = 1, \ldots, m$ and $\omega^{\boldsymbol{\lambda}} = (\omega'_1, \ldots, \omega'_m)$ is the corresponding vector equilibrium constant. This is for any convergent subsequence; since the equilibrium problem does not depend on the sequence of indices Λ' and the solution is unique we obtain the limits in (20). By the same token, the limit in (27) holds true over the whole sequence of indices Λ. Therefore,

$$\lim_{n\in\Lambda}\left|\int Q^2_{\mathbf{n},j}(x)\frac{|\mathcal{H}_{\mathbf{n},j}(x)|d\sigma_j(x)}{|Q_{\mathbf{n},j-1}(x)Q_{\mathbf{n},j+1}(x)|}\right|^{1/2|\mathbf{n}|} = e^{-P_j\omega_j}\,. \tag{33}$$

From (32) it follows that $\omega_1 = \omega_1^\lambda$ when $j = 1$. Suppose that $P_{j-1}\omega_{j-1} = \sum_{k=1}^{j-1}\omega_k^\lambda/P_k$ where $j-1\in\{1,\ldots,m-1\}$. Then, according to (32)

$$P_j\omega_j = \omega_j^\lambda/P_j + P_{j-1}\omega_{j-1} = \sum_{k=1}^{j}\omega_k^\lambda/P_k$$

and (21) immediately follows using (33).

For $j\in\{1,\ldots,m\}$, from (13), we have

$$\Psi_{\mathbf{n},j}(z) = \frac{Q_{\mathbf{n},j+1}(z)}{Q_{\mathbf{n},j}(z)}\int\frac{Q^2_{\mathbf{n},j}(x)}{z-x}\frac{\mathcal{H}_{\mathbf{n},j}(x)d\sigma_j(x)}{Q_{\mathbf{n},j-1}(x)Q_{\mathbf{n},j+1}(x)}, \tag{34}$$

where $Q_{\mathbf{n},0}\equiv Q_{\mathbf{n},m+1}\equiv 1$. Now, (20) implies

$$\lim_{\mathbf{n}\in\Lambda}\left|\frac{Q_{\mathbf{n},j+1}(z)}{Q_{\mathbf{n},j}(z)}\right|^{1/|\mathbf{n}|} = \exp\left(P_jV^{\lambda_j}(z) - P_{j+1}V^{\lambda_{j+1}}(z)\right), \tag{35}$$

uniformly on compact subsets of $\mathbb{C}\setminus(\Delta_j\cup\Delta_{j+1})$ (we also use that the zeros of $Q_{n,j}$ and $Q_{n,j+1}$ lie in Δ_j and Δ_{j+1}, respectively). It remains to find the $|\mathbf{n}|$-th root asymptotic behavior of the integral.

Fix a compact set $\mathcal{K}\subset\mathbb{C}\setminus\Delta_j$. It is not difficult to prove that (for the definition of $K_{n,j}$ see (29))

$$\frac{C_1}{K^2_{\mathbf{n},j}} \le \left|\int\frac{Q^2_{\mathbf{n},j}(x)}{z-x}\frac{\mathcal{H}_{\mathbf{n},j}(x)d\sigma_j(x)}{Q_{\mathbf{n},j-1}(x)Q_{\mathbf{n},j+1}(x)}\right| \le \frac{C_2}{K^2_{\mathbf{n},j}},$$

where

$$C_1 = \frac{\min\{\max\{|u-x|,|v| : z = u+iv\} : z\in\mathcal{K}, x\in\Delta_j\}}{\max\{|z-x|^2 : z\in\mathcal{K}, x\in\Delta_j\}} > 0$$

and

$$C_2 = \frac{1}{\min\{|z-x| : z\in\mathcal{K}, x\in\Delta_j\}} < \infty.$$

Taking into account (21)

$$\lim_{\mathbf{n}\in\Lambda}\left|\int \frac{Q^2_{\mathbf{n},j}(x)}{z-x}\frac{\mathcal{H}_{\mathbf{n},j}(x)\mathrm{d}\sigma_j(x)}{Q_{\mathbf{n},j-1}(x)Q_{\mathbf{n},j+1}(x)}\right|^{1/|\mathbf{n}|} = \exp\left(-2\sum_{k=1}^{j}\omega_k^{\lambda}/P_k\right). \quad (36)$$

From (34), (35), and (36), we obtain (22) and we are done. □

Remark 1 In the case of type I Hermite Padé approximation, asymptotic formulas for the forms $\mathcal{A}_{\mathbf{n},j}$ and the polynomials $A_{\mathbf{n},j}$ can be obtained following arguments similar to those employed above, see [41] or [42]. Basically, all one has to do is replace the use of the formulas in Proposition 4 by the ones in Proposition 2. I recommend doing it as an exercise.

4.3 Application to Hermite-Padé Approximation

The convergence of type II Hermite-Padé approximants for the case of m generating measures and interpolation conditions equally distributed between the different functions was obtained in [12]. When $m = 2$ the result was proved in [40]. Other results for more general sequences of multi-indices and so called multipoint Hermite-Padé approximation were considered in [18, 19]. For type I, the convergence was proved recently in [35]. Here we wish to show how the weak asymptotic of the Hermite-Padé polynomials allows to estimate the rate of convergence of the approximants. We restrict to type II. The presentation follows closely the original result given in [27].

Consider the functions

$$\Phi_{\mathbf{n},j}(z) := (Q_{\mathbf{n}}\widehat{s}_{1,j} - P_{\mathbf{n},j})(z) = O(1/z^{n_j+1}), \qquad z\to\infty, \qquad j = 1,\ldots,m, \quad (37)$$

which are the remainders of the interpolation conditions defining the type II Hermite-Padé approximants with respect to the multi-index $\mathbf{n}$ of the Nikishin system of functions $(\widehat{s}_{1,1},\ldots,\widehat{s}_{1,m})$. Because of (37), $P_{\mathbf{n},j}$ is the polynomial part of the Laurent expansion at ∞ of $Q_{\mathbf{n}}\widehat{s}_{1,j}$. It is easy to check that

$$(Q_{\mathbf{n}}\widehat{s}_{1,j} - P_{\mathbf{n},j})(z) = \int \frac{Q_{\mathbf{n}}(x)ds_{1,j}(x)}{z-x}, \qquad P_{\mathbf{n},j}(z) = \int \frac{Q_{\mathbf{n}}(z)-Q_{\mathbf{n}}(x)}{z-x}ds_{1,j}(x).$$

For example, this follows using Hermite's integral representation of $Q_{\mathbf{n}}\widehat{s}_{1,j} - P_{\mathbf{n},j}$, Cauchy's integral formula, and the Fubini theorem. According to the way in which $s_{1,j}$ is defined, we have

$$\Phi_{\mathbf{n},j}(z) = \int\cdots\int \frac{Q_{\mathbf{n}}(x_1)d\sigma_1(x_1)d\sigma_2(x_2)\cdots d\sigma_j(x_j)}{(z-x_1)(x_1-x_2)\cdots(x_{j-1}-x_j)}.$$

Notice that $\Phi_{\mathbf{n},1} = \Psi_{\mathbf{n},1}$. We wish to establish a connection between the functions $\Phi_{\mathbf{n},j}$ and $\Psi_{\mathbf{n},k}$, $1 \leq j, k \leq m$.

First let us present an interesting formula which connects $(s_{1,1}, \ldots, s_{1,m}) = \mathcal{N}(\sigma_1, \ldots, \sigma_m)$ and $(s_{m,m}, \ldots, s_{m,1}) = \mathcal{N}(\sigma_m, \ldots, \sigma_1)$. When $1 \leq k < j \leq m$, we denote

$$s_{j,k} = \langle \sigma_j, \sigma_{j-1}, \ldots, \sigma_k \rangle.$$

Lemma 6 *For each* $j = 2, \ldots, m$,

$$(\widehat{s}_{1,j} - \widehat{s}_{1,j-1}\widehat{s}_{j,j} + \widehat{s}_{1,j-2}\widehat{s}_{j,j-1} + \cdots + (-1)^{j-1}\widehat{s}_{1,1}\widehat{s}_{j,2} + (-1)^j\widehat{s}_{j,1})(z) \equiv 0, \tag{38}$$

for all $z \in \mathbb{C} \setminus (\Delta_1 \cup \Delta_j)$.

Proof Notice that

$$\widehat{s}_{1,j}(z) + (-1)^j\widehat{s}_{j,1}(z) = \int \cdots \int \frac{(x_1 - x_j)d\sigma_1(x_1)d\sigma_2(x_2)\cdots d\sigma_j(x_j)}{(z - x_1)(x_1 - x_2)\cdots(x_{j-1} - x_j)(z - x_j)}.$$

On the right hand side, use that $x_1 - x_j = (x_1 - x_2) + (x_2 - x_3) + \cdots + (x_{j-1} - x_j)$ to separate the integral in a sum. In each one of the resulting integrals, the numerator cancels one of the factors in the denominator, and the integral splits in the product of two which easily identify with the remaining terms in the formula. □

Now, we can prove the connection formulas.

Lemma 7 *We have* $\Psi_{\mathbf{n},1} = \Phi_{\mathbf{n},1}$ *and for* $j = 2, \ldots, m$

$$\Psi_{\mathbf{n},j}(z) = \sum_{k=2}^{j}(-1)^k\widehat{s}_{j,k}(z)\Phi_{\mathbf{n},k-1}(z) + (-1)^{j+1}\Phi_{\mathbf{n},j}(z), \qquad z \in \mathbb{C} \setminus (\Delta_1 \cup \Delta_j), \tag{39}$$

and

$$\Phi_{\mathbf{n},j}(z) = \sum_{k=2}^{j}(-1)^k\widehat{s}_{k,j}(z)\Psi_{\mathbf{n},k-1}(z) + (-1)^{j+1}\Psi_{\mathbf{n},j}(z), \qquad z \in \mathbb{C} \setminus (\cup_{k=1}^{j}\Delta_k). \tag{40}$$

Proof Obviously, $\Psi_{\mathbf{n},1} = \Phi_{\mathbf{n},1}$. Notice that formula (38) remains valid if the measures $\sigma_1, \ldots, \sigma_m$ are signed and finite. All what is needed is that they are supported on intervals which are consecutively non intersecting. It is easy to see that

$$\Phi_{\mathbf{n},j}(z) = \widehat{\langle \mathcal{Q}_{\mathbf{n}}\sigma_1, \sigma_2, \ldots, \sigma_j \rangle}(z), \qquad j = 2, \ldots, m.$$

The symbol $\widehat{\langle\cdot\rangle}$ means taking the Cauchy transform of $\langle\cdot\rangle$. On the other hand

$$\Psi_{\mathbf{n},j}(z) = \widehat{\langle \sigma_j, \ldots, \sigma_2, Q_{\mathbf{n}}\sigma_1 \rangle}(z), \qquad j = 2, \ldots, m.$$

Taking into consideration the previous remarks, using formula (38) with $d\sigma_1$ replaced with $Q_{\mathbf{n}} d\sigma_1$, after trivial transformations we obtain (39).

The collection of formulas (39) for $j = 2, \ldots, m$ together with $\Phi_{\mathbf{n},1} = \Psi_{\mathbf{n},1}$ can be expressed in matrix form as follows

$$(\Psi_{\mathbf{n},1}, \ldots, \Psi_{\mathbf{n},m})^t = D(\Phi_{\mathbf{n},1}, \ldots, \Phi_{\mathbf{n},m})^t,$$

where $(\cdot)^t$ is the transpose of the vector $(\cdot)$ and D is the $m \times m$ lower triangular matrix given by

$$D := \begin{pmatrix} 1 & 0 & 0 & \cdots & 0 \\ \widehat{s}_{2,2} & -1 & 0 & \cdots & 0 \\ \widehat{s}_{3,2} & -\widehat{s}_{3,3} & 1 & \cdots & 0 \\ \vdots & \vdots & \vdots & \ddots & \vdots \\ \widehat{s}_{m,2} & -\widehat{s}_{m,3} & \widehat{s}_{m,4} & \cdots & (-1)^{m-1} \end{pmatrix}.$$

Obviously D is invertible and

$$(\Phi_{\mathbf{n},1}, \ldots, \Phi_{\mathbf{n},m})^t = D^{-1}(\Psi_{\mathbf{n},1}, \ldots, \Psi_{\mathbf{n},m})^t \tag{41}$$

is the matrix form of the relations which express each function $\Phi_{\mathbf{n},j}$ in terms of $\Psi_{\mathbf{n},k}$, $k = 1, \ldots, j$. The matrix D^{-1} is also lower triangular and it may be proved, using (38) in several ways, that

$$D^{-1} := \begin{pmatrix} 1 & 0 & 0 & \cdots & 0 \\ \widehat{s}_{2,2} & -1 & 0 & \cdots & 0 \\ \widehat{s}_{2,3} & -\widehat{s}_{3,3} & 1 & \cdots & 0 \\ \vdots & \vdots & \vdots & \ddots & \vdots \\ \widehat{s}_{2,m} & -\widehat{s}_{3,m} & \widehat{s}_{4,m} & \cdots & (-1)^{m-1} \end{pmatrix}.$$

Using (41) and the expression of D^{-1} we obtain (40). □

Let $\boldsymbol{\lambda} = (\lambda_1, \ldots, \lambda_m) \in \mathcal{M}_1$ be the vector equilibrium measure determined by the matrix $C_{\mathcal{N}}$ on the system of compact sets $E_j = \operatorname{supp}(\sigma_j)$, $j = 1, \ldots, m$. In the sequel we assume that the hypothesis of Theorem 5 hold. For each $j = 1, \ldots, m$, set

$$U_j^{\boldsymbol{\lambda}} = P_j V^{\lambda_j}(z) - P_{j+1} V^{\lambda_{j+1}}(z) - 2\sum_{k=1}^{j} \omega_k^{\boldsymbol{\lambda}} / P_k\,,$$

($V^{\widehat{\mu}_{m+1}} \equiv 0$). Notice that in a neighborhood of $z = \infty$, we have

$$U_j^{\boldsymbol{\lambda}}(z) = O\left(p_j \log \frac{1}{|z|}\right).$$

The potentials of the components of the equilibrium measure define continuous functions on all $\mathbb{C}$ (see the equilibrium equations). Thus, the functions $U_j^{\boldsymbol{\lambda}}$ are defined and continuous on all $\mathbb{C}$.

Fix $j \in \{1, \ldots m\}$. For $k = 1, \ldots, j$ define the regions

$$D_k^j = \{z \in \mathbb{C} : U_k^{\boldsymbol{\lambda}}(z) > U_i^{\boldsymbol{\lambda}}(z), i = 1, \ldots, j\}.$$

Some D_k^j could be empty. Denote

$$\xi_j(z) = \max\{U_k^{\boldsymbol{\lambda}}(z) : k = 1, \ldots j\}.$$

Corollary 1 *Under the assumptions of Theorem 5, for each $j = 1, \ldots, m$, we have*

$$\lim_{\mathbf{n} \in \Lambda} \left| \widehat{s}_{1,j}(z) - \frac{P_{\mathbf{n},j}(z)}{Q_{\mathbf{n}}(z)} \right|^{1/|\mathbf{n}|} = \exp(V^{\lambda_1} + \xi_j)(z), \qquad z \in (\cup_{k=1}^{j} D_k^j) \setminus (\cup_{k=1}^{j+1} \Delta_k), \tag{42}$$

and

$$\limsup_{\mathbf{n} \in \Lambda} \left| \widehat{s}_{1,j}(z) - \frac{P_{\mathbf{n},j}(z)}{Q_{\mathbf{n}}(z)} \right|^{1/|\mathbf{n}|} \leq \exp(V^{\lambda_1} + \xi_j)(z), \qquad z \in \mathbb{C} \setminus (\cup_{k=1}^{j+1} \Delta_k), \tag{43}$$

uniformly on compact subsets of the indicated regions. Moreover, $(V^{\lambda_1} + \xi_j)(z) < 0, z \in \mathbb{C} \setminus \Delta_1$ which implies that the sequence $(P_{\mathbf{n},j}/Q_{\mathbf{n}}), \mathbf{n} \in \Lambda$, converges to $\widehat{s}_{1,j}$ with geometric rate in $\mathbb{C} \setminus (\cup_{k=1}^{j+1} \Delta_k)$.

Proof By (22) and (40) it follows that the following asymptotic formula takes place (notice that the functions $\widehat{s}_{k,j}$ are different from zero in $\mathbb{C} \setminus (\cup_{k=1}^{j} \Delta_k)$),

$$\lim_{n \in \Lambda} |\Phi_{\mathbf{n},j}(z)|^{1/|\mathbf{n}|} = \exp U_k^{\boldsymbol{\lambda}}(z), \qquad z \in D_k^j \setminus (\cup_{k=1}^{j+1} \Delta_k),$$

uniformly on compact subsets of the specified region. Then

$$\lim_{n \in \Lambda} |\Phi_{\mathbf{n},j}(z)|^{1/|\mathbf{n}|} = \exp \xi_j(z), \qquad z \in (\cup_{k=1}^{j} D_k^j) \setminus (\cup_{k=1}^{j+1} \Delta_k),$$

and

$$\limsup_{n\in\Lambda} |\Phi_{\mathbf{n},j}(z)|^{1/|\mathbf{n}|} \leq \exp \xi_j(z)\,, \qquad z \in \mathbb{C} \setminus (\cup_{k=1}^{j+1}\Delta_k)\,,$$

uniformly on compact subsets of the specified region.

Formulas (42) and (43) follow directly from

$$\widehat{s}_{1,j}(z) - \frac{P_{\mathbf{n},j}(z)}{Q_{\mathbf{n}}(z)} = \frac{\Phi_{\mathbf{n},j}(z)}{Q_{\mathbf{n}}(z)},$$

the asymptotic formulas given for $\Phi_{\mathbf{n},j}$, and (20).

When $j = 1$, we have

$$(V^{\lambda_1} + \xi_1)(z) = 2V^{\lambda_1}(z) - P_2 V^{\lambda_2}(z) - 2\omega_1^{\boldsymbol{\lambda}} = 2(W_1^{\boldsymbol{\lambda}}(z) - \omega_1^{\boldsymbol{\lambda}}).$$

According to (24), $W_1^{\boldsymbol{\lambda}}(z) - \omega_1^{\boldsymbol{\lambda}} \equiv 0, x \in \operatorname{supp}(\lambda_1)$. On the other hand, $W_1^{\boldsymbol{\lambda}}(z) - \omega_1^{\boldsymbol{\lambda}}$ is subharmonic in $\mathbb{C} \setminus \operatorname{supp}(\lambda_1)$ and tends to $-\infty$ as $z \to \infty$. By the maximum principle for subharmonic functions $W_1^{\boldsymbol{\lambda}}(z) - \omega_1^{\boldsymbol{\lambda}} < 0, z \in \mathbb{C} \setminus \operatorname{supp}(\lambda_1)$ (equality cannot occur at any point of this region because it would imply that $W_1^{\boldsymbol{\lambda}}(z) - \omega_1^{\boldsymbol{\lambda}} \equiv 0$ which is impossible).

Let us assume that $(V^{\lambda_1} + \xi_{j-1})(z) < 0, z \in \mathbb{C} \setminus \Delta_1$, where $j \in \{2, \ldots, m\}$ and let us prove that $(V^{\lambda_1} + \xi_j)(z) < 0, z \in \mathbb{C} \setminus \Delta_1$. Obviously,

$$\xi_j(z) = \max\{\xi_{j-1}(z), U_j(z)\}.$$

Consider the difference

$$U_j(z) - U_{j-1}(z) = 2(W_j^{\boldsymbol{\lambda}}(z) - w_j^{\boldsymbol{\lambda}})/P_j = O\left((p_j - p_{j-1})\log(1/|z|)\right), z \to \infty.$$

If $p_j = p_{j-1} = 0$ then $W_j^{\boldsymbol{\lambda}}(z) - w_j^{\boldsymbol{\lambda}}$ is subharmonic in $\overline{\mathbb{C}} \setminus \operatorname{supp}(\lambda_j)$ (at ∞ it is finite) and equals zero on $\operatorname{supp}(\lambda_j)$. Hence, $U_j(z) \leq U_{j-1}(z) \leq \xi_{j-1}(z)$ on $\mathbb{C} \setminus \operatorname{supp}(\lambda_j)$. Therefore, using the equilibrium condition, $U_j(z) = U_{j-1}(z)$ on Δ_j and $U_j(z) < U_{j-1}(z)$ on $\mathbb{C} \setminus \Delta_j$. In this case, $\xi_j(z) = \xi_{j-1}(z), z \in \mathbb{C} \setminus \Delta_1$, and the conclusion follows from the induction hypothesis.

If $p_j < p_{j-1}$, in a neighborhood of ∞ we have $U_j(z) > U_{j-1}(z)$ since $(p_j - p_{j-1})\log(1/|z|) \to +\infty$ as $z \to \infty$. Let $\Gamma = \{z \in \mathbb{C} : U_j(z) = U_{j-1}(z)\}$. This set contains $\operatorname{supp}(\lambda_j)$ and divides $\mathbb{C} \setminus \Delta_1$ in two domains $\Omega_1 = \{z \in \mathbb{C} \setminus \Delta_1 : U_j(z) > U_{j-1}(z)\}$, which contains $z = \infty$, and $\Omega_2 = \{z \in \mathbb{C} \setminus \Delta_1 : U_j(z) < U_{j-1}(z)\}$. Since $U_{j-1}(z) \leq \xi_{j-1}(z)$, on $\Omega_2 \cup \Gamma$ we have $\xi_{j-1}(z) = \xi_j(z)$ and thus $(V^{\lambda_1} + \xi_j) < 0$. On Ω_1 the function $V^{\lambda_1} + U_j$ is subharmonic and on its boundary Γ equals $V^{\lambda_1} + U_{j-1} < 0$. Since $(V^{\lambda_1} + U_j)(z) \to -\infty$ as $z \to \infty$ it follows that on Ω_1 we have $(V^{\lambda_1} + U_j)(z) < 0$. Therefore, $(V^{\overline{\mu}_1} + \xi_j) < 0$ on Ω_1. With this we conclude the proof. □

5 Ratio Asymptotic

In the study of the ratio asymptotic of Hermite-Padé approximants conformal mappings on Riemann surface and boundary problems of analytic functions come into play.

5.1 Preliminaries from Riemann Surfaces and Boundary Value Problems

Let $\Delta_1, \ldots, \Delta_m$ be a collection of intervals contained in the real line as in Definition 2. Consider the $(m+1)$-sheeted Riemann surface

$$\mathcal{R} = \overline{\bigcup_{k=0}^{m} \mathcal{R}_k},$$

formed by the consecutively "glued" sheets

$$\mathcal{R}_0 := \overline{\mathbb{C}} \setminus \Delta_1, \quad \mathcal{R}_k := \overline{\mathbb{C}} \setminus \{\Delta_k \cup \Delta_{k+1}\}, \qquad k = 1, \ldots, m-1, \qquad \mathcal{R}_m = \overline{\mathbb{C}} \setminus \Delta_m,$$

where the upper and lower banks of the slits of two neighboring sheets are identified. Fix $\ell \in \{1, \ldots, m\}$. Let $\psi^{(\ell)}$, $\ell = 1, \ldots, m$, be a single valued rational function on $\mathcal{R}$ whose divisor consists of one simple zero at the point $\infty^{(0)} \in \mathcal{R}_0$ and one simple pole at the point $\infty^{(\ell)} \in \mathcal{R}_\ell$. Therefore,

$$\psi^{(\ell)}(z) = C_1/z + O(1/z^2), \; z \to \infty^{(0)}, \qquad \psi^{(\ell)}(z) = C_2 z + O(1), \; z \to \infty^{(\ell)}, \tag{44}$$

where C_1 and C_2 are constants different from zero. Since the genus of $\mathcal{R}$ equals zero (it is conformally equivalent to $\overline{\mathbb{C}}$), such a single valued function on $\mathcal{R}$ exists and is uniquely determined up to a multiplicative constant. We denote the branches of the algebraic function $\psi^{(\ell)}$, corresponding to the different sheets $k = 0, \ldots, m$ of $\mathcal{R}$ by

$$\psi^{(\ell)} := \{\psi_k^{(\ell)}\}_{k=0}^m \cdot$$

In the sequel, we fix the multiplicative constant in such a way that

$$\prod_{k=0}^{m} |\psi_k^{(\ell)}(\infty)| = 1, \qquad C_1 > 0. \tag{45}$$

Since $\psi^{(\ell)}$ is such that $C_1 > 0$, then

$$\psi^{(\ell)}(z) = \overline{\psi^{(\ell)}(\overline{z})}, \qquad z \in \mathcal{R}.$$

In fact, define $\phi(z) := \overline{\psi^{(\ell)}(\overline{z})}$. Notice that ϕ and $\psi^{(\ell)}$ have the same divisor (same poles and zeros counting multiplicities); consequently, there exists a constant C such that $\phi = C\psi^{(\ell)}$. Comparing the leading coefficients of the Laurent expansion of these two functions at $\infty^{(0)}$, we conclude that $C = 1$.

In terms of the branches of $\psi^{(\ell)}$, the symmetry formula above means that for each $k = 0, 1, \ldots, m$:

$$\psi_k^{(\ell)} : \overline{\mathbb{R}} \setminus (\Delta_k \cup \Delta_{k+1}) \longrightarrow \overline{\mathbb{R}}$$

($\Delta_0 = \Delta_{m+1} = \emptyset$); therefore, the coefficients (in particular, the leading one) of the Laurent expansion at ∞ of the branches are real numbers, and

$$\psi_k^{(\ell)}(x_\pm) = \overline{\psi_k^{(\ell)}(x_\mp)} = \overline{\psi_{k+1}^{(\ell)}(x_\pm)}, \qquad x \in \Delta_{k+1}. \tag{46}$$

Among other things, the symmetry property entails that all the coefficients in the Laurent expansion at infinity of the branches $\psi_k^{(\ell)}$ are real numbers.

Since $\lim_{x\to\infty} x\psi_0^{(\ell)}(x) = C_1 > 0$, by continuity it follows that $\psi_k^{(\ell)}(\infty) > 0, k = 1, \ldots, \ell-1$, $\lim_{x\to\infty} \psi_\ell^{(\ell)}(x)/x = (\psi_\ell^{(\ell)})'(\infty) > 0$, and $\psi_k^{(\ell)}(\infty) < 0, k = \ell+1, \ldots, m$. On the other hand, the product of all the branches $\prod_{k=0}^m \psi_k^{(\ell)}$ is a single valued analytic function on $\overline{\mathbb{C}}$ without singularities; therefore, by Liouville's Theorem it is constant. Due to the previous remark and the normalization adopted in (45), we can assert that

$$\prod_{k=0}^m \psi_k^{(\ell)}(z) \equiv \begin{cases} 1, & m-\ell \text{ is even}, \\ -1, & m-\ell \text{ is odd}. \end{cases} \tag{47}$$

In [6, Lemma 4.2] the following boundary value problem was proved to have a unique solution. In (49) below, we introduce a slight correction to the formula in [6].

Lemma 8 *Let $\ell \in \{1, \ldots, m\}$ be fixed. There exists a unique collection of functions $(F_k^{(\ell)})_{k=1}^m$ which verify the system of boundary value problems*

$$\begin{aligned}
&1)\quad F_k^{(l)}, 1/F_k^{(\ell)} \in \mathcal{H}(\mathbb{C} \setminus \Delta_k), \\
&2)\quad (F_k^{(\ell)})'(\infty) > 0, \quad k = 1, \ldots, l, \\
&2')\quad F_k^{(\ell)}(\infty) > 0, \quad k = l+1, \ldots, m, \\
&3)\quad |F_k^{(\ell)}(x)|^2 \frac{1}{|(F_{k-1}^{(\ell)} F_{k+1}^{(\ell)})(x)|} = 1, \quad x \in \Delta_k,
\end{aligned} \tag{48}$$

where $F_0^{(l)} \equiv F_{m+1}^{(l)} \equiv 1$. *Moreover*

$$F_k^{(l)} = sg\left(\prod_{\nu=k}^{m} \psi_\nu^{(l)}(\infty)\right)\prod_{\nu=k}^{m} \psi_\nu^{(l)}, \tag{49}$$

where $sg\left(\prod_{\nu=k}^{m} \psi_\nu^{(l)}(\infty)\right)$ *denotes the sign of the leading coefficient of the Laurent expansion at* ∞ *of* $\prod_{\nu=k}^{m} \psi_\nu^{(l)}$.

We are ready to state a result on the ratio asymptotic for type II Hermite-Padé polynomials of a Nikishin system. This result was obtained in [6] (see also [33]).

Theorem 6 *Assume that* $\sigma_k' > 0$ *almost everywhere on* $\Delta_k = \operatorname{supp}\sigma_k, k = 1,\ldots,m$. *Let* $\Lambda \subset \mathbb{Z}_+^m(\bullet)$ *be a sequence of multi-indices such that* $n_1 - n_m \leq d$ *for all* $\mathbf{n} \in \Lambda$, *where* d *is some fixed constant. Then for each fixed* $k \in \{1,\ldots,m\}$, *we have*

$$\lim_{\mathbf{n}\in\Lambda} \frac{Q_{\mathbf{n}^l,k}(z)}{Q_{\mathbf{n},k}(z)} = \widetilde{F_k^{(l)}}(z), \tag{50}$$

uniformly on each compact subset of $\mathbb{C} \setminus \Delta_k$, *where* $F_k^{(l)}$ *is given in* (49), *the algebraic functions* $\psi_\nu^{(l)}$ *are defined by* (44)–(45) *and* $\widetilde{F_k^{(l)}}$ *is the result of dividing* $F_k^{(l)}$ *by the leading coefficient of its Laurent expansion at* ∞.

Sketch of the Proof Fix $\ell \in \{1,\ldots,m\}$. Because of the interlacing property of the zeros of the polynomials $Q_{\mathbf{n},k}$ and $Q_{\mathbf{n}^\ell,k}$ it follows that for each fixed $k \in \{1,\ldots,m\}$ the family of functions $(Q_{\mathbf{n}^\ell,k}/Q_{\mathbf{n},k}), n \in \Lambda$, is uniformly bounded on each compact subset of $\mathbb{C} \setminus \Delta_k$. To prove the theorem we need to show that for any $\Lambda' \subset \Lambda$ such that

$$\lim_{\mathbf{n}\in\Lambda'} \frac{Q_{\mathbf{n}^\ell,k}}{Q_{\mathbf{n},k}} = G_k^\ell, \qquad k = 1,\ldots,m,$$

the limiting functions do not depend on Λ'.

To prove this, it is shown that their exist positive constants $c_1,\ldots,c_m$ such that $(c_k G_k^\ell)_{k=1}^m$ verifies the system of boundary value problems (48). Properties 1), 2), and 2') are easily verified by $(G_k)_{k=1}^m$ with 1 on the right hand side of 2) and 2'). Thanks to the orthogonality properties contained in (12), using results on ratio and relative asymptotic of orthogonal polynomials with respect to varying measures contained in [15, Theorem 6] and [7, Theorem 2] one can also prove that $(G_k)_{k=1}^m$ satisfies 3) with a constant different from 1 on the right hand side. Normalizing the functions G_k appropriately one obtains all the boundary conditions and it follows that $c_k G_k^\ell = F_k^\ell, k = 1,\ldots,m$. Then using that $G_k^\ell(\infty) = 1, k = \ell+1,\ldots,m$ and $(G_k^\ell))'(\infty) = 1, k = 1,\ldots,\ell$. one sees that G_k^ℓ has to be the function on the right hand side of (50) independently of Λ'. □

Remark 2 An interesting open question is if one can relax the assumption $n_1 - n_m \leq d$ in the theorem. That restriction is connected with the conditions which have been found to be sufficient for the ratio and relative asymptotic of polynomials orthogonal with respect to varying measures. Those theorems in [7, 15] would have to be improved. Perhaps this could be done without too much difficulty if $n_1 - n_m = o(|\mathbf{n}|)$, $|\mathbf{n}| \to \infty$. Sequences verifying something like (18), as in the case of weak asymptotic, would require a deep consideration and substantial new ideas.

Remark 3 On the basis of (50) and (13) one can also prove ratio asymptotic for the sequences $(K_{\mathbf{n}^\ell,k}/K_{\mathbf{n},k})$, $(\Psi_{\mathbf{n}^\ell,k}/\Psi_{\mathbf{n},k})$, $\mathbf{n} \in \Lambda, k = 1, \ldots, m$ (see [6, 33]). Using Theorem 3 and Proposition 2, it is also possible to prove ratio asymptotic for the polynomials $A_{\mathbf{n},k}$ and the forms $\mathcal{A}_{\mathbf{n},k}$ and it is a good exercise (see also [22]).

Remark 4 The ratio asymptotic of multiple orthogonal polynomials finds applications in the study of asymptotic properties of modified Nikishin systems and the corresponding Hermite-Padé approximants (see, for example, [34] and [36]).

Acknowledgments This work was supported by research grant PGC2018-096504-B-C33 of Ministerio de Ciencias, Innovación y Universidades, Spain.

References

1. Apéry, R.: Irrationalité de $\zeta(2)$ et $\zeta(3)$. Astérisque **61**, 11–13 (1979)
2. Angelesco, M.A.: Sur deux extensions des fractions continues algébriques. C.R. Acad. Sci. Paris **18**, 262–263 (1919)
3. Aptekarev, A.I.: Asymptotics of simultaneously orthogonal polynomials in the Angelesco case. Math. USSR Sb. **64**, 57–84 (1989)
4. Aptekarev, A.I.: Strong asymptotics of multiply orthogonal polynomials for Nikishin systems. Sbornik Math. **190**, 631–669 (1999)
5. Aptekarev, A.I.: Sharp estimates for rational approximations of analytic functions. Sbornik Math. **193**, 3–72 (2002)
6. Aptekarev, A.I., López Lagomasino, G., Rocha, I.A.: Ratio asymptotic of Hermite-Padé orthogonal polynomials for Nikishin systems. Sbornik Math. **196**, 1089–1107 (2005)
7. Barrios Rolanía, D., de la Calle Ysern, B., López Lagomasin, G.: Ratio and relative asymptotic of polynomials orthogonal with respect to varying Denisov-type measures. J. Approx. Theory **139**, 223–256 (2006)
8. Bello Hernández, M., López Lagomasino, G., Mínguez Ceniceros, J.: Fourier-Padé approximants for Angelesco systems. Constr. Approx. **26**, 339–359 (2007)
9. Bleher, P.M., Kuijlaars, A.B.J.: Random matrices with external source and multiple orthogonal polynomials. Int. Math. Res. Notices 2004(3), 109–129 (2004)
10. Bertola, M., Gekhtman, M., Szmigielski, J.: Cauchy biorthogonal polynomials. J. Approx. Theory **162**, 832–867 (2010)
11. Beukers, F.: Padé Approximation in Number Theory. In: Lecture Notes in Math., Vol. 888, pp. 90–99. Springer, Berlin (1981)
12. Bustamante, J., López Lagomasino, G.: Hermite–Padé approximation for Nikishin systems of analytic functions. Russ. Acad. Sci. Sb. Math. **77**, 367–384 (1994)
13. Coates, J.: On the algebraic approximation of functions. I, II, III. Indag. Math. **28**, 421–461 (1966)

14. Daems, E., Kuijlaars, A.B.J.: Multiple orthogonal polynomials of mixed type and non intersecting brownian motions. J. Approx. Theory **146**, 91–114 (2007)
15. de la Calle Ysern, B., López Lagomasino, G.: Weak Convergence of varying measures and Hermite-Padé orthogonal polynomials. Constr. Approx. **15**, 553–575 (1999)
16. Driver, K., Stahl, H.: Simultaneous rational approximants to Nikishin systems. I. Acta Sci. Math. (Szeged) **60**, 245–263 (1995)
17. Driver, K., Stahl, H.: Simultaneous rational approximants to Nikishin systems. II. Acta Sci. Math. (Szeged) **61**, 261–284 (1995)
18. Fidalgo Prieto, U., López Lagomasino, G.: Rate of convergence of generalized Hermite-Padé approximants of Nikishin systems. Constr. Approx. **23**, 165–196 (2006)
19. Fidalgo Prieto, U., López Lagomasino, G.: General results on the convergence of multipoint-Padé approximants of Nikishin systems. Constr. Approx. **25**, 89–107 (2007)
20. Fidalgo Prieto, U., López Lagomasino, G.: Nikishin systems are perfect. Constr Approx. **34**, 297–356 (2011)
21. Fidalgo Prieto, U., López Lagomasino, G.: Nikishin systems are perfect. The case of unbounded and touching supports. J. Approx. Theory **163**, 779–811 (2011)
22. Fidalgo Prieto, U., López, A., López Lagomasino, G., Sorokin, V.N.: Mixed type multiple orthogonal polynomials for two Nikishin systems. Constr. Approx. **32**, 255–306 (2010)
23. Gonchar, A.A., Rakhmanov, E.A.: The equilibrium measure and distribution of zeros of extremal polynomials. Math. USSR Sb. **53**, 119–130 (1986)
24. Gonchar, A.A.: Rational approximations of analytic functions. In: Proc. Internat. Congress Math., Vol. I, pp. 739–748. Amer. Math. Soc., Providence, RI (1987)
25. Gonchar, A.A., Rakhmanov, E.A.: On convergence of simultaneous Padé approximants for systems of functions of Markov type. Proc. Steklov Inst. Math. **157**, 31–50 (1983)
26. Gonchar, A.A., Rakhmanov, E.A.: Equilibrium distributions and degree of rational approximation of analytic functions. Math. USSR Sb. **62**, 305–348 (1989)
27. Gonchar, A.A., Rakhmanov, E.A., Sorokin, V.N.: Hermite–Padé approximants for systems of Markov–type functions. Sb. Math. **188**, 33–58 (1997)
28. Hermite, Ch.: Sur la fonction exponentielle. C. R. Acad. Sci. Paris **77**, 18–24, 74–79, 226–233, 285–293 (1873); Reprinted in his Oeuvres, Tome III, pp. 150–181. Gauthier-Villars, Paris (1912)
29. Jager, H.: A simultaneous generalization of the Padé table. I-VI. Indag. Math. **26**, 193–249 (1964)
30. Krein, M.G., Nudel'man, A.A.: The Markov Moment Problem and Extremal Problems. Transl. Math. Monogr., Vol. 50. Amer. Math. Soc., Providence, RI (1977)
31. Kuijlaars, A.B.J.: Multiple orthogonal polynomial ensembles. In: Arvesú, J., Marcellán, F., Martínez-Finkelshtein, A. (eds.), Recent Trends in Orthogonal Polynomials and Approximation Theory. Contemporary Mathematics, vol. 507, pp. 155–176. Amer. Math. Soc., Providence, RI (2010)
32. Kuijlaars, A.B.J.: Multiple orthogonal polynomials in random matrix theory. In: Proc. Internat. Congress Math., Vol. III, pp. 1417–1432, Hyderabad, India (2010)
33. López García, A., López Lagomasino, G.: Ratio asymptotic of Hermite-Padé orthogonal polynomials for Nikishin systems. II. Adv. Math. **218**, 1081–1106 (2008)
34. López, A., López Lagomasino, G.: Relative asymptotic of multiple orthogonal polynomials of Nikishin systems. J. Approx. Theory **158**, 214–241 (2009)
35. López Lagomasino, G., Medina Peralta, S.: On the convergence of type I Hermite-Padé approximants. Adv. Math. **273**, 124–148 (2015)
36. López Lagomasino, G., Medina Peralta, S.: On the convergence of type I Hermite-Padé approximants for a class of meromorphic functions. J. Comput. Appl. Math. **284**, 216–227 (2015)
37. Mahler, K.: Perfect systems. Compos. Math. **19**, 95–166 (1968)
38. Markov, A.A.: Deux demonstrations de la convergence de certains fractions continues. Acta Math. **19**, 93–104 (1895)

39. Nikishin, E.M.: A system of Markov functions. Vestnik Moskov. Moscow Univ. Math. Bull. **34**, 63–66 (1979)
40. Nikishin, E.M.: On simultaneous Padé approximants. Math. USSR Sb. **41**, 409–425 (1982)
41. Nikishin, E.M.: Asymptotics of linear forms for simultaneous Padé approximants. Sov. Math. (Iz. VUZ) **30**, 43–52 (1986)
42. Nikishin, E.M., Sorokin, V.N.: Rational Approximations and Orthogonality. Transl. Math. Monogr., Vol. 92. Amer. Math. Soc., Providence, RI (1991)
43. Saff, E.B., Totik, V.: Logarithmic Potentials with External Fields. In: Series of Comprehensive Studies in Mathematics, Vol. 316. Springer, New York (1997)
44. Stahl, H.: Best uniform rational approximation of x^α on [0,1]. Acta Math. **90**, 241–306 (2003)
45. Stahl, H., Totik, V.: General Orthogonal Polynomials. Cambridge University Press, Cambridge (1992)
46. Van Assche, W.: Analytic number theory and rational approximation. In: Branquinho, A., Foulquié, A. (eds.), Coimbra Lecture Notes on Orthogonal Polynomials, pp. 197–229. Nova Science Pub., New York, (2008)

Revisiting Biorthogonal Polynomials: An LU Factorization Discussion

Manuel Mañas

Abstract The Gauss–Borel or LU factorization of Gram matrices of bilinear forms is the pivotal element in the discussion of the theory of biorthogonal polynomials. The construction of biorthogonal families of polynomials and its second kind functions, of the spectral matrices modeling the multiplication by the independent variable x, the Christoffel–Darboux kernel and its projection properties, are discussed from this point of view. Then, the Hankel case is presented and different properties, specific of this case, as the three terms relations, Heine formulas, Gauss quadrature and the Christoffel–Darboux formula are given. The classical orthogonal polynomial of Hermite, Laguerre and Jacobi type are discussed and characterized within this scheme. Finally, it is shown who this approach is instrumental in the derivation of Christoffel formulas for general Christoffel and Geronimus perturbations of the bilinear forms.

Keywords Biorthogonal polynomials · Bivariate functionals · Sesquilinear forms · Gram matrix · Gauss–Borel factorization · Jacobi matrix · Very classical orthogonal polynomials · Christoffel formulas · Christoffel perturbations · Geronimus perturbations · Quasi determinants

1 Introduction

These notes correspond to the *orthogonal polynomial* part of the five lectures I delivered during the VII Iberoamerican Workshop in Orthogonal Polynomials and Applications (Seventh EIBPOA). As such, more than presenting new original material, they are intended to give an alternative, but consistent and systematic,

M. Mañas (✉)
Departamento de Física Teórica, Universidad Complutense de Madrid, Madrid, Spain

Instituto de Ciencias Matematicas (ICMAT), Madrid, Spain
e-mail: manuel.manas@ucm.es

F. Marcellán, E. J. Huertas (eds.), *Orthogonal Polynomials: Current Trends and Applications*, SEMA SIMAI Springer Series 22,
https://doi.org/10.1007/978-3-030-56190-1_10

construction of the theory of biorthogonal polynomials using the LU factorization of the Gram matrix of a given bilinear form. We refer the interested reader to the classical texts [23, 36] for a general background on the subject.

At the beginning, when our group started to work in this area, this LU approach was motivated by the connection of the theory of orthogonal polynomials and the theory of integrable systems, and the ubiquity of factorization problems in the description of the later, see [18, 22, 28–31, 33]. Adler and van Moerbeke performed a pioneering work regarding this approach [1–4]. Despite this original motivation, we soon realized that this factorization technique allows for a general and systematic approach to the subject of orthogonal polynomials, giving a unified framework that can be extended to more sophisticated orthogonality scenarios. Given the understandable space constraint of this volume we will not describe the LU description of the Toda/KP integrable hierarchies and refer the reader to the slides of my lectures posted in the web page of Seventh EIBPOA.

With the background given here we hope that the reader could understand further developments applied in other orthogonality situations. Let us now describe what we have done regarding this research in with more general orthogonality frameworks. In [7] we studied the **generalized orthogonal polynomials** [2] and its matrix extensions from the Gauss–Borel view point. In [8] we gave a complete study in terms of this factorization for **multiple orthogonal polynomials of mixed type** and characterized the integrable systems associated to them. Then, we studied Laurent orthogonal polynomials in the unit circle trough the CMV approach in [5] and found in [6] the Christoffel–Darboux formula for generalized orthogonal matrix polynomials. These methods were further extended, for example we gave an alternative Christoffel–Darboux formula for mixed multiple orthogonal polynomials [10] or developed the corresponding theory of **matrix Laurent orthogonal polynomials in the unit circle** and its associated Toda type hierarchy [11]. In [9, 15, 16] a complete analysis, in terms of the spectral theory of matrix polynomials, of Christoffel and Geronimus perturbations and Christoffel formulas was given for **matrix orthogonal polynomials**, while in [17] we gave a complete description for the Christoffel formulas corresponding to Christoffel perturbations for univariate CMV Laurent orthogonal polynomials. We also mention recent developments on **multivariate orthogonal polynomials** in real spaces (MVOPR), the corresponding Christoffel formula and the interplay with algebraic geometry [12, 13]. Similar multivariate situations but for the complex torus and the CMV ordering was analyzed in [14].

During my lectures I learned, with pleasure, that Diego Dominici is also interested in the ways the LU factorization is an useful tool in the theory of orthogonal polynomials. A paper by Dominici related to matrix factorizations and OP will appear in the journal Random Matrices: Theory and Applications.

2 LU Factorization and Gram Matrix

Given a square complex matrix $A \in \mathbb{C}^{N\times N}$, an LU factorization refers to the factorization of A into a lower triangular matrix L and an upper triangular matrix U

2.1 LDU Factorization

An LDU decomposition is a decomposition of the form

$$A = LDU,$$

where D is a diagonal matrix, and L and U are unitriangular matrices, meaning that all the entries on the diagonals of L and U are one.

Existence and Uniqueness

- Any $A \in \mathrm{GL}_N(\mathbb{C})$ admits an LU (or LDU) factorization if and only if all its leading principal minors are nonzero.
- If $A \in \mathbb{C}^{N\times N}$ is a singular matrix of rank r, then it admits an LU factorization if the first r leading principal minors are nonzero, although the converse is not true.
- If $A \in \mathbb{C}^{N\times N}$ has an LDU factorization, then the factorization is unique.

Cholesky Factorization

- If $A \in \mathbb{C}^{N\times N}$ is a symmetric ($A = A^\top$) matrix, we can choose U as the transpose of L

$$A = LDL^\top.$$

- When A is Hermitian, $A = A^\dagger$, we can similarly find

$$A = LDL^\dagger.$$

These decomposition are called Cholesky factorizations. The Cholesky decomposition always exists and is unique—provided the matrix is positive definite.

2.2 Schur Complements

Given a block matrix M

$$M = \left(\begin{array}{c|c} A & B \\ \hline C & D \end{array}\right)$$

with $A \in \mathrm{GL}_p(\mathbb{C})$, $B \in \mathbb{C}^{p\times q}$, $C \in \mathbb{C}^{q\times p}$ and $D \in \mathbb{C}^{q\times q}$, we define its Schur complement with respect to A as

$$M\backslash A := D - CA^{-1}B \in \mathbb{C}^{q\times q}.$$

Proposition 1 (Schur Complements and LU **Factorization)** *If* A *is nonsingular we have for the block matrix* M *the following factorization*

$$M = \begin{pmatrix} I_p & 0_{p\times q} \\ CA^{-1} & I_q \end{pmatrix} \begin{pmatrix} A & 0_{p\times q} \\ 0_{q\times p} & M\backslash A \end{pmatrix} \begin{pmatrix} I_p & A^{-1}B \\ 0_{q\times p} & I_q \end{pmatrix}.$$

For a detailed account of Schur complements see [38].

Given a matrix $M = (M_{i,j})_{i,j=0}^{N-1} \in \mathbb{C}^{N\times N}$ a $\ell+1$, $\ell < N$, truncation is given by

$$M^{[\ell+1]} := \begin{pmatrix} M_{0,0} & M_{0,1} & \dots & M_{0,\ell} \\ M_{1,0} & M_{1,1} & \dots & M_{1,\ell} \\ \vdots & \vdots & \ddots & \vdots \\ M_{\ell,0} & M_{\ell,1} & \dots & M_{\ell,\ell} \end{pmatrix}.$$

Proposition 2 (Nonzero Minors and LU **Factorization)** *Any nonsingular matrix* $M \in \mathrm{GL}_N(\mathbb{C})$ *with all its leading principal minors different from zero, i.e.,* $\det M^{[\ell+1]} \neq 0$, $\ell \in \{0, 1, , \dots, N-1\}$, *has a* LDU *factorization.*

Proof As $\det M^{[N-1]} \neq 0$, we have

$$M = \begin{pmatrix} I_{N-1} & 0_{(N-1)\times 1} \\ v_N & 1 \end{pmatrix} \begin{pmatrix} M^{[N-1]} & 0_{(N-1)\times 1} \\ 0_{1\times(N-1)} & M\backslash M^{[N-1]} \end{pmatrix} \begin{pmatrix} I_{N-1} & w_N \\ 0_{1\times(N-1)} & 1 \end{pmatrix}$$

where v_N and w_N are row and column vectors, respectively, with $N-1$ components.

Now, from $\det M^{[N-2]} \neq 0$ we deduce

$$M^{[N-1]} = \begin{pmatrix} I_{N-2} & 0_{(N-2)\times 1} \\ v_{N-1} & 1 \end{pmatrix} \begin{pmatrix} M^{[N-2]} & 0_{(N-2)\times 1} \\ 0_{1\times(N-2)} & M^{[N-1]}\backslash M^{[N-2]} \end{pmatrix} \begin{pmatrix} I_{N-2} & w_{N-1} \\ 0_{1\times(N-2)} & 1 \end{pmatrix}$$

which inserted in the previous result yields

$$M = \left(\begin{array}{c|cc} I_{N-2} & \multicolumn{2}{c}{0_{(N-2)\times 2}} \\ \hline \multirow{2}{*}{$*$} & 1 & 0 \\ & * & 1 \end{array}\right) \left(\begin{array}{c|cc} M^{[N-2]} & \multicolumn{2}{c}{0_{(N-2)\times 2}} \\ \hline \multirow{2}{*}{$0_{2\times(N-2)}$} & M^{[N-1]}\backslash M^{[N-2]} & 0 \\ & 0 & M\backslash M^{[N-1]} \end{array}\right) \left(\begin{array}{c|cc} I_{N-2} & \multicolumn{2}{c}{*} \\ \hline \multirow{2}{*}{$0_{2\times(N-2)}$} & 1 & * \\ & 0 & 1 \end{array}\right).$$

We finally get an LDU factorization with

$$D = \operatorname{diag}\left(M^{[1]}\backslash M^{[0]}, M^{[2]}\backslash M^{[1]}, \dots, M\backslash M^{[N-1]}\right).$$

□

2.3 *Bilinear Forms, Gram Matrices and LU Factorizations*

A bilinear form $\langle\cdot,\cdot\rangle$ on the ring of complex valued polynomials in the variable x, $\mathbb{C}[x]$, is a continuous map

$$\begin{aligned}\langle\cdot,\cdot\rangle : \mathbb{C}[x]\times\mathbb{C}[x] &\longrightarrow \quad \mathbb{C}\\ (P(x),Q(x)) &\mapsto \langle P(x),Q(y)\rangle\end{aligned}$$

such that for any triple $P(x), Q(x), R(x) \in \mathbb{C}[x]$ the following properties are fulfilled

1. $\langle AP(x)+BQ(x),R(y)\rangle = A\langle P(x),R(y)\rangle + B\langle Q(x),R(y)\rangle, \forall A,B\in\mathbb{C}$,
2. $\langle P(x),AQ(y)+BR(y)\rangle = \langle P(x),Q(y)\rangle A + \langle P(x),R(y)\rangle B, \forall A,B\in\mathbb{C}$.

Observe that we have not chosen the conjugate in one of the variables.

For $P(x)=\sum\limits_{k=0}^{\deg P} p_k x^k$ and $Q(x)=\sum\limits_{l=0}^{\deg Q} q_l x^l$ the bilinear form is

$$\langle P(x),Q(y)\rangle = \sum_{\substack{k=1,\dots,\deg P\\ l=1,\dots,\deg Q}} p_k G_{k,l} q_l, \qquad G_{k,l}=\left\langle x^k,y^l\right\rangle.$$

The corresponding **semi-infinite** matrix

$$G=\begin{pmatrix} G_{0,0} & G_{0,1} & \dots\\ G_{1,0} & G_{1,1} & \dots\\ \vdots & \vdots & \end{pmatrix},$$

is the so called **Gram matrix** of the sesquilinear form.

Examples

- **Borel measures** A first example is given by a complex (or real) Borel measure $\mathrm{d}\mu$ with support $\mathrm{supp}(\mathrm{d}\mu)\subset\mathbb{R}$. Given any pair of matrix polynomials $P(x), Q(x)\in\mathbb{C}[x]$ we introduce the following bilinear form

$$\langle P(x),Q(x)\rangle_\mu = \int_{\mathbb{R}} P(x)\,\mathrm{d}\mu(x)Q(x).$$

- **Example: Linear functionals** We consider the space of polynomials $\mathbb{C}[x]$, with an appropriate topology, as the space of fundamental functions and take the space of generalized functions as the corresponding continuous linear functionals. Take a linear functional $u\in(\mathbb{C}[x])'$ and consider

$$\langle P(x),Q(x)\rangle_u = u(P(x)Q(x)).$$

In both examples the Gram matrix is a Hankel matrix $G_{i+1,j} = G_{i,j+1}$. In these cases, the Gram matrix is also known as moment matrix as we have

$$G_{i,j} = \int_{\mathbb{R}} x^{i+j} \, \mathrm{d}\mu(x) = m_{i+j},$$

where m_j is known as the j-moment, for the measure case, while for the linear functional scenario we have

$$G_{i,j} = u(x^{i+j}).$$

Schwartz Generalized Kernels There are bilinear forms which do not have this particular Hankel type symmetry. Let $u_{x,y} \in (\mathbb{C}[x, y])' \cong \mathbb{C}[\![x, y]\!]$ so that

$$\langle P(x), Q(y)\rangle = \langle u_{x,y}, P(x) \otimes Q(y)\rangle.$$

The Gram matrix of this bilinear form has entries

$$G_{k,l} = \langle u_{x,y}, x^k \otimes y^l\rangle.$$

This gives a continuous linear map $\mathcal{L}_u : \mathbb{C}[y] \to (\mathbb{C}[x])'$ such that $\langle P(x), Q(y)\rangle = \langle \mathcal{L}_u(Q(y)), P(x)\rangle$. See [34] for an introduction to bivariate generalized kernels.

Integrals Kernels A kernel $u(x, y)$ is a complex valued locally integrable function, that defines an integral operator $f(x) \mapsto g(x) = \int u(x, y) f(y) \, \mathrm{d}y$ and

$$\langle P(x), Q(y)\rangle = \int_{\mathbb{R}^2} P(x) u(x, y) Q(y) \, \mathrm{d}x \, \mathrm{d}y$$

There is an obvious way of ordering the monomials $\{x^n\}_{n=0}^{\infty}$ in a semi-infinite vector

$$\chi(x) = (1, x, x^2, \dots)^\top$$

Also, we consider

$$\chi^*(x) = (x^{-1}, x^{-2}, x^{-3}, \dots)^\top, \quad \text{with} \quad (\chi(x))^\top \chi^*(y) = \frac{1}{x - y}, \quad |y| < |x|.$$

The semi-infinite Gram matrix can be written as follows $G = \langle \chi, \chi^\top \rangle$. For a Borel measure it reads $G = \int \chi(x) (\chi(x))^\top \mathrm{d}\mu(x)$ and for a linear functional $G = \left\langle u(x), \chi(x) (\chi(x))^\top \right\rangle$. When dealing with an integral kernel we have $G = \int \chi(x) (\chi(y))^\top u(x, y) \, \mathrm{d}x$ and for a Schwartz kernel $G = \left\langle u(x, y), \chi(x) (\chi(y))^\top \right\rangle$.

3 Orthogonal Polynomials

Definition 1 (Quasi-Definite Bilinear Forms) We say that a bilinear form $\langle\cdot,\cdot\rangle$ is quasi-definite whenever its Gram matrix has all its leading principal minors different from zero.

Proposition 3 (Quasi-Definiteness and LDU Factorization) *The Gram matrix of a quasi-definite bilinear form admits a unique LDU factorization.*

Given a quasi-definite bilinear form in the space of polynomials we consider the LDU factorization of its Gram matrix in the form

$$G = (S_1)^{-1} H (S_2)^{-\top}$$

where S_1 and S_2 are lower unitriangular matrices and H is a diagonal matrix.

When the quasi-definite bilinear form comes from a Borel measure or a linear functional the corresponding Gram matrix, now known as moment matrix, is symmetric: $G = G^{\top}$. Thus, the LDU factorization becomes a Cholesky factorization. Whenever the Borel measure is positive (sign defined will equally do) the moment matrix is a positive definite matrix, i.e., all the principal minors of the moment matrix are strictly positive. Given either a Borel measure or a linear functional, we consider its LDU factorization

$$G = S^{-1} H S^{-\top},$$

where S is a lower unitriangular matrix and H is a diagonal matrix. For a Borel positive measure, the diagonal coefficients H_k are positive, $H_k > 0$.

Definition 2 (Constructing the Polynomials) Given a Gram matrix and its Gauss–Borel factorization we construct the following two families of polynomials

$$P_1(x) := S_1 \chi(x), \qquad\qquad P_2(x) := S_2 \chi(x).$$

Here $P_i = (P_{i,0}, P_{i,1}, \dots,)$ with $P_{i,k}(x) = x^k + \cdots$, monic polynomials of degree k.

Proposition 4 (Orthogonality Relations) *The above families of polynomials satisfy the following orthogonality relations*

$$\left\langle P_{1,k}(x), y^l \right\rangle = \delta_{l,k} H_k, \qquad \left\langle x^l, P_{2,k}(y) \right\rangle = \delta_{l,k} H_k, \qquad 0 \le l < k.$$

Proof We have that $\left\langle P_1(x), \big(\chi(y)\big)^{\top} \right\rangle = S_1 \left\langle \chi(x), (\chi(y))^{\top} \right\rangle = S_1 G = H(S_2)^{-\top}$ is an upper triangular matrix, and the result follows. □

Proposition 5 (Biorthogonality Relations) *The above families of polynomials are biorthogonal*

$$\langle P_{1,k}(x), P_{2,l}(x)\rangle = \delta_{l,k} H_k.$$

Proof We have that $\left\langle P_1(x), \left(P_2(y)\right)^\top\right\rangle = S_1 \left\langle \chi(x), (\chi(y))^\top\right\rangle (S_2)^\top = S_1 G = H.$ □

3.1 *Quasi-Determinants*

We include this brief section here because, despite the fact that for the standard orthogonality the quotient of determinants is enough to describe adequately the results, in more general situations quasi-determinants are needed. Also, even in this situation they give more compact expressions. As we will see we can understand them as an extension of determinants to noncommutative situations and also as Schur complements.

Some History In the late 1920 Archibald Richardson, one of the two responsible of Littlewood–Richardson rule, and the famous logician Arend Heyting, founder of intuitionist logic, studied possible extensions of the determinant notion to division rings. Heyting defined the *designant* of a matrix with noncommutative entries, which for 2×2 matrices was the Schur complement, and generalized to larger dimensions by induction.

The Situation Nowadays 1990 till today, was given by Gel'fand, Rektah and collaborators, see [24]. Quasi-determinants were defined over free division rings and was early noticed that **is not an analog of the commutative determinant but rather of a ratio of determinants**. A cornerstone for quasi-determinants is the *heredity principle*, **quasi-determinants of quasi-determinants are quasi-determinants**; there is no analog of such a principle for determinants.

3.1.1 The Easiest Quasi-Determinant: A Schur Complement

We start with $k = 2$, so that $A = \begin{pmatrix} A_{1,1} & A_{1,2} \\ A_{2,1} & A_{2,2} \end{pmatrix}$. In this case the first quasi-determinant $\Theta_1(A) :- A/A_{1,1}$; i.e., a Schur complement which requires $\det A_{1,1} \neq 0$.

Olver vs Gel'fand The notation of Olver [32] is different to that of the Gel'fand school were $\Theta_1(A) = |A|_{2,2} = \begin{vmatrix} A_{1,1} & A_{1,2} \\ A_{2,1} & \boxed{A_{2,2}} \end{vmatrix}$. There is another quasi-determinant $\Theta_2(A) = A/A_{22} = |A|_{1,1} = \begin{vmatrix} \boxed{A_{1,1}} & A_{1,2} \\ A_{2,1} & A_{2,2} \end{vmatrix}$, the other Schur complement, and we need $A_{2,2}$ to be a non singular matrix. Other quasi-determinants that can be considered for regular square blocks are $\begin{vmatrix} A_{1,1} & A_{1,2} \\ \boxed{A_{2,1}} & A_{2,2} \end{vmatrix}$ and $\begin{vmatrix} A_{1,1} & \boxed{A_{1,2}} \\ A_{2,1} & A_{2,2} \end{vmatrix}$. These last two quasi-determinants are out of the scope of these notes.

Example Consider

$$A = \left(\begin{array}{c|c:c} A_{1,1} & A_{1,2} & A_{1,3} \\ \hline A_{2,1} & A_{2,2} & A_{2,3} \\ \hdashline A_{3,1} & A_{3,2} & A_{3,3} \end{array}\right)$$

and take the quasi-determinant with respect the first diagonal block, which we define as the Schur complement indicated by the non dashed lines

$$\Theta_1(A) = \begin{vmatrix} A_{11,1} & A_{1,2} \; A_{1,3} \\ A_{2,1} & \boxed{\begin{matrix} A_{2,2} & A_{2,3} \\ A_{3,2} & A_{3,3} \end{matrix}} \\ A_{3,1} & \end{vmatrix} = \begin{pmatrix} A_{2,2} & A_{2,3} \\ A_{3,2} & A_{3,3} \end{pmatrix} - \begin{pmatrix} A_{2,1} \\ A_{3,1} \end{pmatrix} A_{1,1}^{-1} \begin{pmatrix} A_{1,2} & A_{1,3} \end{pmatrix}$$

$$= \left(\begin{array}{c:c} A_{2,2} - A_{2,1}A_{1,1}^{-1}A_{1,2} & A_{2,3} - A_{2,1}A_{1,1}^{-1}A_{1,3} \\ \hdashline A_{3,2} - A_{3,1}A_{1,1}^{-1}A_{1,2} & A_{3,3} - A_{3,1}A_{1,1}^{-1}A_{1,3} \end{array}\right).$$

Take the quasi-determinant given by the Schur complement as indicated by the dashed lines

$$\Theta_2(\Theta_1(A)) = \begin{vmatrix} A_{2,2} - A_{2,1}A_{1,1}^{-1}A_{1,2} & A_{2,3} - A_{2,1}A_{1,1}^{-1}A_{1,3} \\ A_{3,2} - A_{3,1}A_{1,1}^{-1}A_{1,2} & \boxed{A_{3,3} - A_{3,1}A_{1,1}^{-1}A_{1,3}} \end{vmatrix}$$

$$= A_{3,3} - A_{3,1}A_{1,1}^{-1}A_{1,3} - \big(A_{3,2} - A_{3,1}A_{1,1}^{-1}A_{1,2}\big)$$

$$\big(A_{2,2} - A_{2,1}A_{1,1}^{-1}A_{1,2}\big)^{-1}\big(A_{2,3} - A_{2,1}A_{1,1}^{-1}A_{1,3}\big)$$

Compute, for the very same matrix

$$A = \left(\begin{array}{cc|c} A_{1,1} & A_{1,2} & A_{1,3} \\ A_{2,1} & A_{2,2} & A_{2,3} \\ \hline A_{3,1} & A_{3,2} & A_{3,3} \end{array}\right)$$

the Schur complement indicated by the non-dashed lines, that is,

$$\Theta_{\{1,2\}}(A) = \begin{vmatrix} A_{1,1} & A_{1,2} & A_{1,3} \\ A_{2,1} & A_{2,2} & A_{2,3} \\ A_{1,3} & A_{2,3} & \boxed{A_{3,3}} \end{vmatrix} = A_{3,3} - \begin{pmatrix} A_{3,1} & A_{3,2} \end{pmatrix} \begin{pmatrix} A_{1,1} & A_{1,2} \\ A_{2,1} & A_{2,2} \end{pmatrix}^{-1} \begin{pmatrix} A_{1,3} \\ A_{2,3} \end{pmatrix}.$$

Now, from

$$\begin{pmatrix} A_{1,1} & A_{1,2} \\ A_{2,1} & A_{2,2} \end{pmatrix}^{-1} = \frac{\begin{pmatrix} A_{1,1}^{-1} + A_{1,1}^{-1}A_{1,2}(A_{2,2} - A_{2,1}A_{1,1}^{-1}A_{1,2})^{-1}A_{2,1}A_{1,1}^{-1} \\ -A_{1,1}^{-1}A_{1,2}(A_{2,2} - A_{2,1}A_{1,1}^{-1}A_{1,2}) \end{pmatrix}}{\begin{pmatrix} -(A_{2,2} - A_{2,1}A_{1,1}^{-1}A_{1,2})^{-1}A_{2,1}A_{1,1}^{-1} \\ (A_{2,2} - A_{2,1}A_{1,1}^{-1}A_{1,2})^{-1} \end{pmatrix}},$$

we deduce $\Theta_2(\Theta_1(A)) = \Theta_{\{1,2\}}(A)$, the simplest case of the heredity principle.

Proposition 6 (Heredity Principle) *Quasi-determinants of quasi-determinants are quasi-determinants.*

Given any set $I = \{i_1, \dots, i_m\} \subset \{1, \dots, k\}$ the heredity principle allows us to define the quasi-determinant

$$\Theta_I(A) = \Theta_{i_1}(\Theta_{i_2}(\cdots \Theta_{i_m}(A) \cdots)).$$

Proposition 7 (Quasi-Determinantal Expressions) *The sequence of biorthogonal polynomials an its squared norms are quasi-determinants*

$$P_{1,k}(x) = \Theta_* \left[\begin{array}{cccc|c} & & & & 1 \\ & G^{[k]} & & & x \\ & & & & \vdots \\ & & & & x^{k-1} \\ \hline G_{k,0} & G_{k,1} & \dots & G_{k,k-1} & x^k \end{array}\right],$$

$$H_k = \Theta_*(G^{[k+1]}),$$

$$P_{2,k}(x) = \Theta_* \left[\begin{array}{cccc|c} & & & & G_{0,k} \\ & G^{[k]} & & & G_{1,k} \\ & & & & \vdots \\ & & & & G_{k,k-1} \\ \hline 1 & x & \dots & x^{k-1} & x^k \end{array}\right].$$

Proof From LU factorization, recalling that $S_{k,k} = 1$, we get

$$\begin{pmatrix} S_{k,0} & S_{k,1} & \dots & S_{k,k-1} \end{pmatrix} G^{[k]} = -\begin{pmatrix} G_{k,0} & G_{k,1} & \dots & G_{k,k-1} \end{pmatrix}.$$

As $G^{[k]}$ is a non singular matrix

$$\begin{pmatrix} S_{k,0} & S_{k,1} & \dots & S_{k,k-1} \end{pmatrix} = -\begin{pmatrix} G_{k,0} & G_{k,1} & \dots & G_{k,k-1} \end{pmatrix} \left(G^{[k]}\right)^{-1}.$$

On the other hand,

$$H_k = \begin{pmatrix} S_{k,0} & S_{k,1} & \dots & S_{k,k-1} & S_{k,k} \end{pmatrix} \begin{pmatrix} G_{0,k} \\ G_{1,k} \\ \vdots \\ G_{k-1,k} \\ G_{k,k} \end{pmatrix} = \begin{pmatrix} S_{k,0} & S_{k,1} & \dots & S_{k,k-1} \end{pmatrix} \begin{pmatrix} G_{0,k} \\ G_{1,k} \\ \vdots \\ G_{k-1,k} \end{pmatrix} + G_{k,k}.$$

□

3.2 Second Kind Functions and LU Factorizations

Definition 3 (Second Kind Functions) For a generalized kernel $u_{x,y}$ we define two families of second kind functions given by

$$C_{1,n}(z) = \left\langle P_{1,n}(x), \frac{1}{z-y} \right\rangle_u, \qquad z \notin \operatorname{supp}_y(u),$$

$$C_{2,n}(z) = \left\langle \frac{1}{z-x}, P_{2,n}(y) \right\rangle_u, \qquad z \notin \operatorname{supp}_x(u).$$

Proposition 8 (LU Factorization Representation of Second Kind Functions) *For z such that $|z| > \sup\big(|y| : y \in \operatorname{supp}_y(u)\big)$ it follows that*

$$C_1(z) = H S_2^{-\top} \chi^*(z),$$

while for z such that $|z| > \sup\big(|x| : x \in \operatorname{supp}_x(u)\big)$ we find

$$C_2(z) = H S_1^{-1} \chi^*(z).$$

Proof Whenever z belongs to an annulus around the origin with no intersection with the y support of the functional

$$\begin{aligned} C_1(z) &= S_1 \left\langle \chi(x), \frac{1}{z-y} \right\rangle_u = S_1 \left\langle \chi(x), \frac{1}{z} \sum_{k=0}^{\infty} \frac{y^k}{z^k} \right\rangle_u \\ &= S_1 \left\langle \chi(x), (\chi(y))^\top \chi^*(z) \right\rangle_u = S_1 G \chi^*(z) = H S_2^{-\top} \chi^*(z). \end{aligned}$$

When z belongs to an annulus around the 0 without intersection with the x support of the functional

$$\begin{aligned}(C_2(z))^\top &= \left\langle \frac{1}{z-x}, (\chi(y))^\top \right\rangle_u (S_2)^\top = \left\langle \frac{1}{z}\sum_{k=0}^{\infty}\frac{x^k}{z^k}, (\chi(y))^\top \right\rangle_u (S_2)^\top \\ &= \left\langle (\chi^*(z))^\top \chi(x), (\chi(y))^\top \right\rangle_u (S_2)^\top \\ &= (\chi^*(z))^\top G (S_2)^\top = (\chi^*(z))^\top S_1^{-\top} H.\end{aligned}$$

□

3.3 Spectral Matrices

Let us introduce the shift or spectral matrix

$$\Lambda := \begin{pmatrix} 0 & 1 & 0 & 0 & \dots \\ 0 & 0 & 1 & 0 & \dots \\ 0 & 0 & 0 & 1 & \dots \\ \vdots & \vdots & \vdots & \vdots & \ddots \end{pmatrix}.$$

The following spectral properties hold

$$\Lambda\chi(x) = x\chi(x), \qquad \Lambda\chi^*(x) = \frac{1}{x}\chi^*(x).$$

If $(E_{i,j})_{s,t} := \delta_{s,i}\delta_{t,j}$ we have

- In the one hand $\Lambda\Lambda^\top = \mathbb{I}$; and, in the other hand, $\Lambda^\top\Lambda = \mathbb{I} - E_{0,0}$.
- $\Lambda^\top\chi(x) = \frac{1}{x}(\mathbb{I} - E_{0,0})\chi(x)$.

We introduce the semi-infinite matrices

$$J_1 := S_1\Lambda(S_1)^{-1}, \qquad J_2 := S_2\Lambda(S_2)^{-1}.$$

Proposition 9 (Spectral Matrices are Hessenberg) *The spectral matrices are lower uni-Hessenberg matrices, i.e., of the form*

$$\begin{pmatrix} * & 1 & 0 & 0 & \dots \\ * & * & 1 & 0 & \dots \\ \vdots & \ddots & \ddots & \ddots & \ddots \end{pmatrix}.$$

Proposition 10 (Spectrality) *The spectral matrices satisfy the eigenvalue property*

$$J_1 P_1(z) = z P_1(z), \qquad J_2 P_2(z) = z P_2(z).$$

Proposition 11 (Eigenvalues of the Truncation and Roots of the Polynomials) *The roots of $P_{i,k}(z)$ and the eigenvalues of the truncation $J_i^{[k]}$ coincide.*

Proof We have

$$J_i^{[k]} \begin{pmatrix} P_{i,0}(x) \\ P_{i,1}(x) \\ \vdots \\ P_{i,k-1}(x) \end{pmatrix} = x \begin{pmatrix} P_{i,0}(x) \\ P_{i,1}(x) \\ \vdots \\ P_{i,k-1}(x) \end{pmatrix} - \begin{pmatrix} 0 \\ 0 \\ \vdots \\ P_{i,k}(x) \end{pmatrix}.$$

For a root α, i.e., $P_{i,k}(\alpha) = 0$ we arrive to

$$\left(J^{[k]}\right) \begin{pmatrix} P_{i,0}(\alpha) \\ P_1(\alpha) \\ \vdots \\ P_{i,k-1}(\alpha) \end{pmatrix} = \alpha \begin{pmatrix} P_{i,0}(\alpha) \\ P_{i,1}(\alpha) \\ \vdots \\ P_{i,k-1}(\alpha) \end{pmatrix}$$

and, therefore, we have the eigenvector $\begin{pmatrix} P_{i,0}(\alpha) \\ P_{i,1}(\alpha) \\ \vdots \\ P_{i,k-1}(\alpha) \end{pmatrix}$ with eigenvalue α. □

3.4 Christoffel–Darboux Kernels

Definition 4 (Christoffel–Darboux Kernels) Given two sequences of matrix biorthogonal polynomials

$$\left\{P_{1,k}(x)\right\}_{k=0}^{\infty} \text{ and } \left\{P_{2,k}(y)\right\}_{k=0}^{\infty},$$

with respect to the sesquilinear form $\langle \cdot, \cdot \rangle_u$, we define the n-th Christoffel–Darboux kernel matrix polynomial

$$K_n(x, y) := \sum_{k=0}^{n} P_{2,k}(y)(H_k)^{-1} P_{1,k}(x),$$

and the mixed Christoffel–Darboux kernel

$$K_n^{\text{mix}}(x,y) := \sum_{k=0}^{n} P_{2,k}(y)(H_k)^{-1}C_{1,k}(x).$$

Proposition 12 (Projection Properties)

(i) For a quasidefinite generalized kernel $u_{x,y}$, the corresponding Christoffel–Darboux kernel gives the projection operator

$$\left\langle K_n(x,z), \sum_{j=0}^{m} \lambda_j P_{2,j}(y)\right\rangle_u = \sum_{j=0}^{n} \lambda_j P_{2,j}(z),$$

$$\left\langle \sum_{j=0}^{m} \lambda_j P_{1,j}(x), K_n(z,y)\right\rangle_u = \sum_{j=0}^{n} C_j P_{1,j}(z).$$

for any $m \in \{0,1,2,\dots\}$.

(ii) In particular, we have

$$\left\langle K_n(x,z), y^l\right\rangle_u = z^l, \qquad l \in\{0,1,\dots,n\}.$$

Proposition 13 (ABC Theorem (Aitken, Berg and Collar) [35]) *We have the following relation*

$$K^{[l]}(x,y) = \left(\chi^{[l]}(y)\right)^{\top}\left(G^{[l]}\right)^{-1}\chi^{[l]}(x).$$

Proof Is a consequence of the following

$$\begin{aligned}\left(\chi^{[l]}(y)\right)^{\top}\left(G^{[l]}\right)^{-1}\chi^{[l]}(x) &= \left(\chi^{[l]}(y)\right)^{\top}\left((S_1^{[l]})^{-1}H^{[l]}(S_2^{[l]})^{-\top}\right)^{-1}\chi^{[l]}(x)\\ &= \left(S_2^{[l]}\chi^{[l]}(y)\right)^{\top}\left(H^{[l]}\right)^{-1}S_1^{[l]}\chi^{[l]}(x)\\ &= \left(P_2^{[l]}(y)\right)^{\top}\left(H^{[l]}\right)^{-1}P_1^{[l]}(x).\end{aligned}$$

□

Proposition 14 (Reproducing Property) *As we are dealing with a projection we find*

$$\langle K^{[l]}(x,z_2), K^{[l]}(z_1,y)\rangle_u = K^{[l]}(z_1,z_2).$$

Proof As an exercise, let us use the ABC theorem

$$\begin{aligned}\langle K^{[l]}(x,z_2), K^{[l]}(z_1,y)\rangle_u &= \Big\langle \left(\chi^{[l]}(z_2)\right)^\top \left(G^{[l]}\right)^{-1}\chi^{[l]}(x),\\ &\quad\times \left(\chi^{[l]}(y)\right)^\top \left(G^{[l]}\right)^{-1}\chi^{[l]}(z_1)\Big\rangle_u\\ &= \left(\chi^{[l]}(z_2)\right)^\top \left(G^{[l]}\right)^{-1}\chi^{[l]}(z_1).\end{aligned}$$

□

4 Standard Orthogonality: Hankel Reduction

Recall that for bilinear forms associated to a Borel measure or a linear functional the Gram matrix is a Hankel matrix, $G_{i,j+1} = G_{i+1,j}$. We will consider in this section some properties that appear in this situation and not in the general scheme.

4.1 Recursion Relations

This property is just a reflection of the self-adjointness of the multiplication operator by x with respect to inner product is reflected in the Hankel structure of the moment matrix

$$\langle xf, h\rangle_\mu = \langle f, xh\rangle_\mu \Rightarrow \Lambda G = G\Lambda^\top \Rightarrow G_{i,j} = G_{i+j}.$$

That leads to the tridiagonal form of the spectral matrices, now named after Jacobi.

Proposition 15 *The spectral matrices are linked by* $J_1 H = H(J_2)^\top$.

Proof From the LU factorization (Cholesky) $G = S^{-1}HS^{-\top}$ and the symmetry $\Lambda G = G\Lambda^\top$ we find

$$S\Lambda S^{-1} = H(S\Lambda S^{-1})^\top H^{-1}.$$

□

This tridiagonal Jacobi matrix $J := J_1$ can be written as follows

$$J := \begin{pmatrix} -S_{10} & 1 & 0 & 0 & 0 & 0 & \dots \\ H_1 H_0^{-1} & S_{10} - S_{21} & 1 & 0 & 0 & 0 & \dots \\ 0 & H_2 H_1^{-1} & S_{21} - S_{32} & 1 & 0 & 0 & \dots \\ 0 & 0 & H_3 H_2^{-1} & S_{32} - S_{43} & 1 & 0 & \dots \\ 0 & 0 & 0 & H_4 H_3^{-1} & S_{43} - S_{54} & 1 & \dots \\ 0 & 0 & 0 & 0 & \ddots & \ddots & \ddots \\ \vdots & \vdots & \vdots & \vdots & & & \end{pmatrix}.$$

Therefore, the spectral properties lead to the well known recursion relations.

Proposition 16 (3-Term Relations) *The orthogonal polynomials and the corresponding second kind functions fulfill*

$$\begin{aligned} JP(x) &= xP(x), & \frac{H_k}{H_{k-1}} P_{k-1} + (S_{k,k-1} - S_{k+1,k}) P_k + P_{k+1} &= x P_k, \\ JC(x) &= xC(x) - H_0 \boldsymbol{e}_0, & \frac{H_k}{H_{k-1}} C_{k-1} + (S_{k,k-1} - S_{k+1,k}) C_k + C_{k+1} &= x C_k, \end{aligned}$$

where $\boldsymbol{e}_0 := (1, 0, 0, \dots, 0, \dots)^\top$.

4.2 Heine Formulas

As the Gram matrix now is a moment matrix we find the well known Heine formulas:

Proposition 17 (Heine Integral Representation) *The orthogonal polynomials can be written as follows*

$$P_k(x) = \frac{1}{k! \det\left[G^{[k]}\right]} \int \prod_{j=1}^{k} (x - x_j) \prod_{1 \leq j < n \leq k} (x_n - x_j)^2 \, \mathrm{d}\mu(x_1) \, \mathrm{d}\mu(x_2) \dots \mathrm{d}\mu(x_k)$$

Proof From the quasi-determinantal expression we get

$$P_k(x) = \frac{1}{\det\left[G^{[k]}\right]} \det \left[\begin{array}{cccc|c} & & & & 1 \\ & & G^{[k]} & & x \\ & & & & \vdots \\ & & & & x^{k-1} \\ \hline G_{k,0} & G_{k,1} & \dots & G_{k,k-1} & x^k \end{array}\right],$$

and using the Vandermonde formula we get the result. □

4.3 Gauss Quadrature Formula

Proposition 18 (Eigenvalues of the Truncation of the Spectral Matrix) *Assume that the measure μ does not change sign in its support (a, b). Then, all eigenvalues of the truncations of the spectral matrices J_1 and J_2 belong to (a, b) and are simple.*

Proof Let $\{a_i\}_{i=1}^m \subset (a, b)$ be the points where the polynomial P_n changes sign in (a, b). Then, as $P_n(z)$ has n roots, $m \leq n$. Therefore, $(x - a_1) \dots (x - a_m)P_n(x)$ does not change sign in (a, b), but from the orthogonality relations we know that P_n is orthogonal to any polynomial of degree less than n and, consequently,

$$\int_a^b (x - a_1) \dots (x - a_m) P_n(x) \, \mathrm{d}\mu(x) = 0,$$

for $m < n$, which is, as μ does not change sign in its support, contradictory. Hence, the only possibility is to have $m = n$. □

Proposition 19 (Powers of Truncations of the Spectral Matrices)

$$\left(J^{[k]}\right)^j = \mathbb{P} \begin{pmatrix} \alpha_{k,1}^j & & & \\ & \alpha_{k,2}^j & & \\ & & \ddots & \\ & & & \alpha_{k,k}^j \end{pmatrix} \mathbb{P}^{-1}, \quad \mathbb{P} := \begin{pmatrix} 1 & 1 & \dots & 1 \\ P_1(\alpha_{k,1}) & P_1(\alpha_{k,2}) & \dots & P_1(\alpha_{k,k}) \\ \vdots & & & \vdots \\ P_{k-1}(\alpha_{k,1}) & P_{k-1}(\alpha_{k,2}) & \dots & P_{k-1}(\alpha_{k,k}) \end{pmatrix}.$$

Proposition 20 *The following identity holds*

$$\int_\Omega x^j \, \mathrm{d}\mu(x) = (J^j)_{0,0} H_0.$$

Proof It follows from 3-term relation

$$x^j P(x) = J^j P(x) \quad \Longrightarrow \quad x^j = (J^j)_{0,0} + (J^j)_{0,1} P_1(x) + \dots + (J^j)_{0,j} P_j(x),$$

and, consequently,

$$\int_\Omega x^j \, \mathrm{d}\mu(x) = (J^j)_{0,0} \langle 1, 1 \rangle + (J^j)_{0,1} \langle P_1, 1 \rangle + \dots + (J^j)_{0,j} \langle P_j, 1 \rangle = (J^j)_{0,0} H_0.$$

□

Proposition 21 (Gaussian Quadrature) *The roots $\{\alpha_{k,l}\}_{l=1}^k$ of the orthogonal polynomials $P_k(x)$ are the k points for the quadrature of μ with precision $2k - 1$. Namely,*

$$\int_\Omega x^j \, \mathrm{d}\mu(x) = \sum_{l=1}^k \lambda_{j,l} \alpha_{k,l}^j, \qquad 0 \leq j \leq 2k - 1,$$

for some coefficients $\lambda_{j,l}$.

Proof From the (2 + 1)-diagonal structure of the Jacobi matrix J it follows that $(J^j)_{0,0} = \left([J^{[k]}]^j\right)_{0,0}$ whenever $0 \leq j \leq 2k-1$.

Now, using the diagonal form of the powers of J and denoting by $\mathbb{P}^{-1}\rfloor_1$ the first column of $\mathbb{P}^{-1}$

$$\int_\Omega x^j \,\mathrm{d}\mu(x) = \left(\left[J^{[k]}\right]^j\right)_{0,0} H_0 = H_0\left(\alpha_{k,1}^j \ \alpha_{k,2}^j \dots \alpha_{k,k}^j\right)\mathbb{P}^{-1}\rfloor_1 = \sum_{l=1}^{k} \lambda_{j,l}\alpha_{k,l}^j,$$

for $0 \leq j \leq 2k-1$.

□

4.4 Christoffel–Darboux Formula

Proposition 22 (Christoffel–Darboux Formula) *The Christoffel–Darboux kernel satisfies*

$$(x-y)K_n(x,y) = P_n(y)(H_n)^{-1}P_{n+1}(x) - P_{n+1}(y)(H_n)^{-1}P_n(x),$$

with confluent version given by

$$\sum_{k=0}^{l-1}\frac{(P_k(x))^2}{H_k} = \frac{1}{H_{l-1}}\left(P_l'(x)P_{l-1}(x) - P_{l-1}'(x)P_l(x)\right)$$

Proof From the eigenvalue property we get

$$\left(H^{-1}J\right)^{[l]}P(y)^{[l]} + \left(H^{-1}J\right)^{[l,\geq l]}P(y)^{[\geq l]} = y\left(H^{-1}\right)^{[l]}P(y)^{[l]},$$

$$[P(x)^\top]^{[l]}\left(H^{-1}J\right)^{[l]} + [P(x)^\top]^{[\geq l]}\left(H^{-1}J\right)^{[\geq l,l]} = x[P(x)^\top]^{[l]}\left(H^{-1}\right)^{[l]}.$$

Left multiply the first eq. by $[P(x)^\top]^{[l]}$ and right multiply the second by $P(y)^{[l]}$, and subtract both results to get

$$[P(x)^\top]^{[l]}\left(H^{-1}J\right)^{[l,\geq l]}P(y)^{[\geq l]} - [P(x)^\top]^{[\geq l]}\left(H^{-1}J\right)^{[\geq l,l]}P(y)^{[l]}$$
$$= (y-x)[P(x)^\top]^{[l]}\cdot\left(H^{-1}\right)^{[l]}P(y)^{[l]}$$

and the result follows. To prove the confluent case just take the limit $y \to x$ in the previous case.

□

Proposition 23 (Christoffel–Darboux Formula for Mixed Kernels) *The mixed Christoffel–Darboux kernel fulfills*

$$(x-y)K_n^{(\text{mix})}(x,y)=P_n(y)(H_n)^{-1}C_{n+1}(x)-P_{n+1}(y)(H_n)^{-1}C_n(x)+1$$

Proof As in the no mixed case consider the expression $P(x)^\top H^{-1}JC(y)$ and the two possible ways of computing them either as $P(x)^\top\left[H^{-1}JC(y)\right]$ or as $\left[P(x)^\top H^{-1}J\right]C(y)$. □

5 Very Classical Orthogonal Polynomials: Hermite, Laguerre and Jacobi Polynomials

Here we study the definite positive case. Neither Bessel polynomials nor Laguerre and Jacobi polynomials with right parameters leading to quasi definite linear functionals are considered. The very classical orthogonal polynomials, Hermite, Laguerre and Jacobi can be understood *à la Bochner* as the eigenfunctions of second order differential operators, associated with corresponding definite positive Borel measures. They also can be characterized by a Rodrigues formula or if you want as those that when derivated preserve the orthogonal character.

Definition 5 (Pearson Equation) The weight u_γ is said very classical whenever we have polynomials $p_2(x)=ax^2+bx+c$ and $p_{1,\gamma}(x)=(A_\gamma-2a)x+(B_\gamma-b)$ with $A_\gamma\neq 0$ such that u_γ satisfies

$$p_2(x)\frac{\mathrm{d}}{\mathrm{d}x}u_\gamma=p_{1,\gamma}(x)u_\gamma$$

The well known very classical weights are:

- Hermite $u(x)=\mathrm{e}^{-x^2}$, $x\in\mathbb{R}$ with $p_1=-2x$, $p_2=1$
- Laguerre $u_\alpha(x)=x^\alpha\,\mathrm{e}^{-x}$, $\alpha>-1$, $x\in\mathbb{R}_+$, with $p_{1,\alpha}=(\alpha-x)$, $p_2=x$
- Jacobi $u_{\alpha,\beta}(x)=(1-x)^\alpha(1+x)^\beta$, $\alpha,\beta>-1$, $x\in(-1,1)$, with $p_{1,\alpha,\beta}=-[(\alpha-\beta)+(\alpha+\beta)x]$, $p_2=1-x^2$

They depend upon: zero ($\gamma=\{\varnothing\}$), one ($\gamma=\{\alpha\}$) and two parameters ($\gamma=\{\alpha,\beta\}$) respectively.

- We denote $u_{\gamma+1}(x)$ the action of increasing by one all the parameters in a classical weight $u_\gamma(x)$. For example, Hermite do not change (no parameter present), in Laguerre we have $\alpha\mapsto\alpha+1$ while in Jacobi the shift is $(\alpha,\beta)\mapsto(\alpha+1,\beta+1)$
- $p_2(x)u_\gamma(x)=u_{\gamma+1}(x)$

Proposition 24 (Properties) *For the very classical weights we have*

$$p_2 u_\gamma\big|_{\partial\Omega} = u_{\gamma+1}\big|_{\partial\Omega} = 0, \quad \langle p_2 f', h\rangle_{u_\gamma} = -\langle f, (p_2' + p_1)h\rangle_{u_\gamma} - \langle f, p_2 h'\rangle_{u_\gamma}.$$

We can matrix represent the action of derivation with the semi-infinite matrix

$$D := \begin{pmatrix} 0\,0\,0\,0 \dots \\ 1\,0\,0\,0 \dots \\ 0\,2\,0\,0 \dots \\ 0\,0\,3\,0 \dots \\ 0\,0\,0\,4 \ddots \\ \vdots\;\vdots\;\vdots\;\vdots \end{pmatrix}, \qquad D\chi(x) = \chi(x)'.$$

Using $f(x) = (f_0, f_1, \dots)\,\chi(x) = \boldsymbol{f}^\top \chi(x)$ we can write

$$x f(x) = \boldsymbol{f}^\top \Lambda \chi(x), \qquad f(x)' = \boldsymbol{f}^\top D\chi(x).$$

Proposition 25 (Symmetry of the Moment Matrix) *The Gram matrices of the classical weights are linked by*

$$DG_{\gamma+1} = -G_\gamma\Big(p_2'(\Lambda) + p_{1,\gamma}(\Lambda) + Dp_2(\Lambda)\Big)^\top.$$

Proposition 26 (*LU* Factorization and Classical Weights) *The LU factorization of the Gram matrix of the classical weights leads to*

$$S_\gamma D S_{\gamma+1}^{-1} = -H_\gamma\Big(S_{\gamma+1}\big(p_2'(\Lambda) + p_{1,\gamma}(\Lambda) + Dp_2(\Lambda)\big)S_\gamma^{-1}\Big)^\top H_{\gamma+1}^{-1}.$$

Equivalently,

$$(S_\gamma)_{n+1,n} = (n+1)\frac{B+nb}{A+2na}, \qquad (H_\gamma)_n = \frac{-n}{A_\gamma + (n-1)a}(H_{\gamma+1})_{n-1}.$$

Proposition 27 (Self-adjoint Differential Operator) *For the classical weights we have a self-adjoint second order differential operator, i.e.,*

$$\left\langle \left[p_2\frac{\mathrm{d}^2}{\mathrm{d}x^2} + (p_2' + p_{1,\gamma})\frac{\mathrm{d}}{\mathrm{d}x}\right] f, h\right\rangle_{u_\gamma} = \left\langle f, \left[p_2\frac{\mathrm{d}^2}{\mathrm{d}x^2} + (p_2' + p_{1,\gamma})\frac{\mathrm{d}}{\mathrm{d}x}\right] h\right\rangle_{u_\gamma}.$$

Proposition 28 (Semi-infinite Matrix Version) *The matrices of classical moments enjoy the following additional symmetry given by the matrix representation* *of a linear second-order differential operator with polynomial coefficients*

$$
\begin{aligned}
&[D^2(a\Lambda^2 + b\Lambda + c) + D(A_\gamma \Lambda + B_\gamma)]G_\gamma \\
&= G_\gamma \left[D^2(a\Lambda^2 + b\Lambda + c) + D(A_\gamma \Lambda + B_\gamma)\right]^\top .
\end{aligned}
$$

Observe that calling $M := SDS^{-1}$, which is a strictly lower triangular matrix with first subdiagonal the sequence of natural numbers fulfilling $[J, M] = I$, the above relation leads to

$$
[M^2(aJ^2 + bJ + c) + M(A_\gamma J + B_\gamma)] = \left[M^2(aJ^2 + bJ + c) + M(A_\gamma J + B_\gamma)\right]^\top .
$$

Proposition 29 (Diagonalizing the Self-adjoint Differential Operator)

$$
\begin{aligned}
N_\gamma &:= S_\gamma \left[D^2(a\Lambda^2 + b\Lambda + c) + D(A_\gamma \Lambda + B_\gamma)\right] S_\gamma^{-1} \\
&= \begin{pmatrix}
0 & 0 & 0 & 0 & 0 & \dots \\
0 & A_\gamma & 0 & 0 & 0 & \dots \\
0 & 0 & 2(A_\gamma + a) & 0 & 0 & \dots \\
0 & 0 & 0 & 3(A_\gamma + 2a) & 0 & \dots \\
0 & 0 & 0 & 0 & \ddots & \\
\vdots & \vdots & \vdots & \vdots & &
\end{pmatrix}.
\end{aligned}
$$

The diagonal coefficients, $(N_\gamma)_n = n(A_\gamma + (n-1)a)$, are the eigenvalues of the sequence of classical orthogonal polynomials, being these one the corresponding eigenfunctions

$$
F_\gamma := p_2 \frac{\mathrm{d}^2}{\mathrm{d}x^2} + (p_2' + p_{1,\gamma}) \frac{\mathrm{d}}{\mathrm{d}x} \qquad \Longrightarrow \qquad F_\gamma\left[P_\gamma(x)\right] = N_\gamma P_\gamma(x).
$$

In particular, for the three classical families, we have

- ***Hermite.*** $\dfrac{\mathrm{d}u}{\mathrm{d}x} = -2xu$; $A = -2$, $B = 0$, $a = 0$, $b = 0$, $c = 1$; $H_0 = \sqrt{\pi}$;

$$
S_{n+1,n} = 0, \qquad H_n = \sqrt{\pi}\frac{n!}{2^n}, \qquad N_n = -n.
$$

- ***Laguerre.*** $x\dfrac{\mathrm{d}u_\alpha}{\mathrm{d}x} = (\alpha - x)u_\alpha$; $A_\alpha = -1$, $B_\alpha = (1+\alpha)$, $a = 0$, $b = 1$, $c = 0$; $(h_\alpha)_0 = \Gamma(\alpha + 1)$;

$$
(S_\alpha)_{n+1,n} = -(n+1)[(n+1)+\alpha], \quad (H_\alpha)_n = n!\Gamma(\alpha + n + 1), \quad (N_\alpha)_n = -2n.
$$

- ***Jacobi***. $(1-x^2)\dfrac{\mathrm{d}u_{\alpha,\beta}}{\mathrm{d}x} = -[(\alpha-\beta)+(\alpha+\beta)x]u_{\alpha,\beta}$; $A_{\alpha,\beta} = -[(\beta+\alpha)+2]$, $B_{\alpha,\beta} = -(\alpha-\beta)$, $a=-1$, $b=0$, $c=1$; $(H_{\alpha,\beta})_0 = \dfrac{\Gamma(\alpha+1)\Gamma(\beta+1)2^{\alpha+\beta+1}}{(\alpha+\beta+1)\Gamma(\alpha+\beta+1)}$;

$$(S_{\alpha,\beta})_{n+1,n} = \frac{(n+1)(\alpha-\beta)}{(\alpha+\beta+2)+2n}$$

$$(H_{\alpha,\beta})_n = n!2^{(\alpha+\beta+2n+1)}\frac{\Gamma(\alpha+\beta+n+1)\Gamma(\alpha+n+1)\Gamma(\beta+n+1)}{(\alpha+\beta+2n+1)\Gamma^2(\alpha+\beta+2n+1)}$$

$$(N_{\alpha,\beta})_n = -n(\beta-\alpha+1+n).$$

6 Christoffel and Geronimus Transformations

6.1 Some History

Three perturbations have attracted the interest of the researchers. Christoffel perturbations, that appear when you consider a new functional $\hat{u} = p(x)u$, where $p(x)$ is a polynomial, were **studied in 1858 by the German mathematician E.B. Christoffel** in [21] in the framework of Gaussian quadrature rules.

Christoffel found explicit formulas relating the corresponding sequences of orthogonal polynomials with respect to two measures, the Lebesgue measure $\mathrm{d}\mu$ supported in the interval $(-1,1)$ and $d\hat{\mu}(x) = p(x)d\mu(x)$, with $p(x) = (x-q_1)\cdots(x-q_N)$ a signed polynomial in the support of $\mathrm{d}\mu$, as well as the distribution of their zeros as nodes in such quadrature rules. Nowadays, these are called **Christoffel formulas**, and can be considered a classical result in the theory of orthogonal polynomials which can be found in a number of textbooks, see for example Chihara, Szegő or Gautschi.

In the theory of orthogonal polynomials, **connection formulas** between two families of orthogonal polynomials allow to express any polynomial of a given degree n as a linear combination of all polynomials of degree less than or equal to n in the second family. A noteworthy fact regarding the Christoffel finding is that in this case the number of terms does not grow with the degree n but remarkably, and on the contrary, remain constant, equal to the degree of the perturbing polynomial.

Geronimus transformation appears when you are dealing with perturbed functionals v defined by $p(x)v = u$, where $p(x)$ is a polynomial. Such a kind of transformations were used by the **Russian mathematician J.L. Geronimus** in [25] in order to have a nice proof of a result by W. Hahn concerning the characterization of classical orthogonal polynomials (Hermite, Laguerre, Jacobi, and Bessel) as those orthogonal polynomials whose first derivatives are also orthogonal polynomials, for an English account of Geronimus' paper see [26].

Again, as happened for the Christoffel transformation, within the Geronimus transformation one can find Christoffel type formulas, now in terms of the second kind functions, relating the corresponding sequences of orthogonal polynomials, for example the work of P. Maroni studied this situation for a perturbation of the type $p(x) = x - a$.

Despite that in the paper by Geronimus no Christoffel type formula was derived, in order to distinguish these Christoffel type formulas from those for Christoffel transformations, we refer to them as Christoffel–Geronimus formulas.

The more general problem related to linear functionals u and v satisfying $p(x)u = q(x)v$, where $p(x), q(x)$ are polynomials has been analyzed the Russian mathematician V.B. Uvarov back in [37] see also [39] where the term linear spectral was given.

Uvarov found Christoffel type formulas, see [37], that allow for any pair of perturbing polynomials $p(x)$ and $q(x)$, to find the new orthogonal polynomials in terms of determinantal expressions of the original unperturbed second kind functions and orthogonal polynomials. On the other hand, the addition of a finite number of Dirac masses to a linear functional appears in the framework of the spectral analysis of fourth order linear differential operators with polynomial coefficients and with orthogonal polynomials as eigenfunctions. Therein you have the so called Laguerre-type, Legendre-type and Jacobi-type orthogonal polynomials introduced by H.L. Krall.

A more general analysis from the point of view of the algebraic properties of the sequences of orthogonal polynomials associated to the linear functionals u and $w = u + \sum_{n=0}^{N} M_n \delta(x - a_n)$, the so-called general Uvarov transformation by Zhedanov, see [39], has been done for the positive definite case by Uvarov.

6.2 Christoffel and Geronimus Transformations

Definition 6 (Christoffel Transformations and the Schwartz Kernel) Given a polynomial $W_C(x)$ of degree N_C, and a generalized kernel $u_{x,y} \in \mathcal{O}'_c$ a linear spectral or Geronimus–Uvarov transformation $\hat{u}_{x,y}$ of $u_{x,y}$ is a matrix of generalized kernels such that

$$\hat{u}_{x,y} = W_C(x) u_{x,y}.$$

Proposition 30 (Christoffel Transformation and the Bilinear Form and Gram Matrix) *The perturbed Gram matrix* $\hat{G} := \langle \chi(x), \chi(y) \rangle_{\hat{u}}$ *and the original one* G *satisfy*

$$\hat{G} = W_C(\Lambda) G.$$

The sesquilinear forms are related by

$$\langle P(x), Q(y) W_G(y) \rangle_{\hat{u}} = \langle P(x) W_C(x), Q(y) \rangle_u .$$

Definition 7 (Geronimus Transformations and the Schwartz Kernel) Given a generalized kernel $u_{x,y} \in \mathcal{O}'_c$ with a given support $\operatorname{supp} u_{x,y}$, and a polynomial $W(y) \in \mathbb{C}[y]$ of degree N, such that $\sigma(W(y)) \cap \operatorname{supp}_y(u) = \varnothing$, a bivariate generalized function $\check{u}_{x,y}$ is said to be a Geronimus transformation of the generalized kernels $u_{x,y}$ if

$$\check{u}_{x,y} W(y) = u_{x,y}.$$

Proposition 31 (Geronimus Transformation and the Bilinear Form and Gram Matrix) *In terms of bilinear forms a Geronimus transformation fulfills*

$$\langle P(x), Q(y) W(y) \rangle_{\check{u}} = \langle P(x), Q(y) \rangle_u ,$$

while, in terms of the corresponding Gram matrices, satisfies

$$\check{G}W(\Lambda^\top) = G.$$

Definition 8 (Linear Spectral Transformations and the Schwartz Kernel) Given two polynomials $W_C(x), W_G(y)$ of degrees N_C, N_G, and a generalized kernel $u_{x,y} \in \mathcal{O}'_c$ such that $\sigma(W_G(y))) \cap \operatorname{supp}_y(u) = \varnothing$, a matrix linear spectral or Geronimus–Uvarov transformation $\hat{u}_{x,y}$ of $u_{x,y}$ is a matrix of generalized kernels such that

$$\hat{u}_{x,y}W_G(y) = W_C(x)u_{x,y}. \tag{1}$$

Proposition 32 (Linear Spectral Transformation and the Bilinear Form and Gram Matrix) *The perturbed Gram matrix* $\hat{G} := \langle \chi(x), \chi(y)\rangle_{\hat{u}}$ *and the original one* G *satisfy*

$$\hat{G}W_G(\Lambda^\top) = W_C(\Lambda)G.$$

The sesquilinear forms are related by

$$\langle P(x), Q(y)W_G(y)\rangle_{\hat{u}} = \langle P(x)W_C(x), Q(y)\rangle_u \,.$$

Proposition 33 (Hankel Case: Christoffel and Geronimus) *Given monic polynomials* $W_C(x) := \prod_{i=1}^d (x-r_i)^{m_i}$ *and* $W_G(x) := \prod_{i=1}^s (x-q_i)^{n_i}$ *with degrees* $\sum_{i=1}^d m_i = M$ y $\sum_{i=1}^s n_i = N$, *respectively, such that* $\{q_1, q_2, \dots q_s\} \cap \Omega = \varnothing$; *the Christoffel* ($W_G(x) = 1$) *and Geronimus* ($W_C(x) = 1$) *transformations of a measure* μ *are*

$$\hat{\mu}(x) := W_C(x)\mu(x), \qquad \hat{\Omega} = \Omega.$$

$$\check{\mu}(x) := \frac{\mu(x)}{W_G(x)} + \sum_{i=1}^{s}\sum_{l=0}^{n_i-1}(-1)^l\frac{\xi_{i,l}}{l!}\delta^{(l)}(x-q_i), \qquad \check{\Omega} = \Omega \cup \{q_1, q_2, \dots q_s\}.$$

Notice that the Geronimus transformation of a measure does not need to be a measure.

Proposition 34 (Hankel Case: Linear Spectral) *Whenever the polynomials* $W_C(x)$ *and* $W_G(x)$ *are coprime the composition of the above transformations is the linear spectral transformation*

$$\tilde{\mu}(x) := \widehat{(\check{\mu})}(x) = \frac{W_C(x)\mu(x)}{W_G(x)} + \sum_{i=1}^{s}\sum_{l=0}^{n_i-1}(-1)^l\frac{\xi_{i,l}}{l!}W_G(x)\delta^{(l)}(x-q_i),$$

$$\tilde{\Omega} = \Omega \cup \{q_1, q_2, \dots q_s\}.$$

The Christoffel formulas found for the three types of transformations hold equally in the more general Schwartz kernel situation. We do not need to have a Hankel symmetry for the Gram matrix.

Why linear spectral? Because the behavior of the Markov function

$$
\begin{aligned}
\hat{C}_0(y) &= \int \frac{(x-r)\,\mathrm{d}\mu}{y-x} = (y-r)\int \frac{\mathrm{d}\mu}{y-x} - \int \mathrm{d}\mu(x) = (y-r)C_0(y) - h_0,\\
\check{C}_0(y) &= \int \frac{\mathrm{d}\mu}{(x-q)(y-x)} + \xi \int \frac{\delta(x-q)\,\mathrm{d}\mu}{y-x}\\
&= \frac{1}{(y-q)}\left[\int \frac{\mathrm{d}\mu}{y-x} - \int \frac{\mathrm{d}\mu(x)}{x-q}\right] + \xi\frac{1}{y-q} = \frac{C_0(y)-C_0(q)+\xi}{(y-q)},\\
\tilde{C}_0(y) &= \int \frac{(x-r)\,\mathrm{d}\mu}{(x-q)(y-x)} + \xi \int \frac{\delta(x-q)(x-r)\,\mathrm{d}\mu}{y-x}\\
&= \frac{(y-r)C_0(y)-(q-r)C_0(q)+(q-r)\xi}{(y-q)}.
\end{aligned}
$$

See [19, 20] for a discussion of perturbations of bilinear forms and [27] for the study of Darboux transformations and orthogonal polynomials.

6.3 Christoffel Perturbations

We began with an example. Take $W_C(x) = x-a$ for $a \in \mathbb{C}$, then the Gram matrices satisfy $\hat{G} = (\Lambda - a)G$ and, therefore, the so called **connector** $\hat{\omega} := \hat{S}_1(\Lambda - a)(S_1)^{-1}$ fulfills $\hat{\omega} := \hat{S}_1(\Lambda - a)(S_1)^{-1} = \hat{H}\left(S_2(\hat{S}_2)^{-1}\right)^{\top} H^{-1}$. The **connector links the original and transformed orthogonal polynomials**

$$
\omega P_1(x) = (x-a)\hat{P}_1(x), \qquad \left(\hat{P}_2(y)\right)^{\top}\hat{H}^{-1}\omega = \left(P_2(y)\right)^{\top} H^{-1},
$$

and is an upper diagonal matrix with only the first superdiagonal non-zero

$$
\omega = \begin{pmatrix}
\omega_{0,0} & 1 & 0 & 0 & \\
0 & \omega_{1,1} & 1 & 0 & \ddots\\
0 & 0 & \omega_{2,2} & 1 & \ddots\\
 & & \ddots & \ddots & \ddots
\end{pmatrix}, \qquad \omega_{k,k} = \frac{\hat{H}_k}{H_k}.
$$

Proposition 35 *The formula*

$$\omega_{n,n} = -\frac{P_{1,n+1}(a)}{P_{1,n}(a)} = \frac{\hat{H}_n}{H_n}$$

holds.

Proof It follows from $\omega_{n,n} P_{1,n}(x) + P_{1,n+1}(x) = (x-a)\hat{P}_{1,n}(x)$. □

Proposition 36 *The CD kernels are connected by*

$$(x-a)\hat{K}_n(x,y) - \hat{P}_{2,n}(y)(\hat{H}_n)^{-1} P_{1,n+1}(x) = K_n(x,y).$$

Proof As ω is upper triangular, from $\big(\hat{P}_2(y)\big)^\top \hat{H}^{-1}\omega = \big(P_2(y)\big)^\top H^{-1}$ we get

$$\begin{aligned}
&\big(\hat{P}_2^{[n+1]}(y)\big)^\top (\hat{H}^{[n+1]})^{-1} \omega^{[n+1]} P_1^{[n+1]}(x)\\
&= \big(P_2^{[n+1]}(y)\big)^\top (H^{[n+1]})^{-1} P_1^{[n+1]}(x) = K_n(x,y).
\end{aligned}$$

Observe also that

$$\begin{aligned}
\omega^{[n+1]} P_1^{[n+1]}(x)) &= (\omega P_1(x))^{[n+1]} - \begin{pmatrix} 0_{n\times 1} \\ P_{1,n+1}(x) \end{pmatrix}\\
&= (x-a)\hat{P}_1^{[n+1]}(x) - \begin{pmatrix} 0_{n\times 1} \\ P_{1,n+1}(x) \end{pmatrix}.
\end{aligned}$$

□

Proposition 37 *The Christoffel formulas are*

$$\begin{aligned}
\hat{P}_{1,n}(x) &= \frac{1}{x-a}\Big(P_{1,n+1}(x) - \frac{P_{1,n+1}(a)}{P_{1,n}(a)} P_{1,n}(x)\Big),\\
\hat{P}_{2,n}(y) &= \frac{K_n(a,y)}{P_{1,n}(a)} H_n, \; \hat{H}_n = -\frac{P_{1,n+1}(a)}{P_{1,n}(a)} H_n.
\end{aligned}$$

The Christoffel formulas in terms of quasi-determinants are

$$\begin{aligned}
\hat{P}_{1,n}(x) &= \frac{1}{x-a}\Theta_* \begin{pmatrix} P_{1,n}(a) & P_{1,n}(x) \\ P_{1,n+1}(a) & P_{1,n+1}(x) \end{pmatrix},\\
\hat{P}_{2,n}(y) &= -\Theta_* \begin{pmatrix} P_{1,n}(a) & H_n \\ K_n(a,y) & 0 \end{pmatrix}, \; \hat{H}_n = \Theta_* \begin{pmatrix} P_{1,n}(a) & H_n \\ P_{1,n+1}(a) & 0 \end{pmatrix}.
\end{aligned}$$

In the Hankel case we have two alternative Christoffel type formulas. Indeed,

$$\hat{P}_n(x)=\frac{1}{x-a}\Theta_*\begin{pmatrix}P_n(a) & P_n(x)\\ P_{n+1}(a) & P_{n+1}(x)\end{pmatrix}=-\Theta_*\begin{pmatrix}P_n(a) & H_n\\ K_n(a,x) & 0\end{pmatrix}.$$

As any polynomial $W_c(x)$ can be factorized in simple factors, the Christoffel formula can be achieved by **iteration** of the previous simple example. We will take another path.

Definition 9 (Jets) Given the set $r=\{(r_i,m_i)\}_{i=1}^d$ determined by the perturbing polynomial $W_C(x):=\prod_{i=1}^d(x-r_i)^{m_i}$ for any function $f(x)$ we define the corresponding jet

$$\mathcal{J}_r[f]:=\left(\frac{f^{(0)}(r_1)}{0!},\frac{f^{(1)}(r_1)}{1!},\dots,\frac{f^{(m_1-1)}(r_1)}{(m_1-1)!};\right.$$
$$\left.\times\frac{f^{(0)}(r_2)}{0!},\dots,\frac{f^{(m_2-1)}(r_2)}{(m_2-1)!};\dots;\frac{f^{(0)}(r_d)}{0!},\dots,\frac{f^{(m_d-1)}(r_d)}{(m_d-1)!}\right).$$

Proposition 38 (General Christoffel Formulas) *The following Christoffel connection formulas hold*

$$\hat{P}_{1,n}(x)=\frac{1}{W_C(x)}\Theta_*\begin{bmatrix}\mathcal{J}P_{1,n} & P_{1,n}(x)\\ \vdots & \vdots\\ \mathcal{J}P_{1,n+N_C} & P_{1,n+N_C}(x)\end{bmatrix},$$

$$\hat{H}_n=\Theta_*\begin{bmatrix}\mathcal{J}P_{1,n} & H_n\\ \mathcal{J}P_{1,n+1} & 0\\ \vdots & \vdots\\ \mathcal{J}P_{1,n+N_C} & 0\end{bmatrix},$$

$$\hat{P}_{2,n}(y)=-\Theta_*\begin{bmatrix}\mathcal{J}P_{1,n} & H_n\\ \mathcal{J}P_{1,n+1} & 0\\ \vdots & \vdots\\ \mathcal{J}P_{1,n+N_C-1} & 0\\ \mathcal{J}K_{n-1}(\cdot,y)(y) & 0\end{bmatrix}.$$

Comments

- The jets appear because the multiplicities bigger than 1 of each root
- The idea of the proof is construct, as previously, a connector ω which is an upper triangular matrix with only the first N_C superdiagonals nonzero

- To get the transformations of the first polynomials $P_{1,n}(x)$ we use the connection formula in terms of the connector, evaluate in its zeroes, taking into account multiplicities and find the connector coefficients
- To get the second polynomials $P_{2,n}$ we derive formulas for the Christoffel–Darboux kernels and apply a similar reasoning as in the example
- For the Hankel reduction we obtain two alternative Christoffel type formulas for the polynomials
  The previous formula reduces to the well known Christoffel formula.

Proposition 39 (Classical Christoffel Formula)

$$\hat{P}_n(x) = \frac{1}{(x-r_1)\cdots(x-r_N)} \frac{\begin{vmatrix} P_n(r_1) & \dots & P_n(r_N) & P_n(x) \\ \vdots & & \vdots & \vdots \\ P_{n+N}(r_1) & \dots & P_{n+N}(r_N) & P_{n+N}(x) \end{vmatrix}}{\begin{vmatrix} P_n(r_1) & \dots & P_n(r_N) \\ \vdots & & \vdots \\ P_{n+N-1}(r_1) & \dots & P_{n+N-1}(r_N) \end{vmatrix}}.$$

6.4 *Geronimus Transformations*

Take $W_G(x) = x - a$ for $a \in \mathbb{C}$, with the perturbed Schwartz kernel given by $\check{u}_{x,y} = \frac{u_{x,y}}{y-a} + \xi_x \delta_{y-a}$ with δ_y being the Dirac delta distribution and ξ_x is a free linear functional. Then, the Gram matrices satisfy $\hat{G}(\Lambda^\top - a) = G$ and, therefore, the so called **connector** $\omega := \check{S}_1(S_1)^{-1}$ fulfills $\omega := \check{S}_1(S_1)^{-1} = \check{H}(\check{S}_2)^{-\top}(\Lambda^\top - a)(S_2)^\top H^{-1}$.

The **connector links the original and transformed orthogonal polynomials**

$$\check{P}_1(x) = \omega P_1(x), \qquad \left(\check{H}^{-1}\omega H\right)^\top \check{P}_2(y) = P_2(y)(y-a),$$

and is a lower diagonal matrix with only the first subdiagonal non-zero

$$\omega = \begin{pmatrix} 1 & 0 & 0 & \dots \\ \omega_{1,0} & 1 & 0 & \\ 0 & \omega_{2,1} & 1 & \ddots \\ & \ddots & \ddots & \ddots \end{pmatrix}, \qquad \omega_{n+1,n} = \check{H}_{n+1}(H_n)^{-1}.$$

Proposition 40 *The Geronimus transformation of the second kind functions satisfies*

$$(x-a)\check{C}_1(x) - \begin{pmatrix} \check{H}_0 \\ 0 \\ \vdots \end{pmatrix} = \omega C_1(x), \qquad \big(\check{C}_2(x)\big)^\top \check{H}^{-1}\omega = \big(C_2(x)\big)^\top H^{-1}.$$

Proof If follows from

$$\begin{aligned}
(z-a)\check{C}_1(z) - \omega C_1(z) &= \left\langle \check{P}_1(x), \frac{1}{z-y}\right\rangle_{\check{u}} (z-a) - \left\langle \check{P}_1(x), \frac{1}{z-y}\right\rangle_{(y-a)\check{u}} \\
&= \left\langle \check{P}_1(x), \frac{z-a-(y-a)}{z-y}\right\rangle_{\check{u}} = \left\langle \check{P}_1(x), 1\right\rangle_{\check{u}} \\
&= \begin{pmatrix} \check{H}_0 \\ 0 \\ \vdots \end{pmatrix}
\end{aligned}$$

and

$$\begin{aligned}
\big(\check{C}_2(x)\big)^\top \check{H}^{-1}\omega - \big(C_2(x)\big)^\top H^{-1} &= \left\langle \frac{1}{z-x}, \check{P}_2(y)\right\rangle_{\check{u}} \check{H}^{-1}\omega - \left\langle \frac{1}{z-x}, P_2(y)\right\rangle_{u} H^{-1} \\
&= \left\langle \frac{1}{z-x}, \big(\check{H}^{-1}\omega\big)^\top \check{P}_2(y)\right\rangle_{\check{u}} - \left\langle \frac{1}{z-x}, H^{-\top} P_2(y)\right\rangle_{u} \\
&= \left\langle \frac{1}{z-x}, (y-a)H^{-\top} P_2(y)\right\rangle_{\check{u}} - \left\langle \frac{1}{z-x}, H^{-\top} P_2(y)\right\rangle_{u} \\
&= 0.
\end{aligned}$$

□

We cannot evaluate directly in $x = a$ as that point belongs to the support of the perturbed functional and the second kind function might not be defined there.

Proposition 41 *For $n > 0$, we have*

$$\omega_{n,n-1} = -\frac{C_{1,n}(a) - \langle \xi_x, P_{1,n}(x)\rangle}{C_{1,n-1}(a) - \langle \xi_x, P_{1,n-1}(x)\rangle}, \qquad \check{H}_0 = -(C_{1,0}(a) - \langle \xi_x, 1\rangle).$$

Proof It follows form

$$
\begin{aligned}
(z-a)\check{C}_1(z) &= (z-a)\left\langle \check{P}_1(x), \frac{1}{z-y}\right\rangle_{\frac{u_{x,y}}{y-a}+\xi_x\delta_{y-a}} \\
&= (z-a)\left\langle \check{P}_1(x), \frac{1}{z-y}\right\rangle_{\frac{u_{x,y}}{y-a}} + \left\langle \xi_x, \check{P}_1(x)\right\rangle \\
&= (z-a)\left\langle \check{P}_1(x), \frac{1}{z-y}\right\rangle_{\frac{u_{x,y}}{y-a}} + \omega\left\langle \xi_x, P_1(x)\right\rangle .
\end{aligned}
$$

Thus, we derive an expression without singularity problems at $z=a$

$$
(z-a)\left\langle \check{P}_1(x), \frac{1}{z-y}\right\rangle_{\frac{u_{x,y}}{y-a}} - \begin{pmatrix} \check{H}_0 \\ 0 \\ \vdots \end{pmatrix} = \omega\big(C_1(z) - \langle \xi_x, P_1(x)\rangle\big).
$$

□

Proposition 42 *For $n>0$, we have the following quasideterminantal expression*

$$
\begin{aligned}
\check{P}_{1,n}(x) &= \Theta_*\begin{pmatrix} C_{1,n-1}(a) - \big\langle \xi_x, P_{1,n-1}(x)\big\rangle & P_{1,n-1}(x) \\ C_{1,n}(a) - \big\langle \xi_x, P_{1,n}(x)\big\rangle & P_{1,n}(x)\end{pmatrix}, \\
\check{H}_n &= \Theta_*\begin{pmatrix} C_{1,n-1}(a) - \big\langle \xi_x, P_{1,n-1}(x)\big\rangle & H_n \\ C_{1,n}(a) - \big\langle \xi_x, P_{1,n}(x)\big\rangle & 0\end{pmatrix}.
\end{aligned}
$$

Proof For $n>0$, the expression for $\omega_{n,n-1}$ implies

$$
\begin{aligned}
\check{P}_{1,n}(x) &= P_{1,n}(x) - \frac{C_{1,n}(a) - \big\langle \xi_x, P_{1,n}(x)\big\rangle}{C_{1,n-1}(a) - \big\langle \xi_x, P_{1,n-1}(x)\big\rangle} P_{1,n-1}(x), \\
\check{H}_n &= -\frac{C_{1,n}(a) - \big\langle \xi_x, P_{1,n}(x)\big\rangle}{C_{1,n-1}(a) - \big\langle \xi_x, P_{1,n-1}(x)\big\rangle} H_{n-1}.
\end{aligned}
$$

□

Proposition 43 *The following connection formulas for Christoffel–Darboux kernels hold*

$$
\check{K}_{n-1}(x,y) = (y-a)K_{n-1}(x,y) - \check{P}_{2,n}(y)\check{H}_n^{-1}\omega_{n,n-1}P_{1,n-1}(x),
$$

and for $n \geq 1$

$$
(x-a)\check{K}_{n-1}^{(\mathrm{mix})}(x,y) = (y-a)K_{n-1}^{(mix)}(x,y) - \check{P}_{2,n}(y)\check{H}_n^{-1}\omega_{n,n-1}C_{1,n-1}(x) + 1.
$$

Proof It follows from the definition of the kernels and the connection formulas

$$\check{P}_1(x) = \omega P_1(x), \qquad \big(\check{P}_2(y)\big)^\top \check{H}^{-1}\omega = \big(P_2(y)\big)^\top (y-a)H^{-1}$$

and $(x-a)\check{C}_1(x) - \big(\check{H}_0, 0, \dots\big)^\top = \omega C_1(x)$. □

Proposition 44 *It also holds that*

$$\check{P}_{2,n}(y) = H_{n-1}\frac{(y-a)(K^{(mix)}_{n-1}(a,y) - \langle \xi_x, K_{n-1}(x,y)\rangle) + 1}{C_{1,n-1}(a) - \big\langle \xi_x, P_{1,n-1}(x)\big\rangle}.$$

Proof The mixed kernel $\check{K}^{(\text{mix})}_{n-1}(x,y)$ will have singularity problems at $x = a$. This issue can be handled as before with the aid of the CD kernel and we get

$$\begin{aligned}
&(y-a)(K^{(\text{mix})}_{n-1}(a,y) - \langle \xi_x, K_{n-1}(x,y)\rangle) + 1\\
&= \check{P}_{2,n}(y)(\check{H}_n)^{-1}\omega_{n,n-1}\big(C_{1,n-1}(a) - \big\langle \xi_x, P_{1,n-1}(x)\big\rangle\big)\\
&= \check{P}_{2,n}(y)(H_{n-1})^{-1}\big(C_{1,n-1}(a) - \big\langle \xi_x, P_{1,n-1}(x)\big\rangle\big).
\end{aligned}$$

□

Proposition 45 *For $n > 0$, the following formulas hold*

$$\check{P}_{2,n}(y) = -\Theta_*\begin{pmatrix} C_{1,n-1}(a) - \big\langle \xi_x, P_{1,n-1}(x)\big\rangle & H_{n-1}\\ (y-a)(K^{(mix)}_{n-1}(a,y) - \langle \xi_x, K_{n-1}(x,y)\rangle) + 1 & 0\end{pmatrix}.$$

Proposition 46 *For a general Geronimus perturbation we find the general Christoffel formulas*

$$\check{P}^{[1]}_n(x) = \Theta_*\begin{pmatrix} \mathcal{J}_{C^{[1]}_{n-N}} - \Big\langle P^{[1]}_{n-N}(x), (\xi)_x\Big\rangle \mathcal{W} & P^{[1]}_{n-N}(x)\\ \vdots & \vdots\\ \mathcal{J}_{C^{[1]}_n} - \Big\langle P^{[1]}_n(x), (\xi)_x\Big\rangle \mathcal{W} & P^{[1]}_n(x)\end{pmatrix},$$

$$\check{H}_n = \Theta_*\begin{pmatrix} \mathcal{J}_{C^{[1]}_{n-N}} - \Big\langle P^{[1]}_{n-N}(x), (\xi)_x\Big\rangle \mathcal{W} & H_{n-N}\\ \mathcal{J}_{C^{[1]}_{n-N+1}} - \Big\langle P^{[1]}_{n-N+1}(x), (\xi)_x\Big\rangle \mathcal{W} & 0\\ \vdots & \vdots\\ \mathcal{J}_{C^{[1]}_n} - \Big\langle P^{[1]}_n(x), (\xi)_x\Big\rangle \mathcal{W} & 0\end{pmatrix},$$

$$\left(\check{P}_n^{[2]}(y)\right)^\top = -\Theta_* \begin{pmatrix} \mathcal{J}_{C_{n-N}^{[1]}} - \left\langle P_{n-N}^{[1]}(x), (\xi)_x \right\rangle \mathcal{W} & H_{n-N} \\ \mathcal{J}_{C_{n-N+1}^{[1]}} - \left\langle P_{n-N+1}^{[1]}(x), (\xi)_x \right\rangle \mathcal{W} & 0 \\ \vdots & \vdots \\ \mathcal{J}_{C_{n-1}^{[1]}} - \left\langle P_{n-1}^{[1]}(x), (\xi)_x \right\rangle \mathcal{W} & 0 \\ W(y)\big(\mathcal{J}_{K_{n-1}^{(pc)}}(y) - \langle K_{n-1}(x,y), (\xi)_x \rangle \mathcal{W}\big) + \mathcal{J}_{\mathcal{V}}(y) & 0 \end{pmatrix}.$$

Here $\mathcal{W}, \mathcal{V} \in \mathbb{C}^{N_G \times N_G}$ are upper triangular matrices determined by $W_G(x)$ and its derivatives, and $\mathcal{V}(x, y)$ is a polynomial constructed in terms of $W_G(x)$ and completely symmetric bivariate polynomials.

Acknowledgments The author thanks the financial support from the Spanish *Agencia Estatal de Investigación*, research project PGC2018-096504-B-C3 entitled *Ortogonalidad y Aproximación: Teoría y Aplicaciones en Física Matemática*. The author acknowledges the exhaustive revision by the anonymous referee that much improved the readability of this contribution.

References

1. Adler, M., van Moerbeke, P.: Group factorization, moment matrices and Toda lattices. Int. Math. Res. Notices **12**, 556–572 (1997)
2. Adler, M., van Moerbeke, P.: Generalized orthogonal polynomials, discrete KP and Riemann–Hilbert problems. Commun. Math. Phys. **207**, 589–620 (1999)
3. Adler, M., van Moerbeke, P.: Darboux transforms on band matrices, weights and associated polynomials. Int. Math. Res. Notices **18**, 935–984 (2001)
4. Adler, M., van Moerbeke, P., Vanhaecke, P.: Moment matrices and multi-component KP, with applications to random matrix theory. Commun. Math. Phys. **286**, 1–38 (2009)
5. Álvarez-Fernández, C., Mañas, M.: Orthogonal Laurent polynomials on the unit circle, extended CMV ordering and 2D Toda type integrable hierarchies. Adv. Math. **240**, 132–193 (2013)
6. Álvarez-Fernández, C., Mañas, M.: On the Christoffel–Darboux formula for generalized matrix orthogonal polynomials. J. Math. Anal. Appl. **418**, 238–247 (2014)
7. Álvarez-Fernández, C., Fidalgo Prieto, U., Mañas, M.: The multicomponent 2D Toda hierarchy: generalized matrix orthogonal polynomials, multiple orthogonal polynomials and Riemann-Hilbert problems. Inverse Probl. **26**, 055009 (15pp.) (2010)
8. Álvarez-Fernández, C., Fidalgo Prieto, U., Mañas, M.: Multiple orthogonal polynomials of mixed type: Gauss–Borel factorization and the multi-component 2D Toda hierarchy. Adv. Math. **227**, 1451–1525 (2011)
9. Álvarez-Fernández, C., Ariznabarreta, G., García-Ardila, J.C., Mañas, M., Marcellán, F.: Christoffel transformations for matrix orthogonal polynomials in the real line and the non-Abelian 2D Toda lattice hierarchy. Int. Math. Res. Notices **2016**, 1–57 (2016)
10. Araznibarreta, G., Mañas, M.: A Jacobi type Christoffel–Darboux formula for multiple orthogonal polynomials of mixed type. Linear Algebra Appl. **468**, 154–170 (2015)
11. Ariznabarreta, G., Mañas, M.: Matrix orthogonal Laurent polynomials on the unit circle and Toda type integrable systems. Adv. Math. **264**, 396–463 (2014)
12. Ariznabarreta, G., Mañas, M.: Multivariate orthogonal polynomials and integrable systems. Adv. Math. **302**, 628–739 (2016)

13. Ariznabarreta, G., Mañas, M.: Christoffel transformations for multivariate orthogonal polynomials. J. Approx. Theory **225**, 242–283 (2018)
14. Ariznabarreta, G., Mañas, M.: Multivariate Orthogonal Laurent Polynomials and Integrable Systems. Publications of the Research Institute for Mathematical Sciences (Kyoto University) (to appear)
15. Ariznabarreta, G., García-Ardila, J.C., Mañas, M., Marcellán, F.: Matrix biorthogonal polynomials on the real line: Geronimus transformations. Bull. Math. Sci. (2018). https://doi.org/10.1007/s13373-018-0128-y
16. Ariznabarreta, G., García-Ardila, J.C., Mañas, M., Marcellán, F.: Non-Abelian integrable hierarchies: matrix biorthogonal polynomials and perturbations. J. Phys. A Math. Theor. **51**, 205204 (2018)
17. Ariznabarreta, G., Mañas, M., Toledano, A.: CMV biorthogonal Laurent polynomials: perturbations and Christoffel formulas. Stud. Appl. Math. **140**, 333–400 (2018)
18. Bergvelt, M.J., ten Kroode, A.P.E.: Partitions, vertex operators constructions and multi-component KP equations. Pac. J. Math. **171**, 23–88 (1995)
19. Bueno, M.I., Marcellán, F.: Darboux transformation and perturbation of linear functionals. Linear Algebra Appl. **384**, 215–242 (2004)
20. Bueno, M.I., Marcellán, F.: Polynomial perturbations of bilinear functionals and Hessenberg matrices. Linear Algebra Appl. **414**, 64–83 (2006)
21. Christoffel, E.B.: Über die Gaussische Quadratur und eine Verallgemeinerung derselben. J. Reine Angew. Math. (Crelle's J.) **55**, 61–82 (1858, in German)
22. Date, E., Jimbo, M., Kashiwara, M., Miwa, T.: Transformation groups for soliton equations. Euclidean Lie algebras and reduction of the KP hierarchy. Publ. Res. Inst. Math. Sci. **18**, 1077–1110 (1982)
23. Gautschi, W.: Orthogonal Polynomials: Computation and Approximation. Oxford University Press, Oxford (2004)
24. Gel'fand, I.M., Gel'fand, S., Retakh, V.S., Wilson, R.: Quasideterminants. Adv. Math.**193**, 56–141 (2005)
25. Geronimus, J.: On polynomials orthogonal with regard to a given sequence of numbers and a theorem by W. Hahn. Izv. Akad. Nauk SSSR **4**, 215–228 (1940, in Russian)
26. Golinskii, L.: On the scientific legacy of Ya. L. Geronimus (to the hundredth anniversary). In: Priezzhev, V.B., Spiridonov, V.P. (eds.) Self-Similar Systems (Proceedings of the International Workshop (July 30–August 7, Dubna, Russia, 1998)), pp. 273–281. Publishing Department, Joint Institute for Nuclear Research, Moscow Region, Dubna (1999)
27. Grünbaum, F.A., Haine, L.: Orthogonal polynomials satisfying differential equations: the role of the Darboux transformation. In: Symmetries and Integrability of Difference Equations (Estérel, 1994). CRM Proceedings Lecture Notes, vol. 9, pp. 143–154. American Mathematical Society, Providence (1996)
28. Kac, V.G., van de Leur, J.W.: The n-component KP hierarchy and representation theory. J. Math. Phys. **44**, 3245–3293 (2003)
29. Mañas, M., Martínez-Alonso, L.: The multicomponent 2D Toda hierarchy: dispersionless limit. Inverse Probl. **25**, 115020 (2009)
30. Mañas, M., Martínez-Alonso, L., Álvarez-Fernández, C.: The multicomponent 2d Toda hierarchy: discrete flows and string equations. Inverse Probl. **25**, 065007 (2009)
31. Mulase, M.: Complete integrability of the Kadomtsev–Petviashvili equation. Adv. Math. **54**, 57–66 (1984)
32. Olver, P.J.: On multivariate interpolation. Stud. Appl. Math. **116**, 201–240 (2006)
33. Sato, M.: Soliton equations as dynamical systems on infinite dimensional Grassmann manifolds (random systems and dynamical systems). Res. Inst. Math. Sci. Kokyuroku **439**, 30–46 (1981)
34. Schwartz, L.: Théorie des noyaux. In: Proceedings of the International Congress of Mathematicians (Cambridge, MA, 1950), vol. 1, pp. 220–230. American Mathematical Society, Providence (1952)

35. Simon, B.: The Christoffel–Darboux kernel. In: Perspectives in Partial Differential Equations, Harmonic Analysis and Applications: A Volume in Honor of Vladimir G. Maz'ya's 70th Birthday. Proc. Sympos. Pure Math., vol. 79, pp. 295–336 (2008)
36. Szego, G.: Orthogonal Polynomials, vol. XXIII of American Mathematical Society Colloquium Publications, American Mathematical Society, New York (1939)
37. Uvarov, V.B.: The connection between systems of polynomials that are orthogonal with respect to different distribution function. USSR Comput. Math. Math. Phys. **9**, 25–36 (1969)
38. Zhang, F. (ed.): The Schur Complement and Its Applications. Springer, New York (2005)
39. Zhedanov, A.: Rational spectral transformations and orthogonal polynomials. J. Comput. Appl. Math. **85**, 67–86 (1997)

Infinite Matrices in the Theory of Orthogonal Polynomials

Luis Verde-Star

Abstract In this chapter we present a brief account of a matrix approach to the study of sequences of orthogonal polynomials. We use an algebra of infinite matrices of generalized Hessenberg type to represent polynomial sequences and linear maps on the complex vector space of all polynomials. We show how the matrices are used to characterize and to construct several sets of orthogonal polynomials with respect to some linear functional on the space of polynomials. The matrices allow us to study several kinds of generalized difference and differential equations, to obtain explicit formulas for the orthogonal polynomial sequences with respect to given bases, and also to obtain formulas for the coefficients of the three-term recurrence relations. We also construct a family of hypergeometric orthogonal polynomials that contains all the families in the Askey scheme and a family of basic hypergeometric q-orthogonal polynomials that contains all the families in the q-Askey scheme.

1 Introduction

In this Chapter we present a brief description of a matrix approach to some aspects of the theory of orthogonal polynomials. We present some basic ideas and a few examples of results that show the usefulness of the approach. We begin here with some remarks about the theory and the applications of infinite matrices.

Infinite matrices of several types have been studied for a long time. They are usually seen as representations of linear operators between vector spaces that have countable bases, or as operators that send sequences to sequences, as it is done in Summability Theory, or that send power series to power series, as in Complex Analysis. Infinite matrices were used often in the development of Quantum Mechanics, considered as linear transformations on a separable Hilbert space. After

L. Verde-Star (✉)
Department of Mathematics, Universidad Autónoma Metropolitana, Mexico City, Mexico
e-mail: verde@xanum.uam.mx

F. Marcellán, E. J. Huertas (eds.), *Orthogonal Polynomials: Current Trends and Applications*, SEMA SIMAI Springer Series 22,
https://doi.org/10.1007/978-3-030-56190-1_11

the appearance of Functional Analysis and Operator Theory, around 1930, the interest in infinite matrices decayed greatly. Another reason for the reduced interest in infinite matrices that persisted for a long time is the fact that most Linear Algebra textbooks only deal with finite dimensional spaces, and most of the material about infinite dimensional spaces is now considered as a part of Functional Analysis.

Some algebraic aspects of the theory of spaces of infinite matrices were studied between 1900 and 1930 by many researchers, such as Hurwitz, Hellinger, Dienes, Riesz, Kothe, and Toeplitz. After that time the analytic aspects of the theory were the main subject of study, mainly due to the interest in Hilbert space operators and Quantum Mechanics. From around 1950 to 1980 infinite matrices were used in very few areas of Mathematics, such as Summability theory and Combinatorics. Infinite Toeplitz matrices have been used for a long time as representations of multiplication operators in some areas of Analysis and Functional Analysis that are very active research areas nowadays. A more recent development is the use of infinite matrices in the theories of Signal and Image processing and Data communication, where the infinite matrices represent transformations of sequences, and are replaced by large finite matrices in the practical applications.

The Pascal triangle is certainly the oldest and most studied infinite matrix, but the study of Pascal-like matrices from the linear algebra perspective started quite recently. There was a renewed interest on linear algebraic properties of finite Pascal matrices and some related matrices around 1980, when some papers on the subject were published. Some of them were motivated by combinatorial problems and others by the group structure of several types of generalized Pascal matrices. There are numerous papers published during the last 30 years where the authors prove, by means of explicit matrix computations, that the matrices $P(x) = [\binom{n}{k} x^{n-k}]$ satisfy $P(x)P(y) = P(x + y)$. The interpretation of Pascal matrices as the matrix representation of translations on the space of polynomials was used by Kalman [12] in 1983. See also [13]. Since the set of translations is obviously a commutative group so is the set of their matrix representations. Some papers about generalizations of Pascal matrices are [9, 26], and [27].

There are some recent papers on the use of matrices and Linear Algebra to study polynomials and polynomial sequences, for example, [2, 3, 6–8], and [30].

In the theory of orthogonal polynomials there are many areas that can be studied using linear algebraic methods, for example, recurrence relations, connection constants, difference and differential equations, moments, and discrete orthogonality. Of course there are also areas that require the use of analytical methods, such as asymptotic properties and perturbation of measures.

It is well-known that the infinite tridiagonal Jacobi matrices associated with orthogonal sequences determine most of the properties of the sequences, since their entries are the coefficients of the three-term recurrence relation, but in the development of the theory of orthogonal polynomials the Jacobi matrices have rarely been studied as elements of an algebra of infinite matrices.

In [20] Maroni proposed an algebraic approach to study orthogonal polynomials based on the use of linear functionals and duality. His approach has been used with great success by numerous researchers.

In our previous papers [28] and [29] we used infinite matrices to study some aspects of the theory of orthogonal polynomials. Our approach can be considered as a complement to Maroni's approach and has been used recently by some researchers. See [10, 11, 19], and [22].

In this chapter we use an algebra of infinite Hessenberg matrices as a setting for the study of polynomial sequences and linear operators on the space of polynomials. Some results that are obtained using our approach, that we don't include in this chapter, can be found in [28] and [29].

The algebra of infinite triangular matrices is the set of infinite matrices that has received more attention from the researchers in the Algebra community. There are some recent articles on the subject, see for example, [23, 24], and [25], and the references therein.

2 The Algebra of Infinite Lower Hessenberg Matrices

We introduce in this section the algebra $\mathcal{L}$ of infinite lower Hessenberg matrices that is an appropriate algebraic setting for the study of polynomial sequences and linear operators such as difference and differential operators on the space of polynomials.

We denote by $\mathcal{L}$ the set of all the infinite matrices $A = [a_{j,k}]$ with complex entries and indices j and k in $\mathbb{N}$ for which there is an integer $m \in \mathbb{Z}$ such that $a_{j,k} = 0$ if $j - k < m$. We say that $a_{j,k}$ lies in the n-th diagonal of A if $j - k = n$. Therefore, if $m > n$ then the m-th diagonal lies below (to the left of) the n-th diagonal. We say that a nonzero element of $\mathcal{L}$ is a diagonal matrix if all of its nonzero elements lie in a single diagonal.

Let A be a nonzero element of $\mathcal{L}$ and let m be the minimum integer such that A has at least one nonzero entry in the m-th diagonal, then we say that A has *index* m and write $i(A) = m$. The index of the zero matrix is defined as infinity.

It is easy to see that $\mathcal{L}$ is a complex vector space with the usual operations of addition and multiplication by scalars. If $A = [a_{j,k}]$ and $B = [b_{j,k}]$ are in $\mathcal{L}$, with $i(A) = m$ and $i(B) = n$, then the usual matrix multiplication of A and B is well defined and the product $C = AB$ is in $\mathcal{L}$, since

$$c_{r,k} = \sum_{j=k+n}^{r-m} a_{r,j} b_{j,k}, \qquad r - k \geq m + n. \tag{2.1}$$

Note that for every pair (r, k) there is a finite number of summands and that $i(AB) \geq i(A) + i(B)$.

If $A \in \mathcal{L}$ then a sufficient, but not necessary, condition for A to have a two-sided inverse is that $i(A) = 0$ and $a_{k,k} \neq 0$ for $k \geq 0$. We denote by $\mathcal{G}$ the set of all matrices that satisfy such condition. $\mathcal{G}$ is a group under matrix multiplication. The unit is the identity matrix I whose entries on the 0-th diagonal are equal to 1 and all other entries are zero.

We define next some particular elements of $\mathcal{L}$ that will be used in the rest of the chapter. Denote by S the diagonal matrix of index 1 with $(S)_{j+1,j} = 1$ for $j \geq 0$. Denote by $\hat{S}$ the transpose of S. Note that $\hat{S}$ is diagonal with index -1 and $\hat{S}S = I$ and $S\hat{S} = J$, where J is the diagonal of index zero with its entry in the position $(0, 0)$ equal to zero and its entries in positions (j, j) equal to 1 for $j \geq 1$. That is, J differs from the identity I only in the position $(0, 0)$. Therefore $\hat{S}$ is a left-inverse for S but not a right-inverse. If m is a positive integer then S^m is the diagonal matrix of index m with all its entries in the m-th diagonal equal to 1. Analogously, $\hat{S}^m$ is diagonal of index $-m$ and all its entries in the $-m$-th diagonal are equal to 1.

We denote by D the diagonal matrix of index 1 with entries $d_{j,j-1} = j$ for $j \geq 1$. Denote by $\hat{D}$ the diagonal matrix of index -1 with entries $\hat{d}_{j-1,j} = 1/j$ for $j \geq 1$. That is,

$$D = \begin{bmatrix} 0 & 0 & 0 & 0 & \dots \\ 1 & 0 & 0 & 0 & \dots \\ 0 & 2 & 0 & 0 & \dots \\ 0 & 0 & 3 & 0 & \dots \\ \vdots & \vdots & \vdots & \ddots & \ddots \end{bmatrix}, \qquad \hat{D} = \begin{bmatrix} 0 & 1 & 0 & 0 & \dots \\ 0 & 0 & 1/2 & 0 & \dots \\ 0 & 0 & 0 & 1/3 & \dots \\ 0 & 0 & 0 & 0 & \ddots \\ \vdots & \vdots & \vdots & \vdots & \ddots \end{bmatrix}. \tag{2.2}$$

It is easy to see that $\hat{D}D = I$ and $D\hat{D} = J$. Let $A \in \mathcal{G}$. Then a simple computation shows that $\hat{S}A^{-1}S$ is the two-sided inverse of $\hat{S}AS$ and also that $\hat{D}A^{-1}D$ is the two-sided inverse of $\hat{D}AD$. Note that $\hat{S}AS$ is obtained from A by shifting the matrix one position to the left and one position upward, so that the 0-th row and the 0-th column of A disappear.

We say that $A \in \mathcal{L}$ is *banded* if there exist a pair of integers (n, m) such that $n \leq m$ and all the nonzero entries of A lie between the diagonals of indices n and m. In such case we say that A is (n, m)-banded. Note that the set of banded matrices is closed under addition and multiplication, but a banded matrix may have an inverse that is not banded. The transpose of a (n, m)-banded matrix is a $(-m, -n)$-banded matrix.

The *Toeplitz matrices* in $\mathcal{L}$ are the elements of the form

$$T = \sum_{k=0}^{\infty} c_k S^k + \sum_{k=1}^{m} c_{-k} \hat{S}^k, \tag{2.3}$$

where the coefficients c_j are complex numbers and m is a positive integer. Note that, if $c_{-m} \neq 0$ then $i(T) = -m$. The 0-th column of T is $[c_0, c_1, c_2, \dots]^{\mathsf{T}}$ and the 0-th row is $[c_0, c_{-1}, c_{-2}, \dots, c_{-m}, 0, 0, \dots]$.

Theorem 1 *Let $A \in \mathcal{L}$ with $i(A) = n$ such that $A\hat{S} = \hat{S}A$, Then we have, if $n \leq 0$ then A is a polynomial in $\hat{S}$ of degree $-n$, and if $n \geq 1$ then $A = 0$.*

Proof Since $\hat{S}S = I$ the equation $A\hat{S} = \hat{S}A$ gives $A = \hat{S}AS$. This equation shows that A must be a Toeplitz matrix. The 0-th column of $\hat{S}A$ is $[a_{1,0}, a_{2,0}, a_{3,0}, \ldots]^{\mathsf{T}}$. It is the 0-th column of A shifted one place upwards. The 0-th column of $A\hat{S}$ is zero and then $a_{j,0} = 0$ for $j \geq 1$. Since A is Toeplitz, all the entries of A below the diagonal of index 0 must be zero. Therefore, if $n > 0$ then $A = 0$ and if $n \leq 0$ we must have

$$A = \sum_{k=0}^{-n} a_{0,k}\hat{S}^k. \tag{2.4}$$

□

If $A \in \mathcal{L}$ has index m, we say that A is *monic* if all the entries in the diagonal of index m are equal to 1.

We say that two elements A and B of $\mathcal{L}$ are *diagonally similar* if and only if there exists an invertible diagonal matrix T of index 0 such that $B = T^{-1}AT$.

We denote by $\mathcal{T}$ the set of $(-1, 1)$-banded matrices whose entries in the diagonals of indices 1 and -1 are all nonzero. The elements of $\mathcal{T}$ are usually called Jacobi matrices and have index equal to -1.

We list some properties of $\mathcal{T}$ in the following proposition. The proofs are straightforward. See [28].

Theorem 2

(a) *If $L \in \mathcal{T}$ then L is diagonally similar to a monic element of $\mathcal{T}$.*
(b) *If $L \in \mathcal{T}$ is monic then there is a unique monic $A \in \mathcal{G}$ such that $A^{-1}LA = \hat{S}$.*
(c) *If L and M are in $\mathcal{T}$ then they are similar to each other.*
(d) *If $L \in \mathcal{T}$ then L is diagonally similar to L^{T}.*
(e) *If $L \in \mathcal{T}$ is monic and has positive entries in the diagonal of index 1 then L is diagonally similar to a symmetric element of $\mathcal{T}$.*

3 Polynomial Sequences and Their Matrix Representation

In this section we present some definitions related with sequences of polynomials and describe their connection with elements of $\mathcal{L}$.

By a *triangular polynomial sequence* we mean a sequence of polynomials $p_0, p_1, p_2, \ldots$ with complex coefficients such that p_n has degree n for $n \geq 0$. Every triangular polynomial sequence is a basis for the complex vector space $\mathcal{P}$ of all polynomials in one variable.

Let $\{p_n\}$ be a triangular polynomial sequence and let A be a nonzero element of $\mathcal{L}$ with $i(A) = m$. We can associate each row of A with a polynomial as follows. Define

$$q_k(t) = \sum_{j=0}^{k-m} a_{k,j}p_j(t), \qquad k \geq 0, \tag{3.1}$$

where empty sums are zero. This means that the k-th row of A is the vector of coefficients of q_k with respect to the basis $\{p_j\}$. If the matrix A is in the group $\mathcal{G}$ then $\{q_k\}$ is a triangular polynomial sequence and thus it is a basis for $\mathcal{P}$.

The sequence of scalar equations (3.1) is equivalent to the vector-matrix equation

$$[q_0, q_1, q_2, \ldots]^{\mathsf{T}} = A\,[p_0, p_1, p_2, \ldots]^{\mathsf{T}}. \tag{3.2}$$

We say that A is the matrix representation of the sequence q_n with respect to the basis $\{p_n\}$.

From (3.2) we see that each matrix in $\mathcal{L}$ can be considered as a linear transformation on the vector space $\mathcal{S}$ of sequences of polynomials $\{u_k\}$ for which the set of numbers $\{\text{degree}(u_k) - k : k \geq 0\}$ has an upper bound in $\mathbb{N}$.

The matrices S and $\hat{S}$, defined in Sect. 2, satisfy

$$\hat{S}[u_0, u_1, u_2, \ldots]^{\mathsf{T}} = [u_1, u_2, u_3, \ldots]^{\mathsf{T}}, \tag{3.3}$$

and

$$[u_0, u_1, u_2, \ldots]\hat{S} = [0, u_0, u_1, u_2, \ldots], \tag{3.4}$$

for any sequence of polynomials $u_0, u_1, u_2, \ldots$. Taking transposes in the previous equations we obtain

$$[u_0, u_1, u_2, \ldots]S = [u_1, u_2, u_3, \ldots], \tag{3.5}$$

and

$$S[u_0, u_1, u_2, \ldots]^{\mathsf{T}} = [0, u_0, u_1, u_2, \ldots]^{\mathsf{T}}. \tag{3.6}$$

Let γ be a linear operator on the space $\mathcal{P}$ of polynomials and let $\{p_k(t) : k \in \mathbb{N}\}$ be a triangular basis for $\mathcal{P}$. The matrix representation of γ with respect to the basis $\{p_k(t)\}$ is the matrix $G = [g_{k,j}]$ where

$$\gamma(p_k(t)) = \sum_{j \geq 0} g_{k,j}\, p_j(t), \qquad k \geq 0. \tag{3.7}$$

Let us note that in the general case G may not be an element of $\mathcal{L}$, but there are many important linear operators on the space $\mathcal{P}$ that have matrix representations in $\mathcal{L}$. If $A \in \mathcal{L}$ is the matrix representation of a sequence of polynomials $\{u_k(t) : k \in \mathbb{N}\}$ in $\mathcal{S}$ with respect to the basis $\{p_k(t)\}$ and $G \in \mathcal{L}$ then AG is the matrix representation of the sequence $\{\gamma(u_k(t)) : k \in \mathbb{N}\}$ with respect to the same basis. Note that the multiplication by G is on the right-hand side. For example, the matrix D is the representation of the usual differential operator with respect to the basis of monomials and then if A is the representation of the sequence $u_k(t)$ with respect to the basis of monomials then AD is the representation of the sequence $\{u'_k(t)\}$ with respect to the same basis.

We define next an important class of triangular bases of $\mathcal{P}$. Let $x_0, x_1, x_2, \ldots$ be a sequence of complex numbers. Define the polynomials $v_0(t) = 1$, and

$$v_k(t) = (t - x_0)(t - x_1)(t - x_2)\cdots(t - x_{k-1}), \qquad k \geq 1. \tag{3.8}$$

The sequence of monic polynomials $\{v_n\}$ is clearly a triangular basis for the space $\mathcal{P}$ of polynomials. It is called the *Newton basis* associated with the sequence x_n. Let V be the infinite matrix that satisfies

$$[v_0(t), v_1(t), v_2(t), \ldots]^{\mathsf{T}} = V[1, t, t^2, \ldots]^{\mathsf{T}}. \tag{3.9}$$

V is in $\mathcal{G}$ and its entries in the n-row are the coefficients of $v_n(t)$ with respect to the basis of monomials. Therefore they are elementary symmetric functions of $x_0, x_1, x_2, \ldots, x_{n-1}$, with appropriate signs. The entries in V^{-1} are complete homogeneous symmetric functions of the x_k. The dual basis of the Newton basis $\{v_n\}$ is the sequence of divided difference functionals $\Delta[x_0, x_1, \ldots, x_{n-1}]$, which give us the coefficients in the representation of any polynomial in terms of the Newton basis.

From (3.8) we see that

$$t v_k(t) = (t - x_k + x_k) v_k(t) = v_{k+1}(t) + x_k v_k(t), \qquad k \geq 0, \tag{3.10}$$

and therefore the matrix X_v that represents the multiplication map $u(t) \to t u(t)$ on the space $\mathcal{P}$ with respect to the Newton basis $\{v_n\}$ is

$$X_v = \begin{bmatrix} x_0 & 1 & 0 & 0 & 0 & \cdots \\ 0 & x_1 & 1 & 0 & 0 & \cdots \\ 0 & 0 & x_2 & 1 & 0 & \cdots \\ 0 & 0 & 0 & x_3 & 1 & \cdots \\ \vdots & \vdots & \vdots & \vdots & \ddots & \ddots \end{bmatrix}. \tag{3.11}$$

When all the x_n are zero the basis $\{v_k\}$ becomes the basis $\{t^k\}$ of monomials and X_v becomes $\hat{S}$.

4 Orthogonal Polynomial Sequences

In this section we show how the matrices in $\mathcal{L}$ can be used to study sequences of orthogonal polynomials.

Let $\mu : \mathcal{P} \to \mathbb{C}$ be a linear functional. A triangular sequence of monic polynomials $\{p_n\}_{n\geq 0}$ is *orthogonal* with respect to μ if and only if

$$\mu(p_k\, p_n) = \gamma_k \delta_{k,n}, \qquad k, n \geq 0, \tag{4.1}$$

where γ_k is a nonzero number for $k \geq 0$ and $\delta_{k,n}$ is Kronecker's function.

Since $\{p_n\}_{n\geq 0}$ is a basis for $\mathcal{P}$, if $n \geq 0$ then the monomial t^n is a linear combination of $p_0(t), p_1(t), p_2(t), \ldots, p_n(t)$ and then it is easy to verify that the orthogonality condition (4.1) is equivalent to

$$\mu(t^n\, p_k(t)) = \begin{cases} 0, & \text{if } 0 \leq n < k \\ \gamma_k, & \text{if } n = k \end{cases}. \tag{4.2}$$

Theorem 3 *Let $\{v_k(t)\}_{k\geq 0}$ be the Newton basis associated with a sequence x_k. Let $A \in \mathcal{G}$ be the matrix representation of a triangular sequence of monic polynomials $\{p_k\}$ with respect to $\{v_k\}$, and let X_v be as defined in the previous section. Then $\{p_k\}$ is orthogonal with respect to some linear functional if and only if the matrix $L = AX_vA^{-1}$ is an element of $\mathcal{T}$.*

Proof Suppose that $\{p_k\}$ is orthogonal with respect to the functional μ. Let $m_k = \mu(v_k)$ for $k \geq 0$ and let $g = [m_0, m_1, m_2, \ldots]^{\mathsf{T}}$. For $j \geq 0$ let e_j denote the infinite column vector with its j-th entry equal to 1 and all other entries equal to zero. From the definition of L we get $L^nA = AX_v^n$ for $n \geq 0$. Therefore $L^nAg = AX_v^ng$ for $n \geq 0$. The k-th entry of AX_v^ng is the product of the k-th row of A and the vector X_v^ng. From (3.10) we get $\mu(tv_k(t)) = \mu(v_{k+1}) + x_k\mu(v_k(t))$, and by (3.11) we see that $X_vg = [\mu(tv_0(t)), \mu(tv_1(t)), \mu(tv_2(t)), \ldots]^{\mathsf{T}}$ and then $X_v^ng = [\mu(t^nv_0(t)), \mu(t^nv_1(t)), \mu(t^nv_2(t)), \ldots]^{\mathsf{T}}$. Therefore the k-th entry of AX_v^ng is

$$\sum_{j=0}^{k} a_{k,j}\mu(t^nv_j(t)) = \mu(t^np_k(t)),$$

which, by (4.2) is zero if $k > n$ and it is γ_n if $k = n$. Therefore $AX_v^ng = L^nAg$ is in the subspace generated by $\{e_0, e_1, \ldots, e_n\}$. By the orthogonality hypothesis $Ag = \gamma_0e_0$ and then $\gamma_0L^ne_0$ is in the subspace generated by $\{e_0, e_1, \ldots, e_n\}$ and there exist coefficients $c_{n,j}$ such that

$$\gamma_0L^ne_0 = \gamma_ne_n + \sum_{j=0}^{n-1} c_{n,j}e_j, \qquad n \geq 0. \tag{4.3}$$

Multiplying on the left by L and using (4.3) with $n+1$ instead of n we get

$$\gamma_0 L^{n+1} e_0 = \gamma_n L e_n + L \sum_{j=0}^{n-1} c_{n,j} e_j = \gamma_{n+1} e_{n+1} + \sum_{j=0}^{n} c_{n+1,j} e_j.$$

Therefore, solving for Le_n we obtain

$$Le_m = \frac{\gamma_{n+1}}{\gamma_n} e_{n+1} + \sum_{j=0}^{n} d_{n,j} e_j,$$

for some coefficients $d_{n,j}$. Therefore the entries of L below the diagonal of index 1 are equal to zero and the entries in the diagonal of index 1 are the quotients γ_{n+1}/γ_n, which are nonzero. On the other hand, since X_v has index -1 and $L = AX_vA^{-1}$ it is clear that L has index -1 and its diagonal of index -1 coincides with $\hat{S}$. Therefore we have proved that L is in $\mathcal{T}$.

Now suppose that $\{p_k\}$ is a sequence of monic polynomials with matrix representation A, and such that $L = AX_vA^{-1}$ is in $\mathcal{T}$. Let α_{n+1} be the $(n+1, n)$ entry of L and let g be the 0-th column of A^{-1}. Since L is $(-1, 1)$-banded then L^n is $(-n, n)$-banded and has nonzero entries in its diagonal of index n. From the equations $AX_v^n g = L^n Ag = L^n e_0$ we see that the orthogonality condition (4.2) is satisfied and therefore the sequence $\{p_k\}$ is orthogonal with respect to the functional μ defined by $\mu(v_j) = (A^{-1})_{j,0}$ for $j \geq 0$.

From $L^n e_0 = AX_v^n g$ we obtain $e_0^{\mathsf{T}} L^n e_0 = (e_0^{\mathsf{T}} A) X_v^n g = e_0^{\mathsf{T}} X_v^n g$ and this means that the $(0, 0)$ entry of L^n is equal to $\mu(v_n)$, for $n \geq 0$. The numbers $\mu(v_n)$ are called the *generalized moments* of μ with respect to the basis $\{v_k\}$. □

The matrix L of the previous theorem has the form

$$L = \begin{bmatrix} \beta_0 & 1 & 0 & 0 & \ldots \\ \alpha_1 & \beta_1 & 1 & 0 & \ldots \\ 0 & \alpha_2 & \beta_2 & 1 & \ldots \\ \vdots & \vdots & \vdots & \vdots & \ddots \end{bmatrix}, \tag{4.4}$$

where

$$\alpha_k = \frac{\gamma_k}{\gamma_{k-1}}, \qquad k \geq 1.$$

The matrix equation $LA = AX_v$ gives us, by equating the k-th rows in both sides, the three-term recurrence relation

$$\alpha_k p_{k-1}(t) + \beta_k p_k(t) + p_{k+1}(t) = t p_k(t), \qquad k \geq 1. \tag{4.5}$$

Notice that $p_0(t) = 1$ and $p_1(t) = t - \beta_0$.

Let $B \in \mathcal{G}$ and $[u_0, u_1, u_2, \ldots] = [v_0, v_1, v_2, \ldots]B$. Then $\{u_k\}$ is also a triangular basis of $\mathcal{P}$ and AB is the matrix representation of $\{p_k\}$ with respect to $\{u_k\}$. Therefore $LAB = ABB^{-1}X_vB$, and it is easy to verify that $B^{-1}X_vB$ is the matrix representation of the multiplication map $w(t) \to tw(t)$ with respect to the basis $\{u_k\}$. This means that the matrix L doesn't change when we use the basis $\{u_k\}$ instead of the basis $\{v_k\}$ to represent the sequence $\{p_k\}$ and the multiplication map.

From Theorem 2, part b) and Theorem 3 we see that each pair of sequences of numbers $(\{\alpha_k\}_{k\geq 1}, \{\beta_k\}_{k\geq 0})$, with $\alpha_k \neq 0$, determines a unique triangular sequence of monic orthogonal polynomials with respect to some linear functional. An important problem is to characterize and to find properties of orthogonal sequences that satisfy certain functional equations, such as differential equations and difference equations, and several generalizations of them. See [1, 4, 15–18], and [21].

We present next an example of a characterization theorem expressed in terms of matrices. Let A be the matrix representation of a sequence of monic orthogonal polynomials $\{p_k\}$ with respect to the basis of monomials. Then $\hat{D}AD$ is the matrix representation of the sequence of monic polynomials $p'_{k+1}(t)/(k+1)$ and it is a monic element of $\mathcal{G}$. The orthogonal sequence $\{p_k\}$ is called *classical* if the sequence $\{p'_{k+1}\}$ is also orthogonal.

Theorem 4 *Let A and $\{p_k\}$ be as defined above. Then $\{p_k\}$ is classical if and only if $A(\hat{D}AD)^{-1}$ is a (0,2)-banded element of $\mathcal{G}$.*

For the proof see [28, Sec. 4].

5 Orthogonal Polynomial Sequences That Satisfy a Generalized First-Order Difference Equation

We introduce a family of matrices that represent operators that generalize the usual differentiation and difference operators. We consider some simple eigenvalue-difference equations and look for orthogonal polynomial solutions of such equations.

From now on by polynomial sequence we mean triangular polynomial sequence and we will consider triangular bases of $\mathcal{P}$. We denote by $\mathcal{D}_0$ the set of diagonal matrices of index 0. For each sequence $s : \mathbb{N} \to \mathbb{C}$ we define the matrix $M(s)$ as the element of $\mathcal{D}_0$ whose (n, n) entry is $s(n)$, for $n \geq 0$.

Definition Let $\{v_k\}$ be a basis of $\mathcal{P}$ and let s be a complex valued sequence. We say that the operator represented by the matrix $M(s(n))S$ with respect to the basis $\{v_k\}$ is a *generalized difference operator of first order*. It is the operator that sends v_n to $s(n)v_{n-1}$, for $n \geq 1$, and sends v_0 to zero.

For example, if $s(n) = n$ for $n \geq 0$ then $M(s(n))S = D$ and represents the usual differentiation operator with respect to the basis of monomials, and represents the usual difference operator with step h with respect to the Newton basis associated with the sequence $x_n = nh$, where h is a fixed nonzero number.

With respect to $\{v_k\}$ the matrix $M(s(n))$ represents the operator that sends v_n to $s(n)v_n$, for $n \geq 0$, and we say that it is a *generalized differential operator of order zero*. It is essentially a rescaling of the sequence $\{v_n\}$.

It is easy to see that $M(s(n))S = SM(\hat{s}(n))$ where $\hat{s}$ is the shifted sequence defined by $\hat{s}(n) = s(n+1)$. In order to simplify the notation we use algebraic expressions in the discrete variable n to denote the corresponding sequences, for example, $M(n^2)$ is the diagonal sequence of index zero with (n, n) entry equal to n^2, and if g is a known sequence then $M(ng(n))$ has its (n, n) entry equal to $ng(n)$.

Let f, g and h be complex valued sequences. Let $\{v_k(t)\}_{k\geq 0}$ be the Newton basis of $\mathcal{P}$ associated with the sequence $x_n = f(n)$. We define the matrix

$$B = M(h(n)) + SM(g(n)). \tag{5.1}$$

B is $(0, 1)$-banded and it is the representation of an operator δ which is the sum of a generalized difference operator of order one and a generalized difference operator of order zero. The operator δ acts on the basis $\{v_k(t)\}$ as follows

$$\delta(v_k(t)) = g(k-1)v_{k-1}(t) + h(k)v_k(t), \qquad k \geq 0, \tag{5.2}$$

where $g(-1) = 0$.

Problem Find all the orthogonal polynomial sequences $\{u_k(t)\}$ that satisfy the generalized difference equation

$$\delta(u_k(t)) = h(k)u_k(t), \qquad k \geq 0, \tag{5.3}$$

for some sequences f, g and h.

This problem can be expressed in terms of matrix equations as follows. Let $C \in \mathcal{G}$ be the matrix representation of the polynomial sequence $\{u_k(t)\}$ with respect to the Newton basis $\{v_k(t)\}$. Let X_v be the matrix representation of the operator of multiplication by t on the space $\mathcal{P}$ with respect to the basis $\{v_k(t)\}$. By Theorem 3 the sequence $\{u_k(t)\}$ is orthogonal if

$$L = CX_vC^{-1} \in \mathcal{T}, \tag{5.4}$$

and the matrix expression of (5.3) is

$$CB = M(h(n))C. \tag{5.5}$$

The problem is now to find sequences f, g, and h such that (5.4) and (5.5) are satisfied. One natural approach is to look for simple sequences, for example, constant sequences. A larger family of simple sequences is the set of linearly recurrent sequences. They are the solutions of linear difference equations with constant coefficients and satisfy an equation of the form

$$s(n+m) = \sum_{j=0}^{m-1} d_j s(n+j), \qquad n \geq 0, \tag{5.6}$$

where $m \geq 1$ and the d_j are numbers. The sequence s is determined by the coefficients d_j and the initial values $s(0), s(1), s(2), \ldots, s(m-1)$ and $s(n)$ can be expressed as a linear combination of functions of the form $p_j(n)\lambda_j^n$, where the p_j are polynomials.

Since B is bidiagonal, the matrix equation (5.5) can be easily solved for $C = [c_{k,j}]$ in terms of g and h. The entries of C are given by

$$c_{k,k} = 1, \qquad k \geq 0, \tag{5.7}$$

and

$$c_{k,j} = \frac{g(j)g(j+1)\cdots g(k-1)}{(h(k)-h(j))(h(k)-h(j+1))\cdots(h(k)-h(k-1))}, \qquad 0 \leq j < k. \tag{5.8}$$

The previous equation shows that if Eq. (5.5) has a solution C then the sequence h must satisfy $h(j) \neq h(k)$ if $j \neq k$.

The inverse matrix $C^{-1} = [m_{k,j}]$ is also easy to determine. We have

$$m_{k,k} = 1, \qquad k \geq 0, \tag{5.9}$$

and

$$m_{k,j} = \frac{g(j)g(j+1)\cdots g(k-1)}{(h(j)-h(j+1))(h(j)-h(j+2))\cdots(h(j)-h(k))}, \qquad 0 \leq j < k. \tag{5.10}$$

If $\{u_k(t)\}$ is orthogonal with respect to a linear functional μ and C is the matrix representation of $\{u_k(t)\}$ with respect to the basis $\{v_k(t)\}$, then the entries $m_{k,0}$ are the generalized moments, that is, $m_{k,0} = \mu(v_k)$ for $k \geq 0$. The standard moments are easily obtained using a change of basis matrix.

6 Hypergeometric Orthogonal Polynomial Sequences

Here we present a solution of (5.4) and (5.5) obtained with sequences f, g, and h that are polynomials in n of small degrees. It turns out that the solution that we obtain in this way is a family of hypergeometric orthogonal polynomial sequences that contains all the families included in the Askey scheme. The coefficients of f, g, and h are the parameters of the family. See [14].

Define the polynomials

$$f(z) = w_0 + w_1 z + w_2 z^2, \quad h(z) = (y_1 + y_2 z)z, \quad g(z) = (1 + e_1 z + e_2 z^2 + e_3 z^3)(z+1). \tag{6.1}$$

Then the sequences $f(n), h(n), g(n)$, for $n \geq 0$, are linearly recurrent. For example, $f(n)$ satisfies

$$f(k+1) - 3f(k) + 3f(k-1) - f(k-2) = 0, \qquad k \geq 2. \tag{6.2}$$

From the Eqs. (5.7) and (5.8) we can find the entries of C in terms of the coefficients y_1, y_2, e_1, e_2, e_3. Then we substitute C in Eq. (5.4) and find that a necessary condition for L to be in $\mathcal{T}$ is

$$e_3 = w_2 y_2, \qquad e_2 = w_1 y_2 + w_2 y_1 + w_2 y_2. \tag{6.3}$$

This shows that the polynomials f, g, h must be related in a particular way. It turns out that the conditions in (6.3) are also sufficient for L to be in $\mathcal{T}$. Therefore there remain 6 parameters: $e_1, y_1, y_2, w_0, w_1, w_2$. A change in the value of w_0 corresponds to a translation of the basis polynomials $v_k(t)$ and, consequently, to a translation of the orthogonal polynomials $u_k(t)$.

The polynomials $u_k(t)$ have a quite simple explicit expression as linear combinations of the polynomials $v_k(t)$.

It is convenient to define the polynomials $\tilde{h}(n) = y_1 + y_2 n$ and $\tilde{g}(n) = 1 + e_1 n + e_2 n^2 + e_3 n^3$. Then we define the coefficients

$$b_{n,k} = \binom{n}{k} \frac{\tilde{h}(n)\tilde{h}(n+1)\cdots\tilde{h}(n+k-1)}{\tilde{g}(0)\tilde{g}(1)\cdots\tilde{g}(k-1)}, \qquad 0 \leq k \leq n. \tag{6.4}$$

The orthogonal polynomials are given by

$$u_n(t) = \frac{1}{b_{n,n}} \sum_{k=0}^{n} b_{n,k} v_k(t), \qquad n \geq 0. \tag{6.5}$$

The matrix $C = [c_{n,k}]$, which is the matrix representation of the sequence $\{u_k(t)\}$ with respect to the basis $\{v_k(t)\}$, is therefore given by

$$c_{n,k} = \frac{b_{n,k}}{b_{n,n}} = \binom{n}{k} \frac{\tilde{g}(k)\tilde{g}(k+1)\cdots\tilde{g}(n-1)}{\tilde{h}(n+k)\tilde{h}(n+k+1)\cdots\tilde{h}(2n-1)}, \qquad 0 \leq k \leq n. \tag{6.6}$$

We also have explicit formulas for the entries of the matrix L, which are the coefficients of the three-term recurrence relation satisfied by the polynomials $u_k(t)$. Let L be as in (4.4) and let

$$\sigma_n = \sum_{j=0}^{n} \beta_j, \qquad n \geq 0. \tag{6.7}$$

Note that $\beta_n = \sigma_n - \sigma_{n-1}$ for $n \geq 1$, and $\beta_0 = \sigma_0$.

Using the explicit formulas for the entries of C and C^{-1} we can compute the entries of L. They turn out to be rational functions of n, and their explicit expressions are

$$\alpha_n = \frac{-ny_2\,\tilde{h}(n-1)\,\tilde{g}(n-1)\,\tilde{g}(-\tilde{h}(n)/y_2)}{\tilde{h}(2n-2)\tilde{h}^2(2n-1)\tilde{h}(2n)}, \qquad n \geq 1. \tag{6.8}$$

and

$$\sigma_n = \frac{-(n+1)\,s(n)}{6\,\tilde{h}(2n+1)}, \qquad n \geq 0, \tag{6.9}$$

where $s(n)$ is the polynomial

$$s(n) = \gamma_3 n^3 + \gamma_2 n^2 + \gamma_1 n + \gamma_0, \qquad n \geq 0, \tag{6.10}$$

and the coefficients are given by

$$\gamma_3 = 2w_2 y_2, \qquad \gamma_2 = 2w_2(2y_1 + y_2),$$

$$\gamma_1 = 6e_1 - 12w_0 y_2 - \tilde{h}(1)(3w_1 + w_2), \qquad \gamma_0 = 6(1 - w_0\tilde{h}(1)). \tag{6.11}$$

From (6.9) we obtain immediately the coefficients $\beta_n = \sigma_n - \sigma_{n-1}$.

The numerator of α_n in (6.8) is a polynomial in n of degree at most 8, and the denominator is a polynomial of degree at most 4. For the sequence σ_n the numerator has degree at most 4 and the denominator has degree at most one.

The generalized moments are given by

$$\mu(v_k) = (C^{-1})_{k,0} = \frac{(-1)^k\tilde{g}(0)\tilde{g}(1)\cdots\tilde{g}(k-1)}{\tilde{h}(1)\tilde{h}(2)\cdots\tilde{h}(k)}, \qquad k \geq 0. \tag{6.12}$$

Note that the generalized moments satisfy a simple two-term recurrence relation and they are normalized so that $\mu(v_0) = 1$.

Giving appropriate values to the 6 parameters we can obtain all the families of hypergeometric orthogonal polynomials in the Askey scheme. We give next some examples.

Since the parameters are the coefficients of f, g, and h, and the equations in (6.3) are satisfied, it is enough to give the polynomials f, $\tilde{h}$, and $\tilde{g}$ to determine all the coefficients. We also list the value of e_1, which is the only free coefficient of $\tilde{g}$.

For the Wilson polynomials, Eq. (9.1.5) in [14], we have

$$f(n) = -(n+a)^2, \qquad \tilde{h}(n) = -\frac{n+a+b+c+d-1}{(a+b)(a+c)(a+d)},$$

$$\tilde{g}(n) = \frac{(n+a+b)(n+a+c)(n+a+d)}{(a+b)(a+c)(a+d)},$$

$$e_1 = \frac{3a^2 + 2(ab+ac+ad) + bc + bd + cd}{(a+b)(a+c)(a+d)}.$$

For the Meixner-Pollaczek polynomials, Eq. (9.7.4) in [14], we have

$$f(n) = -i(n+\lambda), \qquad \tilde{h}(n) = \frac{e^{ix}\sin(x)}{\lambda}, \qquad \tilde{g}(n) = \frac{n+2\lambda}{2\lambda}, \qquad e_1 = \frac{1}{2\lambda}.$$

For the Jacobi polynomials, Eq. (9.8.5) in [14], we have

$$f(n) = 1, \qquad \tilde{h}(n) = \frac{n+\alpha+\beta+1}{2(\alpha+1)}, \qquad \tilde{g}(n) = \frac{n+\alpha+1}{(\alpha+1)}, \qquad e_1 = \frac{1}{(\alpha+1)}.$$

For the Bessel polynomials, Eq. (9.13.4) in [14], we have

$$f(n) = 0, \qquad \tilde{h}(n) = \frac{n+a+1}{2}, \qquad \tilde{g}(n) = 1, \qquad e_1 = 0.$$

7 A Family of q-Orthogonal Polynomial Sequences That Contains All the Families in the q-Askey Scheme

In this section we present another solution of the problem set up in (5.5) and (5.6) that gives us a family of q-orthogonal polynomial sequences that contains all the families in the q-Askey scheme [14]. See [31] for a more detailed account of a related family, using another set of parameters.

Define the Laurent polynomials

$$f(z) = w_0 + w_1 z + w_2 z^{-1}, \qquad h(z) = y_1 z + y_2 z^{-1}, \tag{7.1}$$

and

$$g(z) = a_0 z^{-2} + a_1 z^{-1} + a_2 + a_3 z + a_4 z^2. \tag{7.2}$$

Let q be a complex number that is not a root of unity and define the sequence $x_n = f(q^n)$ for $n \geq 0$, and let $\{v_k(t)\}_{k\geq 0}$ be the Newton basis associated with the sequence x_n.

Define the matrices $G = M(g(q^{n+1}))$ and $B = M(h(q^n)) + SG$ and let C be the unique monic matrix that satisfies $CB = M(h(q^n))C$. Recall that the entries of C are expressed in terms of the sequences $h(q^n)$ and $g(q^{n+1})$ by the formulas (5.7) and (5.8).

Let $L = CX_vC^{-1}$ where X_v is as in (3.11). The equations $L_{4,2} = 0$ and $L_{5,3} = 0$ give us $a_0 = qw_2y_2$ and $a_4 = q^{-1}w_1y_1$, which are necessary conditions for L to be in the set $\mathcal{T}$. The equations $L_{k,0} = 0$, for $k \geq 2$ give 3 values for a_1 for which L becomes $(-1, 1)$-banded, but only when $a_1 = -(qw_2y_2 + a_2 + a_3 + q^{-1}w_1y_1$ we obtain $L \in \mathcal{T}$. For the other values of a_1 some entry in the diagonal of index 1 of L becomes zero. Therefore, if we take

$$g(z) = qw_2y_2z^{-2} - (qw_2y_2 + a_2 + a_3 + q^{-1}w_1y_1)z^{-1} + a_2 + a_3z + q^{-1}w_1y_1z^2, \tag{7.3}$$

then C satisfies $CB = M(h(q^n))C$ and $L = CX_vC^{-1} \in \mathcal{T}$ and therefore C is the matrix representation of an orthogonal polynomial sequence $\{u_k(t)\}_{k\geq 0}$ with respect to the Newton basis $\{v_k(t)\}_{k\geq 0}$.

The sequence $\{u_k(t)\}_{k\geq 0}$ is determined by the coefficients of f, h, and g, so that we have a family with parameters $a_2, a_3, y_1, y_2, w_0, w_1, w_2$. We can obtain another family that has one parameter less by means of the substitutions $y_2 \to pq^{-1}y_1, a_2 \to a_2y_1, a_3 \to a_3y_1$. The parameters y_1, y_2 disappear and the new parameter p is incorporated. In [31] we obtained a similar family that contains all the families in the q-Askey scheme and we present the values of the parameters that correspond to some of the families in the q-Askey scheme.

The generalized moments with respect to the basis $\{v_k(t)\}_{k\geq 0}$ are the coefficients $m_{k,0}$, for $k \geq 0$. They are given by (5.10) and satisfy a simple two-term recurrence relation.

Let $U(z)$ be the rational function

$$U(z) = \frac{q^2y_1y_2z^4g(z)g\left(\dfrac{qy_2}{y_1z}\right)}{(y_1z^2 - q^2y_2)(y_1z^2 - qy_2)^2(y_1z^2 - y_2)}. \tag{7.4}$$

If we write L as in (4.4) then

$$\alpha_k = U(q^k), \qquad k \geq 1. \tag{7.5}$$

Since $g(1) = 0$ we can factor g and then obtain the following expression for $U(z)$

$$U(z) = \frac{(z-1)(y_1z - qy_2)w(z)r(z)}{q(y_1z^2 - q^2y_2)(y_1z^2 - qy_2)^2(y_1z^2 - y_2)}, \tag{7.6}$$

where

$$w(z) = w_1 y_1 z^3 + (qa_3 + w_1 y_1)z^2 + (qa_2 + qa_3 + w_1 y_1)z - q^2 w_2 y_2, \qquad (7.7)$$

and

$$r(z) = qw_2 y_1^2 z^3 - (qa_2 y_1 + qa_3 y_1 + w_1 y_1^2)z^2 - (q^2 a_3 y_2 + qw_1 y_1 y_2)z - q^2 w_1 y_2^2. \qquad (7.8)$$

The rational function $U(z)$ satisfies

$$U(z) = U\left(\frac{qy_2}{y_1 z}\right), \qquad (7.9)$$

and the degrees of its numerator and denominator are at most equal to 8.

Now let

$$s(z) = \frac{(qz-1)((q^2 w_2 y_1 - q^2 a_3 - qw_1 y_1)z - (qw_1 y_2 + qa_2 + qa_3 + w_1 y_1))}{q(q-1)(qy_1 z^2 - y_2)}, \qquad (7.10)$$

and

$$\sigma_k = s(q^k) + (k+1)w_0, \qquad k \geq 0. \qquad (7.11)$$

Then the recurrence coefficients β_k are given by

$$\beta_k = \sigma_k - \sigma_{k-1}, \qquad k \geq 1, \qquad (7.12)$$

and $\beta_0 = \sigma_0$. Let us note that $s(z)$ is a rational function and the degrees of its numerator and denominator are at most equal to 2. It is easy to verify that the sequences $f(q^k)$, $h(q^k)$, and $g(q^{k+1})$ are linearly recurrent.

Giving appropriate values for the parameters $a_2, a_3, y_1, y_2, w_0, w_1, w_2$ we can obtain the coefficients α_k and β_k of the normalized recurrence relation for each one of the families of q-orthogonal polynomial sequences in the q-Askey scheme. For the families for which $w_1 \neq 0$ or $w_2 \neq 0$ it is easy to obtain the discrete orthogonality measure from the sequence of generalized moments by means of simple matrix operations.

8 Final Remarks

We presented in the previous sections two important families of orthogonal polynomial sequences that are solutions of the general problem proposed in Sect. 5. It would be interesting to find other similar families associated with linearly recurrent sequences. Another problem is to determine which polynomial sequences in the families studied here are not included in the Askey scheme or the q-Askey scheme. There are some choices of the sequences associated with f, g, and h that yield a tridiagonal matrix L that has some zero entries in its diagonal of index 1, and thus L is not in $\mathcal{T}$, that should be studied.

The matrix approach seems very well suited to study connection coefficients.

A more general problem is obtained when the matrix B is replaced by a (0, 2)-banded matrix, that represents a generalized difference operator of order 2.

We can also use other bases for the space $\mathcal{P}$ instead of the Newton bases. For example, the bases of Chebyshev polynomials, the Hermite polynomials, or some polynomial sequence that satisfies a simple recurrence relation.

The study of the algebraic properties of the algebra $\mathcal{L}$ and some related algebras is an area that requires further research. In [5] we obtained some results about an extension of $\mathcal{L}$ to an algebra of doubly infinite matrices. The subalgebra of $\mathcal{L}$ of the infinite lower triangular matrices has been studied for a long time and its properties are better known. See [23, 24], and [25] and the references therein.

References

1. Álvarez-Nodarse, R.: On characterizations of classical polynomials. J. Comput. Appl. Math. **196**, 320–337 (2006)
2. Arponen, T.: A matrix approach to polynomials. Linear Algebra Appl. **359**, 181–196 (2003)
3. Arponen, T.: A matrix approach to polynomials II. Linear Algebra Appl. **394**, 257–276 (2005); Classical orthogonal polynomials. Bull. Math. Anal.
4. Al Salam, W.A., Chihara, T.S.: Another characterization of the classical orthogonal polynomials. SIAM J. Math. Anal. **3**, 65–70 (1972)
5. Arenas-Herrera, M.I., Verde-Star, L.: Representation of doubly infinite matrices as non-commutative Laurent series. Spec. Matrices **5**, 250–257 (2017)
6. Costabile, F.A., Longo, E.: An algebraic approach to Sheffer polynomial sequences. Integral Transforms and Special Functions **25**, 295–311 (2014)
7. Costabile, F.A., Gualtieri, M.I., Napoli, A.: Recurrence relations and determinant forms for general polynomial sequences. Application to Genocchi polynomials. Integral Transforms and Special Functions **30** 112–127 (2019)
8. Costabile, F.A., Gualtieri, M.I., Napoli, A.: Polynomial sequences: elementary basic methods and application hints. A survey., Revista de la Real Academia de Ciencias Exactas, Físicas y Naturales. Serie A. Matemáticas, to appear.
9. Ernst, T.: q-Pascal and q-Bernoulli Matrices: An Umbral Approach, Uppsala University, Department of Mathematics, Report 23. Uppsala, Sweden (2008)
10. Garza, L.G., Garza, L.E., Marcellán, F.: A matrix characterization for the D_v-semiclassical and D_v-coherent orthogonal polynomials. Linear Algebra Appl. **487**, 242–259 (2015)
11. Garza, L.G., Garza, L.E., Marcellán, F.: A matrix approach for the semiclassical and coherent orthogonal polynomials. Appl. Math. Comput. **256**, 459–471 (2015)
12. Kalman, D.: Polynomial translation groups. Math. Mag. **56**, 23–25 (1983)
13. Kalman, D., Ungar, A.: Combinatorial and functional identities in one-parameter matrices. Am. Math. Monthly **94**, 21–35 (1987)
14. Koekoek, R., Lesky, P.A., Swarttouw, R.F.: Hypergeometric Orthogonal Polynomials and Their q-Analogues. Springer, Berlin, Heidelberg (2010)

15. Koepf, W., Schmersau, D.: Recurrence equations and their classical orthogonal polynomial solutions, Orthogonal systems and applications. Appl. Math. Comput. **128**, 303–327 (2002)
16. Kwon, K.H., Littlejohn, L.L., Yoon, B.H.: New characterizations of classical orthogonal polynomials. Indag. Math. (N.S.) **7**, 199–213 (1996)
17. Loureiro, A.F.: New results on the Bochner condition about classical orthogonal polynomials. J. Math. Anal. Appl. **364**, (2010) 307–323 (2010)
18. Marcellán, F., Branquinho, A., Petronilho, J.: Classical orthogonal polynomials: a functional approach. Acta Appl. Math. **34** 283–303 (1994)
19. Marcellán, F., Pinzón-Cortés, C.N.: (M, N)-Coherent pairs of linear functionals and Jacobi matrices. Appl. Math. Comput. **232** 76–83 (2014)
20. Maroni, P.: Une théorie algébrique des polynômes orthogonaux. Application aux polynômes orthogonaux semi-classiques. In: Brezinski, C., et al. (eds.), Orthogonal Polynomials and Their Applications. IMACS Ann. Comput. Appl. Math. **9**, 95–130 (1991)
21. Medem, J.C., Álvarez-Nodarse, R., Marcellán, F.: On the q-polynomials: a distributional study. J. Comput. Appl. Math. **135**, 157–196 (2001)
22. Njinkeu Sandjon, M., Branquinho, A., Foupouagnigni, M., Area, I.: Characterization of classical orthogonal polynomials on quadratic lattices. J. Differ. Equ. Appl. **23** 983–1002 (2017)
23. Słowik, R.: Maps on infinite triangular matrices preserving idempotents. Linear and Multilinear Algebra **62**, 938–964 (2014)
24. Słowik, R.: Derivations of rings of infinite matrices. Commun. Algebra **43**, 3433–3441 (2015)
25. Słowik, R.: Every infinite triangular matrix is similar to a generalized infinite Jordan matrix. Linear and Multilinear Algebra **65**, 1362–1373 (2017)
26. Verde-Star, L.: Groups of generalized Pascal matrices. Linear Algebra Appl. **382**, 179–194 (2004)
27. Verde-Star, L.: Infinite triangular matrices, q-Pascal matrices, and determinantal representations. Linear Algebra Appl. **434**, 307–318 (2011)
28. Verde-Star, L.: Characterization and construction of classical orthogonal polynomials using a matrix approach. Linear Algebra Appl. **438**, 3635–3648 (2013)
29. Verde-Star, L.: Recurrence coefficients and difference equations of classical discrete and q-orthogonal polynomial sequences. Linear Algebra Appl. **440**, 293–306 (2014)
30. Verde-Star, L.: Polynomial sequences generated by infinite Hessenberg matrices. Spec. Matrices **5** 64–72 (2017)
31. Verde-Star, L.: A unified construction of all the hypergeometric and basic hypergeometric families of orthogonal polynomial sequences. Submitted